Studies in Systems, Decision and Control

Volume 292

The series "Studies in Systems, Decision and Control" (SSDC) covers both new developments and advances, as well as the state of the art, in the various areas of broadly perceived systems, decision making and control–quickly, up to date and with a high quality. The intent is to cover the theory, applications, and perspectives on the state of the art and future developments relevant to systems, decision making, control, complex processes and related areas, as embedded in the fields of engineering, computer science, physics, economics, social and life sciences, as well as the paradigms and methodologies behind them. The series contains monographs, textbooks, lecture notes and edited volumes in systems, decision making and control spanning the areas of Cyber-Physical Systems, Autonomous Systems, Sensor Networks, Control Systems, Energy Systems, Automotive Systems, Biological Systems, Vehicular Networking and Connected Vehicles, Aerospace Systems, Automation, Manufacturing, Smart Grids, Nonlinear Systems, Power Systems, Robotics, Social Systems, Economic Systems and other. Of particular value to both the contributors and the readership are the short publication timeframe and the world-wide distribution and exposure which enable both a wide and rapid dissemination of research output.

Indexed by SCOPUS, DBLP, WTI Frankfurt eG, zbMATH, SCImago.

All books published in the series are submitted for consideration in Web of Science.

Saliha Marir · Mohammed Chadli

Control of Singular Fractional Order Systems: LMI Approach

Stability, Stabilization and Admissibility

Springer

Saliha Marir
University of Science and Technology
Mohamed Boudiaf
Bir El Djir, Oran, Algeria

Mohammed Chadli
University Paris-Saclay Evry
IBISC Laboratory
Évry, France

ISSN 2198-4182 ISSN 2198-4190 (electronic)
Studies in Systems, Decision and Control
ISBN 978-3-031-87381-2 ISBN 978-3-031-87382-9 (eBook)
https://doi.org/10.1007/978-3-031-87382-9

This Springer imprint is published by the registered company Springer Nature Switzerland AG
The registered company address is: Gewerbestrasse 11, 6330 Cham, Switzerland

Introduction

In the understanding and development of large class of systems it is now a well realized and accepted fact that the researchers have taken their initiation from nature. Natural things can be well understood in two possible ways, quantitative and qualitative. Mathematics plays a central role in this direction. It is the science of patterns and relationships. When we go back to understand the quantitative and qualitative behaviour of nature, it seems that evolution is from integer to fraction. Quantitative behaviour can be well explained using number theory, which started from integer and reached to fractional due to division operation and finally converged to real numbers. Calculus is a branch of mathematics describing how things change. It provides a framework for modelling systems undergoing change, and a way to deduce the predictions of such models. All these resulted in pointing a fact that integer order calculus is a subset of fractional calculus.

Fractional calculus can be defined as the generalization of classical calculus to orders of integration and differentiation not necessarily integer, goes back to the initiative of the philosopher and creator of modern calculus G. W. Leibniz, who made some remarks on the meaning and possibility of fractional derivative of order $\frac{1}{2}$ in the late 17th century. However a rigorous investigation was first carried out by Liouville in a series of papers from 1832 to 1837, where he defined the first outcast of an operator of fractional integration. Surveys of the history of the fractional theory derivative can be found in [1–5].

For three centuries, the theory of fractional derivatives developed mainly as a pure theoretical field of mathematics useful only for mathematicians. Starting from the sixties, the researches in this domain pointed out that the non-integer order derivative revealed to be more adequate tool for the description of properties of various real materials as polymers. Various types of physical phenomena, in favour of the use of models with the help of fractional derivative, that is, fractality, recursivity, diffusion and/or relaxation phenomena are given in [6]. Recent books [7–11] provide a rich source of information on fractional-order calculus and its applications. The book by M. Caputo [12], published in 1969, in which he systematically used his original definition of fractional differentiation for formulating and solving problems of viscoelasticity and his lectures on seismology [13] must also added to this gallery as well as a series of A. Oustaloup's books on applications of fractional derivatives in control theory [14–16].

Nowadays, interest in fractional differentiation keeps growing. Fractional tools also appear in automatic, particularly in control of dynamic systems where the system to be controlled and/or the controller are governed by fractional differential equations. The introduction of these instruments is driven by the robust character that provides Crone control (robust-control of non-integer order), introduced by A. Oustaloup in 1983.

Due to the potentialities of the fractional-order systems to improve classical systems results, an extensive research has been devoted to the controller synthesis of systems involving a fractional order derivative [17–26] and the references therein.

The main advantage of fractional derivative is that it provides an excellent tool for the description of memory and hereditary properties of various materials and processes in comparison with classical integer-order models.

As we all know, the problem of stability is very essential and crucial issue in control theory, especially on control of fractional-order systems. Very recently, stability and robustness of such class of systems have been investigated extensively both from an algebraic and an analytic point of view [27–34] and references therein. In spite of intensive researches, many challenging and unsolved problems related to control theory of fractional-order systems remain an open problems. The monograph includes the analysis of the admissibility and stabilization condition for singular fractional-order linear time-invariant systems.

The monograph consists of seven chapters, four Appendices and a list of references.

Chapter 1: "Overview on Fractional Calculus" This chapter is dedicated to provide an introduction to the basic definitions of special functions (the Gamma function, the Mittag-Leffler function) which play the most important role in the theory of fractional calculus. Some linear operators of fractional integration and fractional differentiation are introduced. The representation of fractional state in the state space is given (in fact, it should rather speak of pseudo-state). The choice of the approach of Caputo for further developments in this manuscript is justified. Particular attention is devoted to the technique of Laplace transforms for solving linear fractional differential equations and some applications of fractional calculus in various fields is also surveyed.

Chapter 2: "Stability Analysis of Fractional-Order Systems" The most important results existing in the literature concerning the stability of linear fractional-order systems are recalled, namely the famous result on the localization poles of the system and those obtained by the resolution of the linear matrix inequalities in a convex region in the complex plane for the case where the fractional-order derivative α satisfies $1 \leq \alpha < 2$ and in a non-convex region for the case $0 < \alpha < 1$. These results are used in the discussion concerning the stabilization for such class of systems in Chap. 3. This part is enriched by numerical examples and simulations.

Chapter 4: "H_∞ Control for Fractional-Order Systems" This chapter concerns with the $\mathcal{H}_\infty$ control problem for fractional-order systems. The purpose is to design a controller such that the closed-loop system is stable and its transfer function satisfies a prescribed $\mathcal{H}_\infty$ performance level. To solve this problem, firstly versions of the bounded real lemma for fractional-order systems are established. Based on these results, the existence condition and explicit expression of a desirable $\mathcal{H}_\infty$ controller are presented. All required conditions are characterized in the form of linear matrix inequalities whose feasible solutions can be obtained easily and directly.

Chapter 5: "Admissibility Analysis of Singular Fractional-Order Systems"

This chapter covers a new class of dynamic systems, it is about singular fractional-order linear systems. In first time, the model is presented and the solution to the state equation is derived with some examples. As we all know that for singular systems, we need to consider not only stability but also the regularity and the non-impulsiveness. Specifically, regularity guarantees the existence and the uniqueness of a solution to a given singular system, while non-impulsiveness ensures no infinite dynamical modes in such system. Analysis and synthesis for singular fractional-order systems were investigated in some papers. This chapter is dedicated to the admissibility of singular linear systems of fractional order. The various results existing in the literature characterizing the admissibility are recalled with detailed demonstrations. Almost the derived results are in the form of $\mathcal{LMI}$s conditions and obtained either by using the decomposition of the original system with Weierstrass canonical form or involving the matrices of the original system.

Chapter 6: "Stabilization of Singular Fractional-Order Systems" This chapter is devoted to the stabilization of singular linear systems with non-integer derivation. We are particularly interested in the following types of commands: static feedback, static output feedback and an observer-based control. The stabilization is based either on the approach by pole placement, or on the resolution of linear matrix inequalities ($\mathcal{LMI}$) applied to the two cases: case where the fractional-order derivative α is between 0 and 1 and the second case where alpha is between 1 and 2. The concept of the normalization to check the admissibility of the closed-loop system is also used. At the end of the chapter, we also discussed the robust stabilization in the presence of uncertainties in the model parameters.

Chapter 7: "$\mathcal{H}_\infty$ Control for Singular Fractional-Order Systems" This chapter investigates the problem of $\mathcal{H}_\infty$ control for descriptor fractional-order continuous-time systems. Firstly, formulations of a bounded real lemma corresponding to this class of dynamic systems are proposed. Using these results, the existence condition and explicit expression of a proper controller that satisfies a prescribed $\mathcal{H}_\infty$ norm condition as well as stabilizing the closed-loop system are derived. All required conditions are characterized in the form of linear matrix inequalities to which we can get solutions easily and directly.

References

1. Dugowson, S.: Les différentielles métaphysiques: histoire de philosophie de la généralisation de l'ordre de dérivation. Thèse de doctorat, University of Paris13 (1994)
2. Miller, K.S., and Ross, B.: An introduction to the fractional calculus and fractional differential equations. John Wiley and Sons Inc., New York (1993)
3. Oldham, K.B., and Spanier, J.: The fractional calculus. Academic Press, New York-London (1974)

4. Ross, B.: A brief history and exposition of fundamental theory of the fractional calculus. Lecture notes in mathematics, **457**, 1–36, Springer Vergla, New York (1975)
5. Samko, S.G., Kilbas, A.A., and Marichev, O.I.: Integrals and derivatives of the fractional order and some of their applications. Nauka i Tekhnika, Minsk, Russian (1987)
6. Caputo, M., and Mainardi, F.: A new dissipation model based on memory mechanism. Pure and Applied Geophysics, **91**, 134–147 (1971)
7. Gorenflo, R., and Mainardi, F.: Fractional calculus, integral and differential equations of fractional order. CISM, Lectures Notes, International center of mechanical sciences, Udine, Italy (2000)
8. Kilbas, A.A., Srivastava, H.M., and Trujillo, J.J.: Theory and applications of fractional differential equations. North-Holland Mathematics Studies. Elsevier, Amsterdam, **204** (2006)
9. Podlubny, I.: Fractional differential equations. Academic Press, New York (1999)
10. Samko, S.G., Kilbas, A.A., and Marichev, O.I.: Fractional integrals and derivatives: Theory and applications. Gordon and Breach Science Publisher, Amsterdam, the Netherlands (1993)
11. Kaczorek, T.: Selected problems of fractional systems theory. Springer-Verlag, Lecture Notes in Control and Information Sciences: 10.1007/978-3-642-20502-6, Heidelberg, Berlin, Germany (2011)
12. Caputo, M.: Elasticità e dissipazione. Zanichelli, Bologna (1969)
13. Caputo, M.: Lectures on seismology and rheological tectonics. Univ. degli. Studi di Roma "La Sapienza", 1992–1993
14. Oustaloup, A.: Systémes asservis linéaires d'ordre fractionnaire. Masson, Paris (1983)
15. Oustaloup, A.: La dérivation non entiére: Théorie, synthése et applications.Hermes, Paris (1995)
16. Oustaloup, A., Sabatier, J., and Lanusse, P., and Melchior, P.: Fractional differentiation and robust control design. Springer May 06, 2015
17. Bingi, R., and Ibrahim,K.,Karsiti,M.N., Hassan,A.M., andHarindran, V.R.: Fractionalorder systems and pid controllers. Springer October 31, 2019
18. Petras, I., Fractional-order nonlinear systems. Springer May 30, 2011
19. Valério, D., and Costa, J.: An introduction to fractional control. The Institution of Engineering and Technology, January 01 2012
20. Xue, D.: Fractional-order control systems. Walter De Gruyter, July 10, 2017
21. Xuefeng, Z.: Stability and stabilization of singular systems: Strict lmi sufficient conditions.Proceedings of the 10th World Congress on Intelligent Control and Automation, July 6–8, Beijing, China (2012)
22. Xuefeng, Z., and Chen, Y.Q.: Admissibility and robust stabilization of continuous linear singular fractional order systems with the fractional order α: The $0 < \alpha < 1$ case. ISA Transactions, **82**, 42–50 (2018)
23. Marir, S., Chadli, M., Basin, M.V.: Stabilization and H∞ control for descriptor fractional order systems. In: 2023 IEEE International Conference on Networking, Sensing and Control,Marseille, France, pp. 1–6, Oct 2023. https://doi.org/10.1109/ICNSC58704.2023.10319004
24. Marir, S.,Chadli,M.,Basin,M.V.:Necessary and sufficient admissibility conditions of dynamic output feedback for singular linear fractional–order systems. Asian J. Control **25**, 2439–2450 (2022). https://doi.org/10.1002/asjc.3001
25. Marir, S., Chadli, M., Basin, M.V.: Bounded real lemma for singular linear continuous-time fractional-order systems. Automatica 135 (2022).https://doi.org/10.1016/j.automatica.2021.109962
26. Marir, S., Chadli, M., Basin, M.V.: H∞ static output feedback controller design for singu lar fractional-order systems. In: 2022 European Control Conference (ECC), London, United Kingdom, July 2022, pp. 1–6. https://doi.org/10.23919/ECC55457.2022.9838112
27. Abusaksaka, A.B., and Partington, J.R.: Bibo stability of some classes of delay systems and fractional systems. Systems and Control Letters, **64**, 43–46 February (2014)
28. Bonnet, C.: and Partington, J.R.: Analysis of fractional delay systems of retarded and neutral type. Automatica, **38**, 1133–1138 (2002)

29. Lu, J.G., and Chen, Y.Q.: Robust stability and stabilization of fractional-order interval systems with the fractional order α: the $0 < \alpha < 1$ case. IEEE Transactions on Automatic Control, **55**, 152–158 (2010)
30. Matignon,D.: Stability results for fractional differential equations with applications to control. Processing, Computational Engineering in Systems and Application Multi-conference, IMACS, IEEE-SMC, **2**, 963–968, Lille, France, July (1996)
31. Trigeassou, J.C., and Maamri, N.: Analysis, modeling and stability of fractional order differential systems 1, 2. Wiley, ISTE, (2019)
32. Marir, S., Chadli, M., Bouagada, D.: New admissibility conditions for singular linear continuous-time fractional-order systems. J. Franklin Inst. **354**, 752–766 (2017)
33. Marir, S., Chadli, M., Bouagada, D.: A novel approach of admissibility for singular linear continuous-time fractional-order systems. Int. J. Control Autom. Syst. (2017).https://doi.org/10.1007/s12555-016-0003-0
34. Marir, S., Chadli, M.: Robust admissibility and stabilization of uncertain singular fractional order linear time-invariant systems. IEEE/CAA J. Automatica Sinica **6** (3), 685–692 (2019).https://doi.org/10.1109/JAS.2019.1911480

Contents

Symbols

$\mathbb{R}$	Field of real numbers
$\mathbb{C}$	Field of complex numbers
$\mathbb{C}^-$	$\{s/s \in \mathbb{C},\ \mathrm{Re}(s) < 0\}$
$\mathbb{R}^{n\times m}$	Space of $n \times m$ real matrices
$\mathbb{R}^n$	Space of n-dimensional real vectors
$\mathbb{C}^n$	Space of n-dimensional complex vectors
$\mathbb{C}^{n\times m}$	Space of $n \times m$ complex matrices
$\mathbb{H}_n$	The set of $n \times n$ Hermitian matrices
$\in$	Belongs to
$\times$	Inner product
$*$	Convolution product
$\otimes$	Kronecker product
$p \Leftrightarrow q$	Statements p and q are equivalent
$p \Rightarrow q$	Statement p implies statement q
rank(.)	Rank of a matrix
det(.)	Determinant of a matrix
deg	Degree of a polynomial
Re	Real part of a complex number
Im	Imaginary part of a complex number
I_n	Identity matrix of the size $n \times n$
$0_{n\times m}$	Zero matrix of the size $n \times m$
$X^\top$	Transpose of matrix X
$X_\perp$	Orthogonal of matrix X
X^{-1}	Inverse matrix of X, X must be square with $\det(X) \neq 0$
X^*	Conjugate transpose of matrix X
$\mathrm{sym}\{X\}$	$X + X^T$, X real matrix
$\mathrm{Her}\{X\}$	$X + X^*$, X complex matrix
$\mathrm{diag}\{X_1,\ldots,X_m\}$	Block diagonal matrix with blocks $X_1,\ldots,X_m$
$X \geq 0$	X is real symmetric (or hermitian) positive semi-definite matrix
$X > 0$	X is real symmetric (or hermitian) positive definite matrix
$\mathrm{spec}(X)$	Spectrum of a matrix X: set of eigenvalues of X

spec(E, A)	$\{s/s \in \mathbb{C},\ s \text{ finite},\ \det(sE - A) = 0\}$
$\overline{\sigma}(X)$	The maximum singular value of matrix X
$\begin{bmatrix} X & Y \\ * & Z \end{bmatrix}$	$\begin{bmatrix} X & Y \\ Y^{\mathsf{T}} & Z \end{bmatrix}$
$\Vert \cdot \Vert$	Modulus of a complex number
arg(.)	Argument of a complex number
$n!$	Factorial (n), $n \in \mathbb{N}$: The product of all the integers from 1 to n
[.]	The integer part of a real number
D^{α}	Fractional order derivative
$\mathcal{LMI}$	Linear matrix inequality
$\mathcal{GLMI}$	Generalized linear matrix inequality
SISO	Single input single output
BIBO	Bounded input bounded output
LTI	Linear time invariant
FOS	Fractional order system
$FO-LTI$	Fractional order-Linear time invariant
SFOS	Singular fractional order system
$\mathcal{L}(.)$	Laplace transform of an argument
SVD	Singular value decomposition

Chapter 1
Overview on Fractional Calculus

1.1 Introduction

Fractional calculus can be defined as the generalization of classical calculus to order of integration and differentiation not necessary integer. The theory of derivatives of non-integer order goes back to the Leibniz's note in his letter to l'Hospital [1], dated 30 September 1695, in which the meaning of the derivative of order one half is discussed. The question raised by Leibniz for a non-integer-order derivative was an ongoing topic for more than 300 years. Thereafter, the theory of derivatives and integrals of arbitrary order has appeared, which by the end of XIX century took more or less form due primarily to Liouville (1832, 1837), Riemann (1847), Grünwald (1867), Letnikov (1868) [2]. For three centuries the theory of fractional derivatives was developed mainly as a pure theoretical field of mathematics useful only for mathematicians.

However, during the past few years, the fractional calculus has aroused a growing interest taking benefit of the fractional operator compacity for description of memory and hereditary properties of various materials and physical systems which are often neglected in the classical integer-order models and plays a significant role in modeling of real-word phenomena as electromagnetic systems [3], dielectric polarization [4], viscoelastic systems [5]. In this Chapter we will give in first time a brief overview of the fractional order calculus, then the fractional order dynamical systems and their behavior will be also discussed.

S. Marir and M. Chadli, *Control of Singular Fractional Order Systems: LMI Approach*, Studies in Systems, Decision and Control 292,
https://doi.org/10.1007/978-3-031-87382-9_1

1.2 Fractional Order Operators and Properties

1.2.1 Special Functions of the Fractional Calculus

Here, we give some information on the Euler's gamma and the Mittag-Leffler functions which play the most important role in the theory of the differentiation of arbitrary order.

- **The gamma function**
  One of the basic function of the fractional calculus is Euler's gamma function $\Gamma(z)$, which generalizes the factorial $n!$ and allows n to take also non-integer and even complex values.

Definition 1.2.1 (*gamma function*) The gamma function $\Gamma(z)$ is defined by the integral

$$\Gamma(z) = \int_0^{\infty} e^{-t} t^{z-1} dt \tag{1.1}$$

which converges in the right half of the complex plane $\mathrm{Re}(z) > 0$.

The Fig. 1.1 represents the gamma function.

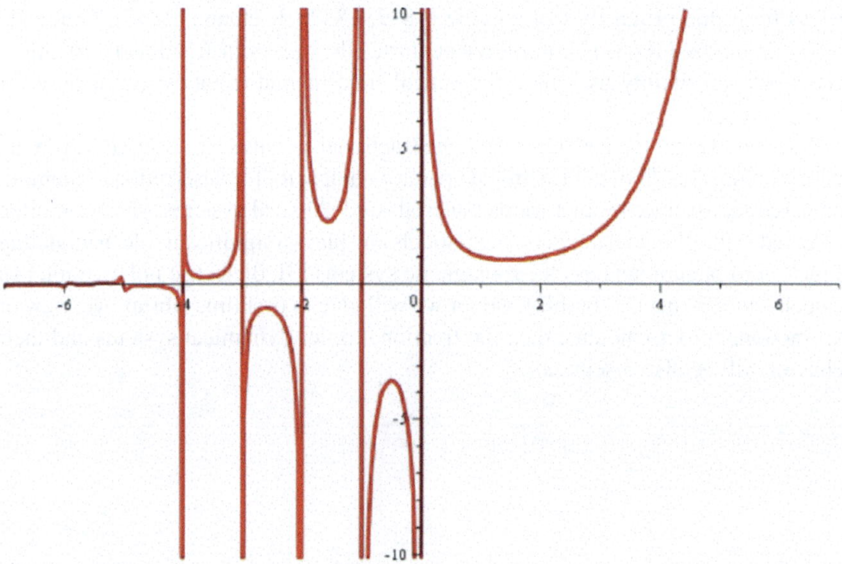

Fig. 1.1 Representation of the gamma function on $\mathbb{R}$

One of the basic properties of the gamma function is that it satisfies the following function equation

$$\Gamma(z+1) = z\Gamma(z)$$

which can be easily proved by integrating by parts

$$\begin{aligned}\Gamma(z+1) &= \int_0^{\infty} e^{-t} t^z dt \\ &= \left[-e^{-t} t^z\right]_0^{\infty} + \int_0^{\infty} z e^{-t} t^{z-1} dt \\ &= z\Gamma(z)\end{aligned}$$

- **The Mittag-Leffler Function**
  The exponential function, e^z, plays a very important role in the theory of integer-order differential equations. Its one parameter generalization, called the Mittag-Leffler function, is the function which is now denoted by [6]

$$E_\alpha(z) = \sum_{k=0}^{\infty} \frac{z^k}{\Gamma(\alpha k + 1)}$$

  It was introduced by G. M. Mittag-Leffler [7–9] and studied also by A. Wiman [10]. For $\alpha = 1$, we obtain

$$E_1(z) = \sum_{k=0}^{\infty} \frac{z^k}{\Gamma(k+1)} = \sum_{k=0}^{\infty} \frac{z^k}{k!} = e^z$$

  The two-parameter function of the Mittag-Leffler type, which plays a very important role in the fractional calculus, was introduced by Agarwal in [11]. This function could have been called the Agarwal function. However, Humbert and Agarwal obtain a number of relationships for this function and generously left the same notation as for the one parameter Mittag-Leffler function, and that is the reason that now the two-parameter function is also called the Mittag-Leffler function.

Definition 1.2.2 A two-parameter function of the Mittag-Leffler type is defined by the series expansion [6]

$$E_{\alpha,\beta}(z) = \sum_{k=0}^{\infty} \frac{z^k}{\Gamma(\alpha k + \beta)}, \alpha > 0, \beta > 0$$

It follows from the definition

for $\beta = 1, \alpha > 0, \ E_{\alpha,1}(z) = \sum\limits_{k=0}^{\infty} \frac{z^k}{\Gamma(\alpha k+1)} = E_\alpha(z)$

for $\beta = \alpha = 1, \ E_{1,1}(z) = \sum\limits_{k=0}^{\infty} \frac{z^k}{\Gamma(k+1)} = \sum\limits_{k=0}^{\infty} \frac{z^k}{k!} = e^z$

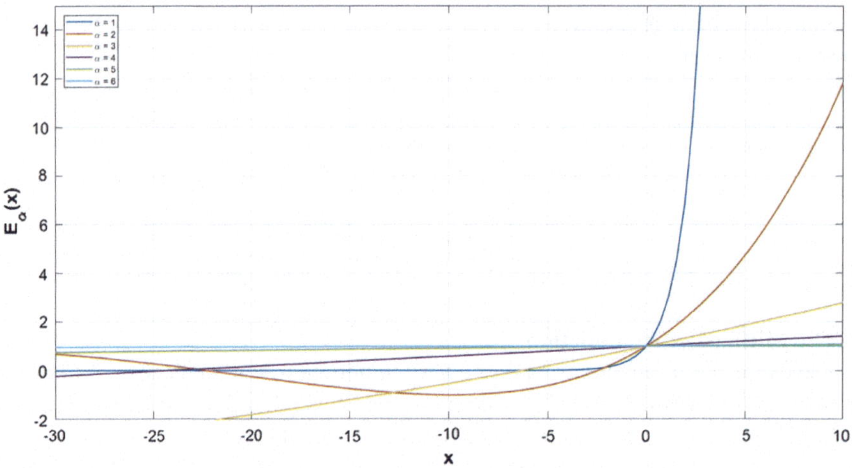

Fig. 1.2 Representation of the Mittag-Leffler for different values of α

$$\text{for } \beta = 2, \alpha = 1,\ E_{1,2}(z) = \sum_{k=0}^{\infty} \frac{z^k}{\Gamma(k+2)} = \sum_{k=0}^{\infty} \frac{z^k}{(k+1)!} = \frac{1}{z} \sum_{k=0}^{\infty} \frac{z^{k+1}}{(k+1)!} = \frac{e^z - 1}{z}$$

$$\text{for } \beta = 3, \alpha = 1,\ E_{1,3}(z) = \sum_{k=0}^{\infty} \frac{z^k}{\Gamma(k+3)} = \sum_{k=0}^{\infty} \frac{z^k}{(k+2)!} = \frac{1}{z^2} \sum_{k=0}^{\infty} \frac{z^{k+2}}{(k+2)!} = \frac{e^z - z - 1}{z^2}$$

The Fig. 1.2 represents the Mittag-Leffler function.

1.2.2 *Fractional-Order Integration*

According to the Riemann-Liouville approach to fractional calculus, the notion of fractional integral of order α ($\alpha > 0$) is a natural consequence of the well known formula (usually attributed to Cauchy), that reduces the calculation of the n-fold primitive of a continuous function f(t) to a single integral of convolution type. In our notation the Cauchy formula reads

$$J^n f(t) = \frac{1}{(n-1)!} \int_0^t (t-\tau)^{n-1} f(\tau) d\tau, t > 0, n \in \mathbb{N}^* \tag{1.2}$$

In a natural way we are led to extend the above formula from positive integer values of the index to any positive real values by using the gamma function and introducing the arbitrary positive real number α, we define the fractional integral of order $\alpha \geq 0$:

$$J^\alpha f(t) = \begin{cases} \frac{1}{\Gamma(\alpha)} \int_0^t (t-\tau)^{\alpha-1} f(\tau) d\tau, \text{t} > 0, & \alpha > 0 \\ f(t), & \alpha = 0 \end{cases} \tag{1.3}$$

We note the semi-group property

$$J^{\alpha} J^{\beta} f(t) = J^{\alpha+\beta} f(t), \alpha, \beta \geq 0 \tag{1.4}$$

which implies the commutative property

$$J^{\alpha} J^{\beta} f(t) = J^{\beta} J^{\alpha} f(t), \alpha, \beta \geq 0 \tag{1.5}$$

and the effect of the operator J^{α} on the power functions is (see [12]):

$$J^{\alpha} t^{\gamma} = \frac{\Gamma(\gamma+1)}{\Gamma(\gamma+1+\alpha)} t^{\gamma+\alpha}, \alpha > 0, \gamma > -1, t > 0 \tag{1.6}$$

For $\alpha = 1$ and $\gamma = n \in \mathbb{N}$, the usual result is recovered, namely

$$J t^{n} = \frac{t^{n+1}}{n+1} \tag{1.7}$$

1.2.3 Riemann-Liouville Fractional Derivative

Denoting by D^n with $n \in \mathbb{N}$, the operator of the derivative of order n , we first note that

$$D^n J^n = I, \; J^n D^n \neq I, n \in \mathbb{N} \tag{1.8}$$

i.e. D^n is left-inverse (and not right-inverse) to the corresponding integral operator J^n. As a consequence we expect that D^{α} is defined as left-inverse to J^{α}. For this purpose, introducing the smallest positive integer m such that $0 \leq m-1 < \alpha < m$, the Riemann-Liouville Fractional Derivative of order $\alpha > 0$ is defined as [12, 13]

$$^{rl}D^{\alpha} f(t) = D^m J^{m-\alpha} \tag{1.9}$$

namely,

$$^{rl}D^{\alpha} f(t) = \begin{cases} \frac{1}{\Gamma(m-\alpha)} \frac{d^m}{dt^m} \int_0^t (t-\tau)^{m-\alpha-1} f(\tau) d\tau, & m-1 \leq \alpha < m \\ \frac{d^m}{dt^m} f(t), & \alpha = m \end{cases} \tag{1.10}$$

Defining for complement $D^0 f(t) = f(t)$, then we easily recognize that

$$^{rl}D^{\alpha} J^{\alpha} f(t) = f(t) \tag{1.11}$$

and

$$ {}^{rl}D^{\alpha}t^{\gamma} = \frac{\Gamma(\gamma+1)}{\Gamma(\gamma+1-\alpha)}t^{\gamma-\alpha}, \alpha > 0, \gamma > -1, t > 0 \tag{1.12} $$

Note the remarkable fact that the fractional derivative $D^{\alpha} f(t)$ is not zero for the constant function $f(t) \equiv 1$ if α is not integer. indeed, if we replace γ by zero in (1.12), we obtain

$$ D^{\alpha}1 = \frac{t^{-\alpha}}{\Gamma(1-\alpha)} \tag{1.13} $$

1.2.4 *Caputo Fractional Derivative*

Applied problems require definitions of fractional derivatives allowing the utilization of physically interpretable initial conditions, which contain $f(0)$, $f'(0)$, $f"(0)$, etc. Unfortunately, the Riemann-Liouville approach leads to initial conditions containing the limit values of the Riemann-Liouville derivative at the lower terminal $\lim_{t\to 0} D^{\alpha-1} f(t), \lim_{t\to 0} D^{\alpha-2} f(t), \cdots \lim_{t\to 0} D^{\alpha-n} f(t)$ and there is no physical interpretation for such types of initial conditions.

A certain solution to this conflict was proposed by the so-called Caputo fractional derivative definition for $\alpha > 0$ [12, 13]

$$ {}^{c}D^{\alpha} f(t) = J^{(m-\alpha)} D^{m} f(t), \quad m-1 < \alpha < m \tag{1.14} $$

namely,

$$ {}^{c}D^{\alpha} f(t) = \begin{cases} \frac{1}{\Gamma(m-\alpha)} \int_0^t (t-\tau)^{m-\alpha-1} f^{(m)}(\tau)d\tau, & m-1 < \alpha < m \\ \frac{d^{(m)}}{dt^m} f(t), & \alpha = m \end{cases} \tag{1.15} $$

This definition is of course more restrictive than Riemann-Liouville definition, since it requires the absolute integrability of the derivative of order m. The main advantage of Caputo's approach is that the initial conditions for fractional differential equation with Caputo derivative take the same form of integer-order.

In particular, according to this definition, the relevant property for which the fractional derivative of a constant is still zero, i.e.

$$ {}^{c}D^{\alpha}1 = 0, \ \alpha > 0 \tag{1.16} $$

1.2.5 Grünwald-Letnikov Fractional Derivative

The derivative of a continuous function proposed by Grünwald-Letnikov can be obtained intuitively from the definition of the usual derivative (the derivative of integer order):

$$f^{(n)}(t) = \frac{d^n}{dt^n} f(t) = \lim_{h \to 0} \frac{1}{h^n} \sum_{k=0}^{n} (-1)^k \binom{n}{k} f(t - kh)$$

where $\binom{n}{k} = \frac{n!}{(n-k)!k!} = \frac{\Gamma(n+1)}{\Gamma(n-k+1)\Gamma(k+1)}$ is the usual binomial coefficients. A generalization of the backward difference by allowing the derivative order to be an arbitrary positive real was proposed by Grünwald-Letnikov [12, 13]

$${}_0^{gl}D_t^{\alpha} f(t) = \lim_{h \to 0} \frac{1}{h^{\alpha}} \sum_{k=0}^{[\frac{t}{h}]} (-1)^k \binom{\alpha}{k} f(t - kh), \alpha > 0 \tag{1.17}$$

where $[\frac{t}{h}]$ denotes the integer part and $\binom{\alpha}{k}$ represents binomial coefficient generalized to real numbers. Namely,

$$\binom{\alpha}{k} = \frac{\Gamma(\alpha + 1)}{\Gamma(\alpha - k + 1)\Gamma(k + 1)} \tag{1.18}$$

1.2.6 Laplace Transforms of Fractional Order Derivatives

The Laplace transform of the function $f(t)$ is the function $F(s)$ of the complex variable s defined by

$$F(s) = \mathcal{L}(f(t)) = \int_0^{\infty} e^{-st} f(t) dt \tag{1.19}$$

For the existence of the integral (1.19), we need to define the exponential order function.

Definition 1.2.3 A function f has exponential order a if there exist constants $M > 0$ such that for some $T \geq 0$,

$$|f(t)| \leq Me^{at}, t \geq T.$$

For the existence of the integral (1.19), we have this theorem.

Theorem 1.2.1 *If f is piecewise continuous on $[0, \infty)$ and of exponential order a, then the Laplace transform $\mathcal{L}(f(t))$ exists for $\text{Re}(s) > a$ and converges absolutely.*

In order to apply the Laplace transform to physical problems, it is necessary to invoke the inverse transform. If $\mathcal{L}(f(t)) = F(s)$, then the inverse Laplace transform is denoted by

$$\mathcal{L}^{-1}(F(s)) = f(t), t \geq 0$$

which maps the Laplace transform of a function back to the original function and is defined by

$$f(t) = \mathcal{L}^{-1}(F(s)) = \int_{c-i\infty}^{c+i\infty} e^{st} F(s)ds, c = \text{Re}(s) > c_0 \tag{1.20}$$

where c_0 lies in the right half plane of the absolute convergence of the Laplace integral.

We mention above some properties of the Laplace transform which we will need later.

- The Laplace transform is a linear mapping, i.e. for all functions f and g admitting Laplace transforms and for all real α and β:

$$\mathcal{L}(\alpha f(t) + \beta g(t)) = \alpha\mathcal{L}(f(t)) + \beta\mathcal{L}(g(t)) \tag{1.21}$$

- Under the same conditions we have

$$\mathcal{L}((f * g)(t)) = \mathcal{L}(f(t))\mathcal{L}(g(t)) \tag{1.22}$$

  where $*$ denotes the convolution product defined by

$$(f * g)(t) = \int_0^t f(t-\tau)g(\tau)d\tau = \int_0^t g(t-\tau)f(\tau)d\tau$$

  with f and g are continuous functions on $[0, \infty)$.
- An other useful property which we need is the Laplace transform of the derivative of an integer order n of a function $f(t)$:

$$\mathcal{L}(f^{(n)}(t)) = s^n F(s) - \sum_{k=0}^{n-1} s^{n-k-1} f^{(k)}(0) = s^n F(s) - \sum_{k=0}^{n-1} s^k f^{(n-k-1)}(0) \tag{1.23}$$

  where $F(s) = \mathcal{L}(f(t))$.

At present we will give the Laplace transform of fractional derivative of order $\alpha > 0$ view previously

- The writing of the fractional integral of order $\alpha > 0$ (1.3) as a convolution product

$$J^{\alpha} f(t) = \frac{t^{\alpha-1}}{\Gamma(\alpha)} * f(t) \tag{1.24}$$

allows to calculate its Laplace transform [14]:

$$\mathcal{L}(J^{\alpha} f(t)) = \frac{F(s)}{s^{\alpha}} \tag{1.25}$$

Indeed,

$$\mathcal{L}(J^{\alpha} f(t)) = \mathcal{L}(\frac{t^{\alpha-1}}{\Gamma(\alpha)})\mathcal{L}(f(t))$$

let $F(s) = \mathcal{L}(f(t))$ and

$$\begin{aligned}\mathcal{L}(\tfrac{t^{\alpha-1}}{\Gamma(\alpha)}) &= \tfrac{1}{\Gamma(\alpha)} \int_0^{\infty} t^{\alpha-1} e^{-st} dt \\ &= \tfrac{1}{\Gamma(\alpha)} \int_0^{\infty} \tfrac{x^{\alpha-1}}{s^{\alpha}} e^{-x} dx \\ &= \tfrac{1}{\Gamma(\alpha)} \tfrac{\Gamma(\alpha)}{s^{\alpha}} \\ &= \tfrac{1}{s^{\alpha}}\end{aligned}$$

then we get the following property

$$\mathcal{L}(\frac{t^{\alpha-1}}{\Gamma(\alpha)}) = \frac{1}{s^{\alpha}} \tag{1.26}$$

Thus the formula (1.25) is derived.

- The expression of the Laplace transform of the Riemann-Liouville Fractional Derivative of order $\alpha > 0$ is given by

$$\mathcal{L}({}^{rl}D^{\alpha} f(t)) = s^{\alpha} F(s) - \sum_{k=0}^{m-1} s^k \left[D^{(\alpha-k-1)} f(t)\right]_{t=0} \tag{1.27}$$

where $m - 1 \leq \alpha < m, m \in \mathbb{N}^*$.
Indeed, according to the Riemann-Liouville Fractional Derivative definition in the form (1.9) and by putting

$$g(t) = J^{m-\alpha} f(t)$$

we obtain

$${}^{rl}D^{\alpha} f(t) = D^m (g(t))$$

then the application of the Laplace transform of an integer order derivative (1.23) gives

$$\mathcal{L}({}^{rl}D^{\alpha} f(t)) = s^m G(s) - \sum_{k=0}^{m-1} s^k \left[\frac{d^{(m-k-1)}}{dt^{(m-k-1)}} g(t) \right]_{t=0}$$

with $G(s) = \mathcal{L}(g(t))$. Taking in to account the Laplace transform of the fractional integral (1.25) with $F(s) = \mathcal{L}(f(t))$ then

$$G(s) = \frac{F(s)}{s^{m-\alpha}}$$

In the other hand, we have

$$\begin{aligned} \frac{d^{(m-k-1)}}{dt^{(m-k-1)}} g(t) &= \frac{d^{(m-k-1)}}{dt^{(m-k-1)}} (J^{m-\alpha} f(t)) \\ &= \frac{d^{(m-k-1)}}{dt^{(m-k-1)}} J^{m-k-1-(\alpha-k-1)} f(t) \\ &= {}^{rl}D^{(\alpha-k-1)}(f(t)) \end{aligned}$$

Finally, we obtain

$$\mathcal{L}({}^{rl}D^{\alpha} f(t)) = s^{\alpha} F(s) - \sum_{k=0}^{m-1} s^k \left[{}^{rl}D^{(\alpha-k-1)} f(t) \right]_{t=0}$$

As can be seen there is the limit values of fractional derivatives at the lower terminate $t = 0$ which have not physical interpretation, that is why the practical applicability of the Laplace transform of the Riemann-Liouville is limited.

- By writing the Caputo fractional derivative as

$$ {}^{c}D^{\alpha} f(t) = J^{m-\alpha}(f^m(t)), m-1 < \alpha < m \tag{1.28}$$

and by using the Laplace transform of the fractional order integral (1.25) and the integer-order derivative (1.21), we get the Laplace transform of the Caputo fractional derivative for $m - 1 < \alpha \leq m$

$$\mathcal{L}({}^{c}D^{\alpha} f(t)) = s^{\alpha} F(s) - \sum_{k=0}^{m-1} s^{\alpha-k-1} \left[D^{(k)} f(t) \right]_{t=0} \tag{1.29}$$

- Assuming that $m - 1 < \alpha < m$ and using the Laplace transform of the power function (1.26), the formula of the Laplace transform of the convolution and the Laplace transform of the integer-order derivative (1.21), we obtain the Laplace transform of the Grünwald-Letnikov derivative as

$$\mathcal{L}({}^{gl}D^{\alpha} f(t)) = s^{\alpha} F(s) \tag{1.30}$$

Remark 1.2.1 It can be noticed that Laplace transformation of the RL derivative requires information of initial conditions of the function i.e. $D^{\alpha-k-1}f(0)$ (see (1.27)). From the mathematical point of view initial value problem is rigorous and elegant. However, This is found to be difficult to measure in term of physical quantities. In the Caputo' s definition initial conditions $x(0), \dot{x}(0), \ldots, x^{(n-1)}(0)$ are the same as for the integer derivatives of f, which has well known physical interpretation.For example, if one can interprets $x(t)$ as a position, then $\dot{x}(t)$ stands for speed and $\ddot{x}(t)$ expresses as an acceleration. This gives a strong support for acceptance of Caputo' s derivative to researchers and practising engineers.

In the remaining work, the fractional derivative of Caputo will be used and simply denoted by D.

1.2.7 Applications of Fractional Calculus

the basics of fractional calculus (integral and differential operations of non integer order) were treated and improved long ago by the mathematicians Leibniz (1695), Liouville (1834), Riemann (1892), and others and attracted the attention in various fields of science and engineering by Oliver Heaviside in the 1890s, it was not until 1974 that the first book on the topic was published by Oldham and Spanier. Recent monographs and symposia proceedings have highlighted the application of fractional calculus in physics, robotics, signal processing, and electromagnetics. Here we state some of applications.

- **Electric transmission lines**
  During the last decades of the nineteenth century, Heaviside successfully developed his operational calculus without rigorous mathematical arguments. In 1892 he introduced the idea of fractional derivatives in his study of electric transmission lines. Based on the symbolic operator form solution of heat equation due to Gregory(1846), Heaviside introduced the letter p for the differential operator $\frac{d}{dt}$ and gave the solution of the diffusion equation

$$\frac{\partial^2 u}{\partial x^2} = a^2 p$$

  for the temperature distribution $u(x,t)$ in the symbolic form

$$u(x,t) = Ae^{(ax\sqrt{p})} + Be^{(-ax\sqrt{p})}$$

  in which $p = \frac{d}{dx}$ was treated as constant, where a, A and B are also constant.
- **Application of Fractional Calculus to Fluid Mechanics**
  V. Kulish and José L. Lage [15]
  Application of fractional calculus to the solution of time-dependent, viscous-diffusion fluid mechanics problems are presented. Together with the Laplace

transform method, the application of fractional calculus to the classical transient viscous-diffusion equation in a semi-infinite space is shown to yield explicit analytical (fractional) solutions for the shearstress and fluid speed anywhere in the domain. Comparing the fractional results for boundary shear-stress and fluid speed to the existing analytical results for the first and second Stokes problems, the fractional methodology is validated and shown to be much simpler and more powerful than existing techniques.

- **Wave propagation in viscoelastic horns using a fractional calculus rheology model**
  Margulies, Timothy [16]
  The complex mechanical behavior of materials are characterized by fluid and solid models with fractional calculus differentials to relate stress and strain fields. Fractional derivatives have been shown to describe the viscoelastic stress from polymer chain theory for molecular solutions. Here the propagation of infinitesimal waves in one dimensional horns with a small cross-sectional area change along the longitudinal axis are examined. In particular, the linear, conical, exponential, and catenoidal shapes are studied. The wave amplitudes versus frequency are solved analytically and predicted with mathematical computation. Fractional rheology data from Bagley are incorporated in the simulations. Classical elastic and fluid "Webster equations" are recovered in the appropriate limits. Horns with real materials that employ fractional calculus representations can be modeled to examine design trade-offs for engineering or for scientific application.
- **Fractional differentiation for edge detection**
  B. Mathieu, P. Melchior, A. Oustaloup, Ch. Ceyral [17]
  In image processing, edge detection often makes use of integer-order differentiation operators, especially order 1 used by the gradient and order 2 by the Laplacian. This paper demonstrates how introducing an edge detector based on non-integer (fractional) differentiation can improve the criterion of thin detection, or detection selectivity in the case of parabolic luminance transitions, and the criterion of immunity to noise, which can be interpreted in term of robustness to noise in general.
- **Using Fractional Calculus for Lateral and Longitudinal Control of Autonomous Vehicles**
  J.I. Suárez, B.M. Vinagre , A.J. Calderón , C.A. Monje and Y.Q. Chen [18]
  Here it is presented the use of Fractional Order Controllers (FOC) applied to the path-tracking problem in an autonomous electric vehicle. A lateral dynamic model of an industrial vehicle has been taken into account to implement conventional and Fractional Order Controllers. Several control schemes with these controllers have been simulated and compared.

1.3 Models and Representations of Fractional-Order Systems

In the following, we will focus only on the continuous-time representation.

The equations for a continuous-time dynamic system of fractional order can be written as follows

$$H(D^{\alpha_0\alpha_1\alpha_2\ldots\alpha_m})(y_1, y_2, \ldots, y_p) = G(D^{\beta_0\beta_1\beta_2\ldots\beta_n})(u_1, u_2, \ldots, u_k) \tag{1.31}$$

where y_i, u_j functions of time representative the outputs and the input respectively of the system described by (1.31). $H(.)$, $G(.)$are combinations (not necessarily linear) laws of the fractional-order derivative operator and $\alpha_i, \beta_j > 0$ are fractional order derivatives relating respectively to the output and input to the system.

Let us now consider a SISO LTI FOS. By means of its dynamic input-output relation, we can derive its continuous-time models. In all what follows, we use Caputo' s definition of a fractional derivative with initial time $t = 0$. The derived differential equation is then expressed by [13]

$$\sum_{i=0}^{n} a_i D^{\alpha_i} y(t) = \sum_{j=0}^{m} b_j D^{\beta_j} u(t) \tag{1.32}$$

where a_i, b_j are real constants. In Eq. (1.32), which describes the dynamics of a LTI system mono-variable of fractional order, two cases arise and lead to two types of systems: commensurate systems (systems with commensurate orders) and non-commensurate systems (systems with non commensurate orders). A system is with commensurate order if all the differentiation orders of the fractional differential equation are integer multiples of a basic order, α, that is, $\alpha_i = i\alpha$, $\beta_j = j\alpha$, the Eq. (1.32) becomes

$$\sum_{i=0}^{n} a_i D^{i\alpha} y(t) = \sum_{j=0}^{m} b_j D^{j\alpha} u(t) \tag{1.33}$$

If in (1.33), $\alpha = \frac{1}{q}, q \in \mathbb{N}^*$, the system will be of rational order. The Eq. (1.32), in which α_i, β_j are the fractional orders that can be either commensurate or noncommensurate, can be written as

$$D^{\alpha_n} y(t) = -\frac{a_{n-1}}{a_n} D^{\alpha_{n-1}} y(t) - \frac{a_{n-2}}{a_n} D^{\alpha_{n-2}} y(t) - \cdots \frac{a_0}{a_n} D^{\alpha_0} y(t) + \sum_{j=0}^{m} \frac{b_j}{a_n} D^{\beta_j} u(t) \tag{1.34}$$

Assume that $\alpha_0 \leq \alpha_1 \leq \alpha_2 \leq \ldots \leq \alpha_n$. The procedure for obtaining a state-space representation from (1.34) is as follows: firstly, let us introduce an intermediate variable

$$x_i(t) = D^{\alpha_{i-1}} y(t), i = 1, 2, \ldots, n \tag{1.35}$$

Then we obtain, with the help of the semi-group property of the fractional-order derivative, successively by induction

$$
\begin{aligned}
x_1(t) &= D^{\alpha_0} y(t) \\
x_2(t) &= D^{\alpha_1} y(t) = D^{\alpha_1-\alpha_0+\alpha_0} y(t) = D^{\alpha_1-\alpha_0}(D^{\alpha_0} y(t)) = D^{\alpha_1-\alpha_0} x_1(t) \\
x_3(t) &= D^{\alpha_2} y(t) = D^{\alpha_2-\alpha_1+\alpha_1} y(t) = D^{\alpha_2-\alpha_1}(D^{\alpha_1} y(t)) = D^{\alpha_2-\alpha_1} x_2(t) \\
&\vdots \\
x_n(t) &= D^{\alpha_{n-1}} y(t) = D^{\alpha_{n-1}-\alpha_{n-2}+\alpha_{n-2}} y(t) = D^{\alpha_{n-1}-\alpha_{n-2}}(D^{\alpha_{n-2}} y(t)) \\
&= D^{\alpha_{n-1}-\alpha_{n-2}} x_{n-1}(t) \\
D^{\alpha_n} y(t) &= D^{\alpha_n-\alpha_{n-1}+\alpha_{n-1}} y(t) = D^{\alpha_n-\alpha_{n-1}}(D^{\alpha_{n-1}} y(t)) = D^{\alpha_n-\alpha_{n-1}} x_n
\end{aligned}
\tag{1.36}
$$

With (1.34) and (1.36), it is easy to build the following group of equations

$$
\begin{aligned}
D^{\alpha_1-\alpha_0} x_1(t) &= x_2(t) \\
D^{\alpha_2-\alpha_1} x_2(t) &= x_3(t) \\
&\vdots \\
D^{\alpha_i-\alpha_{i-1}} x_i(t) &= x_{i+1}(t) \\
&\vdots \\
D^{\alpha_n-\alpha_{n-1}} x_n &= -\frac{a_{n-1}}{a_n} x_n - \frac{a_{n-2}}{a_n} x_{n-1} - \ldots - \frac{a_0}{a_n} x_1 + \sum_{j=0}^{m} \frac{b_j}{a_n} D^{\beta_j} u(t)
\end{aligned}
\tag{1.37}
$$

which can be expressed in matrix form

$$
D^{\alpha} x(t) = A x(t) + B D^{\beta} u(t) \tag{1.38}
$$

with $\alpha = (\alpha_1 - \alpha_0, \alpha_2 - \alpha_1, \ldots, \alpha_n - \alpha_{n-1})$, $\beta = (\beta_0, \beta_1, \ldots, \beta_m)$, $x(t) = \left[x_1(t)\ x_2(t)\ \ldots\ x_n(t) \right]^{\mathrm{T}}$ and

$$
A = \begin{bmatrix}
0 & 1 & 0 & 0 & \ldots & 0 \\
0 & 0 & 1 & 0 & \ldots & 0 \\
\vdots & \vdots & \vdots & \vdots & \vdots & 0 \\
-\frac{a_{n-1}}{a_n} & -\frac{a_{n-2}}{a_n} & -\frac{a_{n-3}}{a_n} & -\frac{a_{n-4}}{a_n} & \ldots & -\frac{a_0}{a_n}
\end{bmatrix},
\quad B = \begin{bmatrix}
0 & \cdots & 0 \\
\vdots & \ddots & \vdots \\
0 & \ldots & 0 \\
\frac{b_0}{a_n} & \ldots & \frac{b_m}{a_n}
\end{bmatrix}
\tag{1.39}
$$

With the given of the input $u(t)$, we can determine the second term of the right member in (1.32), leading to a new state representation where the matrix B would be a simple column matrix of the same size as the state vector $x(t)$. Indeed, by posing

$$
\sum_{j=0}^{m} b_j D^{\beta_j} u(t) = e(t) \tag{1.40}
$$

the representation (1.39) becomes

$$D^{\alpha}x(t) = Ax(t) + Be(t),\ B = \begin{bmatrix}0\,0\ldots0\,1\end{bmatrix}^{\mathrm{T}} \in \mathbb{R}^{n\times 1} \tag{1.41}$$

For a fractional order system described by n-term fractional differential equation

$$\sum_{i=0}^{n} a_i D^{\alpha_i} y(t) = e(t) \tag{1.42}$$

The legitimate choice of the derivative order $\alpha_0 = 0$ of the output corresponding to the coefficient a_0 allows us to establish a state equation augmented by an output equation of the form $y(t) = Cx(t)$, hence the following realization is obtained

$$\begin{aligned} D^{\alpha}x(t) &= Ax(t) + Be(t) \\ y(t) &= Cx(t) \end{aligned} \tag{1.43}$$

In the case of commensurate systems, the differentiation fractional-order $\alpha_i - \alpha_{i-1}$ are equal to a unique value, say α. In this case, (1.43) becomes

$$\begin{aligned} D^{\alpha}x(t) &= \begin{bmatrix} D^{\alpha}x_1(t) \\ D^{\alpha}x_2(t) \\ \vdots \\ D^{\alpha}x_n(t) \end{bmatrix} = Ax(t) + Be(t) \\ y(t) &= Cx(t) \end{aligned} \tag{1.44}$$

Applying the Laplace transform to (1.32) with zero initial conditions, the input-output representations of fractional-order systems can be obtained. In the case of continuous models, a fractional-order system will be given by a transfer function of the form

$$G(s) = \frac{Y(s)}{U(s)} = \frac{b_m s^{\beta_m} + b_{m-1}s^{\beta_{m-1}} \ldots + b_0 s^{\beta_0}}{a_n s^{\alpha_n} + a_{n-1}s^{\alpha_{n-1}} \ldots + a_0 s^{\alpha_0}} \tag{1.45}$$

As can be seen in the previous equations, a fractional-order system has an irrational-order transfer function in Laplace' s domain. Because of this, it can be said that a fractional-order system has an unlimited memory, and obviously the systems of integer-order are just particular cases.

In the case of a commensurate-order system, the continuous-time transfer function is given by

$$G(s) = \frac{\sum_{k=0}^{m} b_k (s^{\alpha})^k}{\sum_{k=0}^{n} a_k (s^{\alpha})^k} \tag{1.46}$$

which can be considered as a pseudo-rational function, $H(\lambda)$, of the variable $\lambda = s^\alpha$,

$$H(\lambda) = \frac{\sum_{k=0}^{m} b_k \lambda^k}{\sum_{k=0}^{n} a_k \lambda^k} \tag{1.47}$$

In this monograph, the behavior of the fractional systems studied will be approached by commensurate models with the properties of linearity and invariance in continuous-time where the Caputo's fractional derivative is suitable.

1.3.1 Temporal Response of Fractional Continuous-Time Linear System

Consider the continuous-time linear system described by the equation

$$D^\alpha x(t) = Ax(t) + Bu(t), 0 < \alpha \leq 1 \tag{1.48a}$$

$$y(t) = Cx(t) + Du(t) \tag{1.48b}$$

where $x(t) \in \mathbb{R}^n$, $u(t) \in \mathbb{R}^m$, $y(t) \in \mathbb{R}^p$ are the state, input and output vectors and $A \in \mathbb{R}^{n\times n}$, $B \in \mathbb{R}^{n\times m}$, $C \in \mathbb{R}^{p\times n}$, $D \in \mathbb{R}^{p\times m}$.

Theorem 1.3.1 *[19] The solution of the Eq.* (1.48a) *has the form*

$$x(t) = \Phi_0(t)x_0 + \int_0^t \Phi(t-\tau)Bu(\tau)d\tau,\ x(0) = x_0 \tag{1.49}$$

where

$$\Phi_0(t) = E_\alpha(At^\alpha) = \sum_{k=0}^{\infty} \frac{A^k t^{k\alpha}}{\Gamma(k\alpha + 1)} \tag{1.50}$$

$$\Phi(t) = \sum_{k=0}^{\infty} \frac{A^k t^{(k+1)\alpha - 1}}{\Gamma((k+1)\alpha)} \tag{1.51}$$

$E_\alpha(At^\alpha)$ *is the Mittag-Leffler function to one parameter and* $\Gamma(x)$ *is the Euler gamma function.*

Proof Applying the Laplace transform (1.19) to (1.48a) and taking in to account (1.29) and that $0 < \alpha \leq 1$, we obtain

$$s^\alpha X(s) - s^{\alpha-1}x_0 = AX(s) + BU(s) \tag{1.52}$$

where

$$X(s) = \mathcal{L}(x(t)),\ U(s) = \mathcal{L}(u(t))$$

Then we get

$$(s^{\alpha} I_n - A)X(s) = s^{\alpha-1} x_0 I_n + BU(s) \tag{1.53}$$

Since the pencil of matrices $s^{\alpha} I_n - A$ is regular then (1.53) implies that

$$X(s) = (s^{\alpha} I_n - A)^{-1}(s^{\alpha-1} x_0 I_n + BU(s)) \tag{1.54}$$

As

$$\begin{aligned}(s^{\alpha} I_n - A)(\sum_{k=0}^{\infty} \tfrac{A^k}{s^{(k+1)\alpha}}) &= \sum_{k=0}^{\infty} \tfrac{A^k s^{\alpha}}{s^{(k+1)\alpha}} - \sum_{k=0}^{\infty} \tfrac{A^{k+1}}{s^{(k+1)\alpha}} \\ &= \sum_{k=0}^{\infty} \tfrac{A^k}{s^{k\alpha}} - \sum_{k=0}^{\infty} \tfrac{A^{k+1}}{s^{(k+1)\alpha}} \\ &= \tfrac{A^0}{s^0} + \sum_{k=1}^{\infty} \tfrac{A^k}{s^{k\alpha}} - \sum_{h=1}^{\infty} \tfrac{A^h}{s^{h\alpha}} \\ &= I_n\end{aligned}$$

then

$$(s^{\alpha} I_n - A)^{-1} = \sum_{k=0}^{\infty} \frac{A^k}{s^{(k+1)\alpha}} \tag{1.55}$$

by substitution of (1.55) in (1.54), we get

$$X(s) = \sum_{k=0}^{\infty} \frac{A^k}{s^{k\alpha+1}} x_0 + \sum_{k=0}^{\infty} \frac{A^k}{s^{(k+1)\alpha}} BU(s) \tag{1.56}$$

Applying the inverse Laplace transform and taking into account its linearity, the Eq. (1.56) is written as

$$\mathcal{L}^{-1}(X(s)) = \sum_{k=0}^{\infty} A^k \mathcal{L}^{-1}\left(\frac{1}{s^{k\alpha+1}}\right) x_0 + \sum_{k=0}^{\infty} A^k B \mathcal{L}^{-1}\left(\frac{U(s)}{s^{(k+1)\alpha}}\right) \tag{1.57}$$

According to to the effect of the Laplace transform on the power function in (1.26) as well as on the fractional integral in (1.25), the Eq. (1.57) becomes

$$\begin{aligned}x(t) &= \sum_{k=0}^{\infty} A^k \tfrac{t^{k\alpha}}{\Gamma(k\alpha+1)} x_0 + \sum_{k=0}^{\infty} A^k B J^{(k+1)\alpha} u(t) \\ &= \sum_{k=0}^{\infty} \tfrac{A^k t^{k\alpha}}{\Gamma(k\alpha+1)} x_0 + \sum_{k=0}^{\infty} A^k B \tfrac{1}{\Gamma((k+1)\alpha)} \int_0^t (t-\tau)^{(k+1)\alpha-1} u(\tau) d\tau \\ &= \sum_{k=0}^{\infty} \tfrac{A^k t^{k\alpha}}{\Gamma(k\alpha+1)} x_0 + \int_0^t \sum_{k=0}^{\infty} \tfrac{A^k (t-\tau)^{(k+1)\alpha-1}}{\Gamma((k+1)\alpha)} Bu(\tau) d\tau\end{aligned} \tag{1.58}$$

Defining

$$\Phi_0(t) = \sum_{k=0}^{\infty} \frac{A^k t^{k\alpha}}{\Gamma(k\alpha + 1)}$$

and

$$\Phi(t) = \sum_{k=0}^{\infty} \frac{A^k (t)^{(k+1)\alpha-1}}{\Gamma((k+1)\alpha)}$$

the state response in (1.49) is derived. □

In the same way, the state response is obtained for $m-1 < \alpha \le m$.

Theorem 1.3.2 *[20] The solution of the Eq.* (1.48a) *for* $m-1 < \alpha \le m$ *has the form*

$$x(t) = \sum_{h=1}^{m} \Phi_h(t) x^{(m-h)}(0^+) + \int_0^t \Phi(t-\tau) B u(\tau) d\tau \tag{1.59}$$

where

$$\Phi_h(t) = \sum_{k=0}^{\infty} \frac{A^k t^{(k\alpha+h)-1}}{\Gamma(k\alpha + h)}, \quad \Phi(t) = \sum_{k=0}^{\infty} \frac{A^k t^{(k+1)\alpha-1}}{\Gamma((k+1)\alpha)}$$

Proof Applying the Laplace transform to (1.48a) while considering (1.29) for $m-1 < \alpha \le m$ yields

$$s^\alpha X(s) - \sum_{h=1}^{m} s^{\alpha-h} x^{(m-h)}(0^+) = AX(s) + BU(s) \tag{1.60}$$

where

$$X(s) = \mathcal{L}(x(t)), \ U(s) = \mathcal{L}(u(t))$$

Using (1.55), the Eq. (1.60) can be written as

$$\begin{aligned} X(s) &= \sum_{k=0}^{\infty} \frac{A^k}{s^{(k+1)\alpha}} \left(\sum_{h=1}^{m} s^{\alpha-h} x^{(m-h)}(0^+) + BU(s) \right) \\ &= \sum_{k=0}^{\infty} \sum_{h=1}^{m} \frac{A^k}{s^{(k\alpha+h)}} x^{(m-h)}(0^+) + \sum_{k=0}^{\infty} \frac{A^k}{s^{(k+1)\alpha}} BU(s) \\ &= \sum_{h=1}^{m} \sum_{k=0}^{\infty} \frac{A^k}{s^{(k\alpha+h)}} x^{(m-h)}(0^+) + \sum_{k=0}^{\infty} \frac{A^k}{s^{(k+1)\alpha}} BU(s) \end{aligned} \tag{1.61}$$

Applying the inverse Laplace transform enables us to obtain

$$x(t) = \sum_{h=1}^{m} \Phi_h(t) x^{(m-h)}(0^+) + \int_0^t \Phi(t-\tau) B u(\tau) d\tau \tag{1.62}$$

with

$$\Phi_h(t) = \sum_{k=0}^{\infty} A^k \mathcal{L}^{-1}\left(\frac{1}{s^{(k\alpha+h)}}\right) = \sum_{k=0}^{\infty} \frac{A^k t^{(k\alpha+h)-1}}{\Gamma(k\alpha+h)}$$

and

$$\Phi(t) = \sum_{k=0}^{\infty} \frac{A^k (t)^{(k+1)\alpha-1}}{\Gamma((k+1)\alpha)}$$

□

Remark 1.3.1 As the fractional-order derivative of x depends on the "history" of x from the lower limit of the integral t_0 that defines this operator until the present instant t, the knowledge of $x(t_0)$ is not sufficient to determine the future behavior of the system [21]. Consequently, vector $x(t)$ does not strictly represent the state of the system. Then we should denote $x(t)$ as a pseudo-state. However, we call this vector a state only for lexical simplifying purposes.

1.4 Summary

The aim of this chapter has been to provide the reader with the essentials of the fractional calculus. Some basic theory of special functions which play the most important role in the theory of differentiation of arbitrary order and in the theory of fractional differential equations. Then, several approaches to generalization of the notion of differentiation and integration are considered. The choice has been reduced to those definitions which are related to applications. As an introduction to the fractional-order control, a brief study of the input-output representation of the fractional continuous-time linear systems is established, then the temporal solutions of these systems are derived by using the Laplace transform.

References

1. Leibniz, G.W.: Mathematische shiften. Georg Olms Verlagsbuchhandlung, Hildsheim (1962)
2. Letnikov, A.V.: Theory of differentiation of an arbitrary order. Mat. Sb. **3**, 1–68 (1968)
3. Ehgheta, N.: On fractional calculus and fractional multipoles in electromagnetism. IEEE Trans. Antennas Propag. **44**, 554–566 (1996)
4. Sun, H., Abdelwahad, A., Onaral, B.: Linear approximation of transfer function with a pole of fractional order. IEEE Trans. Autom. Control **29**, 441–444 (1984)
5. Bagley, R., Calico, R.: Fractional order state equations for control of viscoelastically damped structures. J. Guidance, Control Dyn. **14**, 304–311 (1991)
6. Erdélyi, A. (Ed.).: Higher Transcendental Functions, vol. 3. McGraw Hill, New York (1955)

7. Mittag-Leffler, G.M.: Sur la nouvelle fonction $e_\alpha(x)$. C. R. Académie des Sciences **137**, 554–558 (1903)
8. Mittag-Leffler, G.M.: Sopra la funzione $e_\alpha(x)$. Red. Acc. Lincei **13**, 3–5 (1904)
9. Mittag-Leffler, G.M.: Sur la representation analytique d'une branche uniforme d'une fonction homogene. Acta Mathematica **29**, 101–182 (1905)
10. Wiman, A.: Über. den fundamentalsatz in der teorie der funktionen $e_\alpha(x)$. Acta Math. **29**, 191–201 (1905)
11. Agarwal, R.P.: A propos de notes de m. pierre humbert. c. r. Séances Acd. Sci. **236**, 2031–2032 (1953)
12. Samko, S.G., Kilbas, A.A., Marichev, O.I.: Fractional Integrals and Derivatives: Theory and Applications. Gordon and Breach Science Publisher, Amsterdam, The Netherlands (1993)
13. Podlubny, I.: Fractional Differential Equations. Academic Press, New York (1999)
14. Oldham, K.B., Spanier, J.: The Fractional Calculus. Academic Press, New York-London (1974)
15. Kulish, V.V., José, L.: Application of fractional calculus to fluid mechanics. J. Fluids Eng. **124**, 803–806 (2002)
16. Margulies, T.: Wave propagation in viscoelastic horns using a fractional calculus rheology model. Acoust. Soc. Am. J. **114**, 2442–2443 (2003)
17. Mathieu, B., Melchior, P., Oustaloup, A., Ceyral, C.: Fractional differentiation for edge detection. Fract. Signal Process. Appl. **83**, 2285–2480 (2003)
18. Suárez, J.I., Vinagre, B.M., Calderón, A.J., Monje, C.A., Chen, Y.Q.: Using fractional calculus for lateral and longitudinal control of autonomous vehicles. Lecture Notes in Computer Science, p. 2809. Springer, Berlin/Heidelberg (2004)
19. Kaczorek, T.: Singular fractional linear systems and electrical circuits. Int. J. Math. Comput. Sci. **21**, 379–384 (2011)
20. Kaczorek, T.: Selected problems of fractional systems theory. Lecture Notes in Control and Information Sciences. Springer, Heidelberg, Berlin, Germany (2011). https://doi.org/10.1007/978-3-642-20502-6
21. Lorenzo, D.F., Hartley, T.T.: Initialization in fractional order systems. In: European Control Conference, Porto, Portugal, pp. 1471–1476 (2001)

Chapter 2
Stability Analysis of Fractional-Order Systems

2.1 Introduction

The stability of fractional-order linear systems is a fundamental topic in control theory, given their ability to model complex dynamical behaviors more accurately than integer-order systems. Fractional-order systems are characterized by differential equations with non-integer derivatives, which provide a more flexible framework to describe processes such as viscoelasticity, diffusion, and memory effects in various physical and engineering applications.

Stability analysis for these systems, however, poses unique challenges. Unlike classical linear systems, where stability is determined by the eigenvalues of the system matrix, fractional-order systems require additional considerations, such as the order of the derivatives and their impact on the system's dynamic response. This makes the study of stability conditions critical for understanding and designing robust fractional-order systems in both theoretical and practical contexts.

2.2 Stability Region of Fractional-Order Systems ($0 < \alpha < 2$)

As in classical calculus, stability analysis is a central task in the study of fractional differential system and fractional control. As for linear time invariant integer order systems, it is now well known that stability of a linear fractional order system depends on the location of the system poles in the complex plane.

Theorem 2.2.1 *[1] A fractional order system defined by its transfer function*

$$G(s) = \frac{Q(s)}{P(s)} \tag{2.1}$$

S. Marir and M. Chadli, *Control of Singular Fractional Order Systems: LMI Approach*, Studies in Systems, Decision and Control 292,
https://doi.org/10.1007/978-3-031-87382-9_2

for $\mathrm{Re}(s) \geq 0$*, where*

$$P(s) = \sum_{k=0}^{p} P_k s^{\alpha_k}$$

with $\alpha_{k+1} > \alpha_k \geq 0$ *and*

$$Q(s) = \sum_{l=0}^{q} q_l s^{\beta_l}$$

with $\beta_{l+1} > \beta_l \geq 0$ *are no longer polynomials.*

The system has the main property of:

$$BIBO\ stability \Leftrightarrow \exists M > 0, \|G(s)\| \leq M, \forall s, \mathrm{Re}(s) \geq 0$$

Moreover, in the case where no simplification occurs between P *and* Q*, that is all the roots of* $P(s) = 0$ *not being roots of* $Q(s) = 0$*, the stability property then reads:*

$$BIBO\ stability \Leftrightarrow P(s) \neq 0, \forall s, \mathrm{Re}(s) \geq 0$$

This statement appears for the first time in [1] as Theorem 2.24 (conjecture), which was fully proved, and solved later in [2]. However such a theorem does not permit to conclude to system stability without system poles computation, which constitutes a tedious work. However, poles location analysis remains a difficult task in the general case. For commensurate fractional order systems, powerful criteria have been proposed. Using commensurate order hypothesis, the system (2.1) also admits a state space like representation:

$$\begin{aligned} D^{\alpha}x(t) &= Ax(t) + Bu(t) \\ y(t) &= Cx(t) \end{aligned} \tag{2.2}$$

In [1], the given result permits to check the system stability through the location in the complex plane of the dynamic matrix eigenvalues. This work is in fact the starting point of several results in the field. For $1 \leq \alpha < 2$, stability condition is given in [3].

Theorem 2.2.2 *[1, 4, 5] Autonomous system:*

$$D^{\alpha}x(t) = Ax(t), x(0) = x_0, \ 0 < \alpha < 2 \tag{2.3}$$

is asymptotically stable if and only if

$$|\arg(\mathrm{spec}(A))| > \alpha\frac{\pi}{2} \tag{2.4}$$

where $\mathrm{spec}(A)$ *is the spectrum (set of all eigenvalues) of* A*.*

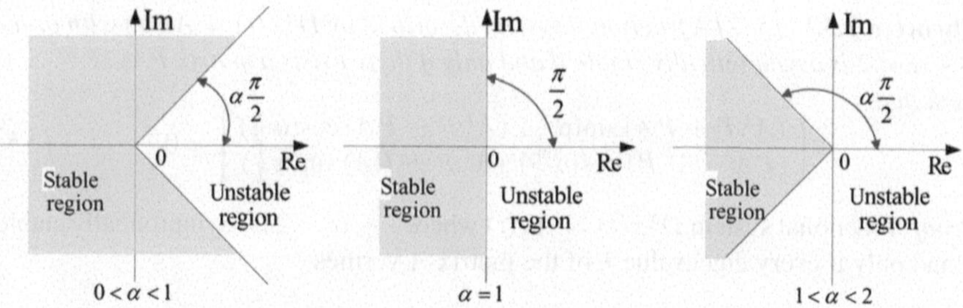

Fig. 2.1 Stable regions D_s^α for $0 < \alpha < 1$, $\alpha = 1$ and $1 < \alpha < 2$

For a minimal realization[1] of (2.2), in [4], the following result has been also demonstrated.

Theorem 2.2.3 *[4] If the triplet* (A, B, C) *is minimal,*[2] *system* (2.2) *is BIBO stable if and only if* $|\arg(\text{spec}(A))| > \alpha\frac{\pi}{2}$

Based on the previous results, stable regions $D_s^\alpha = \left\{\lambda/\lambda \in \mathbb{C}, |\arg(\lambda)| > \alpha\frac{\pi}{2}\right\}$ for a fractional system depending on its differentiation order α and on the value of $|\arg(\text{spec}(A))|$ are illustrated in Fig. 2.1.

The location of the eigenvalues of a matrix in a particular region of the complex plane can be solved using the formalism of $\mathcal{LMI}$ regions introduced in [6] which are the basis of the stability criteria in terms of $\mathcal{LMI}$ for non-integer models presented in this section. A recall on the $\mathcal{LMI}$ areas is presented in Annexe B.

In the first paragraph, the case of order belonging $1 \leq \alpha < 2$ is considered. The stability domain associated is then convex and can be described using $\mathcal{LMI}$ regions. The case $0 < \alpha \leq 1$ is then discussed. In this case, the stability domain is not convex and can not be described by $\mathcal{LMI}$ areas.

2.3 $\mathcal{LMI}$ Stability Conditions: Case $1 \leq \alpha < 2$

As can be shown in Fig. 2.1, the stability domain of a fractional system with order $1 \leq \alpha < 2$ is a convex set, $\mathcal{LMI}$ methods for defining such a region can be used. Hence a $\mathcal{LMI}$-based sufficient and necessary condition for the stability can be formulated in the following theorem.

---

[1] A state space model (A, B, C, D) is a realization of a transfer function $G(s)$ if its transfer function coincides with G(s).

[2] A realization (A, B, C, D) of a transfer function/matrix $G(s)$ is said to be minimal if no other realization of $G(s)$ has smaller dimension.

Theorem 2.3.1 *[5, 7] A fractional system described by $D^\alpha x(t) = Ax(t)$ with order $1 \leq \alpha < 2$ is asymptotically stable if and only if there exists a matrix $P \in \mathbb{R}^{n \times n} > 0$ such that*

$$\begin{bmatrix} (A^T P + PA)\sin(\alpha\frac{\pi}{2}) & (A^T P - PA)\cos(\alpha\frac{\pi}{2}) \\ (PA - A^T P)\cos(\alpha\frac{\pi}{2}) & (A^T P + PA)\sin(\alpha\frac{\pi}{2}) \end{bmatrix} < 0 \tag{2.5}$$

Proof Fractional system $D^\alpha x(t) = Ax(t)$ where $1 \leq \alpha < 2$ is asymptotically stable if and only if every eigenvalue λ of the matrix A verifies

$$|\arg(\lambda)| > \alpha\frac{\pi}{2}$$

that is if and only if λ belongs to $D_s^\alpha = D_{s_1} \cap D_{s_2}$, where D_{s_1} and D_{s_2} are defined from rotation of λ of angles $(\alpha - 1)\frac{\pi}{2}$ and $(1 - \alpha)\frac{\pi}{2}$ respectively and given by

$$D_{s_1} = \left\{\lambda \in \mathbb{C} : \mathrm{Re}(\lambda e^{j(\alpha-1)\frac{\pi}{2}}) < 0\right\} \tag{2.6}$$

$$D_{s_2} = \left\{\lambda \in \mathbb{C} : \mathrm{Re}(\lambda e^{j(1-\alpha)\frac{\pi}{2}}) < 0\right\} \tag{2.7}$$

As for $\lambda \in \mathrm{spec}(A)$, $\lambda^* \in \mathrm{spec}(A)$ and D_{s_1} and D_{s_2} are symmetric with respect to the real axis of the complex plane,

$$(\exists\lambda_1 \in \mathrm{spec}(A), \lambda_1 \in D_{s_1}) \Leftrightarrow (\exists\lambda_2 \in \mathrm{spec}(A), \lambda_2 \in D_{s_2})$$

only relation (2.6) is necessary for our purpose.

As for any $z \in \mathbb{C}$, $\mathrm{Re}(z) = \frac{z+z^*}{2}$, relation (2.6) reduces to

$$\lambda e^{j(1-\alpha)\frac{\pi}{2}} + \lambda^* e^{-j(1-\alpha)\frac{\pi}{2}} < 0 \tag{2.8}$$

or to

$$\lambda(\cos((1-\alpha)\frac{\pi}{2}) + j\sin((1-\alpha)\frac{\pi}{2})) + \lambda^*(\cos((\alpha-1)\frac{\pi}{2}) + j\sin((\alpha-1)\frac{\pi}{2})) < 0 \tag{2.9}$$

The inequality (2.9) can be rewritten as

$$\mathrm{Re}(\lambda)\sin(\alpha\frac{\pi}{2}) - \mathrm{Im}(\lambda)\cos(\alpha\frac{\pi}{2}) < 0 \tag{2.10}$$

From [8] (see Appendix B.4, conic sector), relation (2.10) is verified if and only if the following $\mathcal{LMI}$ feasibility problem is verified:

$$\exists P \in \mathbb{R}^{n\times n} > 0, (A^{\mathrm{T}}P + PA)\sin(\alpha\frac{\pi}{2}) + j(A^{\mathrm{T}}P - PA)\cos(\alpha\frac{\pi}{2}) < 0 \tag{2.11}$$

As an $\mathcal{LMI}$ involving real terms can be verified from a complex one, the problem becomes
$\exists P \in \mathbb{R}^{n\times n} > 0$:

$$\begin{bmatrix} (A^{\mathrm{T}}P + PA)\sin(\alpha\frac{\pi}{2}) & (A^{\mathrm{T}}P - PA)\cos(\alpha\frac{\pi}{2}) \\ (PA - A^{\mathrm{T}}P)\cos(\alpha\frac{\pi}{2}) & (A^{\mathrm{T}}P + PA)\sin(\alpha\frac{\pi}{2}) \end{bmatrix} < 0 \tag{2.12}$$

This completes the proof. □

Remark 2.3.1 In what follows, variants of the condition (2.12) will be presented.

- Using the Kronecker product, (2.12) can be written as

$$\mathrm{sym}\left\{\Theta \otimes (A^{\mathrm{T}}P)\right\} < 0 \tag{2.13}$$

  where $\Theta = \begin{bmatrix} \sin(\alpha\frac{\pi}{2}) & \cos(\alpha\frac{\pi}{2}) \\ -\cos(\alpha\frac{\pi}{2}) & \sin(\alpha\frac{\pi}{2}) \end{bmatrix}$

- By pre- and post-multiplying the formula (2.12) by $\begin{bmatrix} P^{-1} & 0 \\ 0 & P^{-1} \end{bmatrix}$, an equivalent condition for the stability is then obtained
  $\exists P \in \mathbb{R}^{n\times n} > 0$

$$\begin{bmatrix} (PA^{\mathrm{T}} + AP)\sin(\alpha\frac{\pi}{2}) & (PA^{\mathrm{T}} - AP)\cos(\alpha\frac{\pi}{2}) \\ (AP - PA^{\mathrm{T}})\cos(\alpha\frac{\pi}{2}) & (PA^{\mathrm{T}} + AP)\sin(\alpha\frac{\pi}{2}) \end{bmatrix} < 0 \tag{2.14}$$

- The formula (2.14) can also be written as

$$\mathrm{sym}\left\{\Theta^{\mathrm{T}} \otimes (AP)\right\} < 0 \tag{2.15}$$

- Since for any complex matrix $M < 0$, its conjugate verifies $\bar{M} < 0$, (2.11) is equivalent to

$$(A^{\mathrm{T}}P + PA)\sin(\alpha\frac{\pi}{2}) - j(A^{\mathrm{T}}P - PA)\cos(\alpha\frac{\pi}{2}) < 0 \tag{2.16}$$

  which can be expressed as

$$\mathrm{sym}\left\{\Theta^{\mathrm{T}} \otimes (A^{\mathrm{T}}P)\right\} < 0 \tag{2.17}$$

2.4 $\mathcal{LMI}$ Stability Conditions: Case $0 < \alpha \leq 1$

Figure 2.1 shows that the stability domain D_s^α of a linear fractional system is not convex when $0 < \alpha < 1$. Due to the absence of the convexity property, the $\mathcal{LMI}$ conditions can not be derived directly as in the case of integer order or fractional order with $1 \leq \alpha < 2$. However, in literature different approaches are suggested to by pass this problem and $\mathcal{LMI}$ condition are derived indirectly. Some of the well recognized results which exist in literature are discussed here. An analysis via stability domain decomposition was proposed in [7] This analysis is based on the fact that the stability domain D_s^α given by (2.4) can be viewed as the union of two half planes, denoted D_{s_1} and D_{s_2}. They result from the rotation of the left-half plane with angles $\varphi_1 = \varphi$ and $\varphi_2 = -\varphi$ respectively, where $\varphi = (1-\alpha)\frac{\pi}{2}$, as shown in Fig. 2.2.

Consequently, stability region D_s^α can be defined by

$$D_s^\alpha = D_{s_1} \cup D_{s_2} \tag{2.18}$$

where

$$D_{s_i} = \left\{z \in \mathbb{C} : \operatorname{Re}(ze^{j\varphi_i}) < 0\right\}, \forall i \in \{1, 2\} \tag{2.19}$$

Since D_{s_1} and D_{s_2} are not symmetrical with respect to the real axis, they do not constitute $\mathcal{LMI}$ regions. The formalism introduced in [6] and developed in [9] permits to deal with this case using the concept of generalized $\mathcal{LMI}$ regions $\mathcal{GLMI}$, initially extended in the case of fractional order systems in [7]. In order not overburdening this section, a recall on the $\mathcal{GLMI}$ region, the application of the D-stability in such region have been developed in Appendix C. A necessary and sufficient condition in terms of $\mathcal{LMI}$ for stability of a non-integer system with order $0 < \alpha < 1$ can thus be obtained by using Theorem C.2.1, see Appendix C.2, on $\mathcal{GLMI}$ region D of the complex plane,

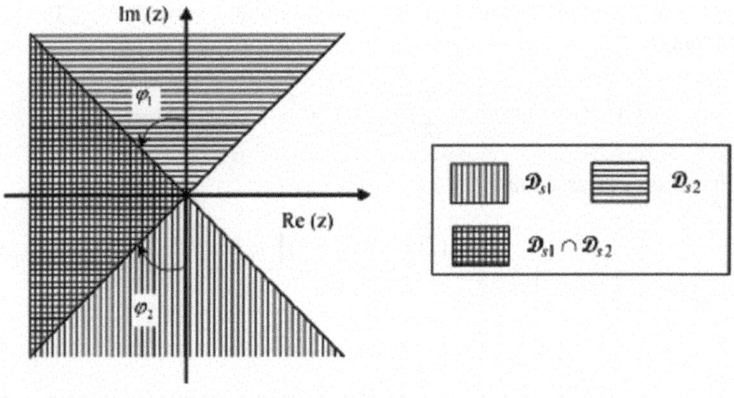

Fig. 2.2 Stability region as a union of two half planes

Lemma 2.4.1 *[10] Let $A \in \mathbb{R}^{n\times n}$, $0 < \alpha \leq 1$ and $\theta = (1-\alpha)\frac{\pi}{2}$. The fractional-order system $D^\alpha x(t) = Ax(t)$ is asymptotically stable if and only if there exist two positive definite Hermitian matrices $X_1 = X_1^* \in \mathbb{C}^{n\times n}$, $X_2 = X_2^* \in \mathbb{C}^{n\times n}$ such that*

$$e^{-j\theta}X_1A^T + e^{j\theta}AX_1 + e^{j\theta}X_2A^T + e^{-j\theta}AX_2 < 0 \tag{2.20}$$

or

$$e^{-j\theta}X_1A + e^{j\theta}A^TX_1 + e^{j\theta}X_2A + e^{-j\theta}A^TX_2 < 0 \tag{2.21}$$

Proof As each domain $\mathcal{D}_{s_i}$, $i \in \{1, 2\}$ has the form (C.10), then it is a $\mathcal{GLMI}$ region of first order with $m = 1$, $\alpha_i = 0$, $\beta_i = e^{j\varphi_i}$ and $\mathcal{D}_s$ has the form (C.13), based on the recall made in Appendix C.3, we deduce that $\mathcal{D}_s$ is a $\mathcal{GLMI}$ region of order $l = m + 1 = 3$ characterized by

$$\theta_1 = \frac{1}{2}\begin{bmatrix}\alpha_1 & 0 & 0\\ 0 & -1 & 0\\ 0 & 0 & 0\end{bmatrix}, \ \psi_1 = \begin{bmatrix}\beta_1 & 0 & 0\\ 0 & 0 & 0\\ 0 & 0 & 0\end{bmatrix}, \ H_1 = -J_1 = \begin{bmatrix}0 & 0 & 0\\ 0 & 1 & 0\\ 0 & 0 & 0\end{bmatrix}$$

$$\theta_2 = \frac{1}{2}\begin{bmatrix}\alpha_2 & 0 & 0\\ 0 & 0 & 0\\ 0 & 0 & -1\end{bmatrix}, \ \psi_2 = \begin{bmatrix}\beta_2 & 0 & 0\\ 0 & 0 & 0\\ 0 & 0 & 0\end{bmatrix}, \ H_2 = -J_2 = \begin{bmatrix}0 & 0 & 0\\ 0 & 0 & 0\\ 0 & 0 & 1\end{bmatrix}$$

and

$$\omega = \begin{bmatrix}1\\ 1\\ 1\end{bmatrix}$$

It is easy to see that conditions (C.1), (C.2) and (C.3) are verified. This allows us to apply the Theorem C.2.1, to get necessary and sufficient condition for the stability of the matrix A in the region $\mathcal{D}_s$. Therefore, there exist two matrices $X_1 \in \mathbb{C}^{n\times n}$ and $X_2 \in \mathbb{C}^{n\times n}$ such that (C.4) and (C.5) hold.

Condition (C.4) becomes

$$\begin{aligned}&\theta_1 \otimes X_1 + \theta_1^* \otimes X_1^* + \psi_1 \otimes (AX_1) + \psi_1^* \otimes (AX_1)^*\\ &+\theta_2 \otimes X_2 + \theta_2^* \otimes X_2^* + \psi_2 \otimes (AX_2) + \psi_2^* \otimes (AX_2)^* < 0\end{aligned}$$

After few computations and taking into account that $\varphi_1 = \varphi$ and $\varphi_2 = -\varphi$ we obtain

$$\begin{bmatrix} e^{j\varphi}AX_1 + e^{-j\varphi}(AX_1)^* + e^{-j\varphi}AX_2 + e^{j\varphi}(AX_2)^* & 0 & 0\\ 0 & -\frac{1}{2}(X_1 + X_1^*) & 0\\ 0 & 0 & -\frac{1}{2}(X_2 + X_2^*)\end{bmatrix} < 0$$

Then we deduce

$$e^{j\varphi}AX_1 + e^{-j\varphi}(AX_1)^* + e^{-j\varphi}AX_2 + e^{j\varphi}(AX_2)^* < 0$$

We can also deduce that

$$X_1 + X_1^* > 0$$

and

$$X_2 + X_2^* > 0$$

In the other hand, Condition (C.5) becomes

$$H_1 \otimes X_1 + J_1 \otimes X_1^* + H_2 \otimes X_2 + J_2 \otimes X_2^* = 0_{3n}$$

After computation, we get

$$\begin{bmatrix} 0_n & 0_n & 0_n \\ 0_n & X_1 - X_1^* & 0_n \\ 0_n & 0_n & X_2 - X_2^* \end{bmatrix} = 0_{3n}$$

which means that $X_1 = X_1^*$ and $X_2 = X_2^*$. Finally, we obtain that the matrices X_1 and X_2 are hermitian definite positive and (2.20) is derived. The proof is then complete. □

The following Lemma proposed in [11] presents another version of Lemma 2.4.1, where the complex $\mathcal{LMI}$ (2.20) is replaced by a real linear matrix inequality through the use of the Kronecker product.

Lemma 2.4.2 *[11] Let* $A \in \mathbb{R}^{n\times n}$ *and* $0 < \alpha < 1$. *The fractional-order system* $D^{\alpha}x(t) = Ax(t)$ *is asymptotically stable if and only if there exist two real symmetric positive definite matrices* $X_{i1} \in \mathbb{R}^{n\times n}, i = 1, 2,$ *and two skew-symmetric matrices* $X_{i2} \in \mathbb{R}^{n\times n}, i = 1, 2,$ *such that*

$$\begin{bmatrix} X_{11} & X_{12} \\ -X_{12} & X_{11} \end{bmatrix} > 0, \begin{bmatrix} X_{21} & X_{22} \\ -X_{22} & X_{21} \end{bmatrix} > 0 \tag{2.22}$$

$$\sum_{i=1}^{2}\sum_{j=1}^{2} \text{sym}\left\{\Theta_{ij} \otimes (AX_{ij})\right\} < 0 \tag{2.23}$$

where

$$\Theta_{11} = \begin{bmatrix} \sin(\alpha\frac{\pi}{2}) & -\cos(\alpha\frac{\pi}{2}) \\ \cos(\alpha\frac{\pi}{2}) & \sin(\alpha\frac{\pi}{2}) \end{bmatrix}$$

$$\Theta_{12} = \begin{bmatrix} \cos(\alpha\frac{\pi}{2}) & \sin(\alpha\frac{\pi}{2}) \\ -\sin(\alpha\frac{\pi}{2}) & \cos(\alpha\frac{\pi}{2}) \end{bmatrix}$$

$$\Theta_{21} = \begin{bmatrix} \sin(\alpha\frac{\pi}{2}) & \cos(\alpha\frac{\pi}{2}) \\ -\cos(\alpha\frac{\pi}{2}) & \sin(\alpha\frac{\pi}{2}) \end{bmatrix} \tag{2.24}$$

$$\Theta_{22} = \begin{bmatrix} -\cos(\alpha\frac{\pi}{2}) & \sin(\alpha\frac{\pi}{2}) \\ -\sin(\alpha\frac{\pi}{2}) & -\cos(\alpha\frac{\pi}{2}) \end{bmatrix}$$

To prove the Lemma 2.4.2, we need the following result.

Lemma 2.4.3 *For complex matrix $M \in \mathbb{C}^{n\times n}$, define*

$$M = A + jB,\ A, B \in \mathbb{R}^{n\times n}.$$

Then $M > 0$ if and only if

$$\begin{bmatrix} A & B \\ -B & A \end{bmatrix} > 0, \tag{2.25}$$

or equivalently

$$\begin{bmatrix} A & -B \\ B & A \end{bmatrix} > 0. \tag{2.26}$$

Proof see Appendix A.2. □

Proof From Lemma 2.4.1, it follows that the fractional-order system $D^\alpha x(t) = Ax(t)$ with $0 < \alpha < 1$ is asymptotically stable if and only if there exist two hermitian positive definite matrices X_i, $i = 1, 2$, such that the $\mathcal{LMI}$ (2.20) holds. Define $X_{i1} = \mathrm{Re}\,(X_i)$, $X_{i2} = \mathrm{Im}\,(X_i)$, $i = 1, 2$. From $X_i = X_{i1} + jX_{i2}$, $i = 1, 2$ and$X_i = X_i^* > 0$, it follows that the matrices X_{i1}, $i = 1, 2$ are symmetric definite positive and the matrices X_{i2}, $i = 1, 2$ are skew matrices.[3] According to Lemma 2.4.3, we have

$$\begin{bmatrix} X_{11} & X_{12} \\ -X_{12} & X_{11} \end{bmatrix} > 0, \begin{bmatrix} X_{21} & X_{22} \\ -X_{22} & X_{21} \end{bmatrix} > 0,$$

so the inequality (2.22) holds. Substituting in the inequality (2.20).

$$X_i = X_{i1} + jX_{i2}, i = 1, 2 \text{ and } e^{j\theta} = \sin(\alpha\frac{\pi}{2}) + j\cos(\alpha\frac{\pi}{2}),$$

[3] A skew-symmetric (or antisymmetric) matrix is a square matrix whose transpose is its negation; that is, it satisfies the condition $A^{\mathrm{T}} = -A$.

we obtain

$$
\begin{aligned}
&(\sin(\alpha\frac{\pi}{2}) + j\cos(\alpha\frac{\pi}{2}))(X_{11} + jX_{12})A^{\mathrm{T}} \\
&+(\sin(\alpha\frac{\pi}{2}) - j\cos(\alpha\frac{\pi}{2}))A(X_{11} + jX_{12}) \\
&+(\sin(\alpha\frac{\pi}{2}) - j\cos(\alpha\frac{\pi}{2}))(X_{21} + jX_{22})A^{\mathrm{T}} \\
&+(\sin(\alpha\frac{\pi}{2}) + j\cos(\alpha\frac{\pi}{2}))A(X_{21} + jX_{22}) < 0,
\end{aligned}
\tag{2.27}
$$

which can be rewritten as follows

$$
\begin{aligned}
&(\sin(\alpha\frac{\pi}{2})X_{11}A^{\mathrm{T}} - \cos(\alpha\frac{\pi}{2})X_{12}A^{\mathrm{T}}) + j(\sin(\alpha\frac{\pi}{2})X_{12}A^{\mathrm{T}} + \cos(\alpha\frac{\pi}{2})X_{11}A^{\mathrm{T}}) \\
&+(\sin(\alpha\frac{\pi}{2})AX_{11} + \cos(\alpha\frac{\pi}{2})AX_{12}) + j((\sin(\alpha\frac{\pi}{2})AX_{12} - \cos(\alpha\frac{\pi}{2})AX_{11}) \\
&+(\sin(\alpha\frac{\pi}{2})X_{21}A^{\mathrm{T}} + \cos(\alpha\frac{\pi}{2})X_{22}A^{\mathrm{T}}) + j(\sin(\alpha\frac{\pi}{2})X_{22}A^{\mathrm{T}} - \cos(\alpha\frac{\pi}{2})X_{21}A^{\mathrm{T}}) \\
&+(\sin(\alpha\frac{\pi}{2})AX_{21} - \cos(\alpha\frac{\pi}{2})AX_{22}) + j(\sin(\alpha\frac{\pi}{2})AX_{22} + \cos(\alpha\frac{\pi}{2})AX_{21}) < 0.
\end{aligned}
\tag{2.28}
$$

Using the fact that $X_{12}^{\mathrm{T}} = -X_{12}$ and $X_{22}^{\mathrm{T}} = -X_{22}$, (2.28) is equivalent to

$$
\begin{aligned}
&(\sin(\alpha\frac{\pi}{2})X_{11}A^{\mathrm{T}} + \cos(\alpha\frac{\pi}{2})X_{12}^{\mathrm{T}}A^{\mathrm{T}}) + j(-\sin(\alpha\frac{\pi}{2})X_{12}^{\mathrm{T}}A^{\mathrm{T}} + \cos(\alpha\frac{\pi}{2})X_{11}A^{\mathrm{T}}) \\
&+(\sin(\alpha\frac{\pi}{2})AX_{11} + \cos(\alpha\frac{\pi}{2})AX_{12}) + j(\sin(\alpha\frac{\pi}{2})AX_{12} - \cos(\alpha\frac{\pi}{2})AX_{11}) \\
&+(\sin(\alpha\frac{\pi}{2})X_{21}A^{\mathrm{T}} - \cos(\alpha\frac{\pi}{2})X_{22}^{\mathrm{T}}A^{\mathrm{T}}) + j(-\sin(\alpha\frac{\pi}{2})X_{22}^{\mathrm{T}}A^{\mathrm{T}} - \cos(\alpha\frac{\pi}{2})X_{21}A^{\mathrm{T}}) \\
&+(\sin(\alpha\frac{\pi}{2})AX_{21} - \cos(\alpha\frac{\pi}{2})AX_{22}) + j(\sin(\alpha\frac{\pi}{2})AX_{22} + \cos(\alpha\frac{\pi}{2})AX_{21}) < 0,
\end{aligned}
\tag{2.29}
$$

which is the same as

$$
\begin{aligned}
&\mathrm{sym}\left\{\sin(\alpha\frac{\pi}{2})AX_{11} + \cos(\alpha\frac{\pi}{2})AX_{12} + \sin(\alpha\frac{\pi}{2})AX_{21} - \cos(\alpha\frac{\pi}{2})AX_{22}\right\} \\
&+j(-\sin(\alpha\frac{\pi}{2})X_{12}^{\mathrm{T}}A^{\mathrm{T}} + \cos(\alpha\frac{\pi}{2})X_{11}A^{\mathrm{T}} + \sin(\alpha\frac{\pi}{2})AX_{12} - \cos(\alpha\frac{\pi}{2})AX_{11}) \\
&+j(-\sin(\alpha\frac{\pi}{2})X_{22}^{\mathrm{T}}A^{\mathrm{T}} - \cos(\alpha\frac{\pi}{2})X_{21}A^{\mathrm{T}} + \sin(\alpha\frac{\pi}{2})AX_{22} + \cos(\alpha\frac{\pi}{2})AX_{21}) < 0.
\end{aligned}
\tag{2.30}
$$

or identically

$$
\begin{aligned}
&\operatorname{sym}\left\{\sin(\alpha\frac{\pi}{2})AX_{11}\right\} + \operatorname{sym}\left\{\cos(\alpha\frac{\pi}{2})AX_{12}\right\} \\
&+ \operatorname{sym}\left\{\sin(\alpha\frac{\pi}{2})AX_{21}\right\} + \operatorname{sym}\left\{-\cos(\alpha\frac{\pi}{2})AX_{22}\right\} \\
&+ j\cos(\alpha\frac{\pi}{2})(X_{11}A^{\mathrm{T}} - AX_{11}) + j\sin(\alpha\frac{\pi}{2})(AX_{12} - X_{12}^{T}A^{\mathrm{T}}) \\
&+ j\cos(\alpha\frac{\pi}{2})(AX_{21} - X_{21}A^{\mathrm{T}}) + j\sin(\alpha\frac{\pi}{2})(AX_{22} - X_{22}^{\mathrm{T}}A^{\mathrm{T}}) < 0. \qquad (2.31)
\end{aligned}
$$

According to Lemma 2.4.3, inequality (2.31) can be expressed in real terms as follows

$$
\Xi = \begin{bmatrix} \Phi & \Psi \\ -\Psi & \Phi \end{bmatrix} < 0, \qquad (2.32)
$$

with

$$
\begin{aligned}
\Phi = {} & \operatorname{sym}\left\{\sin(\alpha\tfrac{\pi}{2})AX_{11}\right\} + \operatorname{sym}\left\{\cos(\alpha\tfrac{\pi}{2})AX_{12}\right\} \\
& + \operatorname{sym}\left\{\sin(\alpha\tfrac{\pi}{2})X_{21}AX_{21}\right\} + \operatorname{sym}\left\{-\cos(\alpha\tfrac{\pi}{2})AX_{22}\right\},
\end{aligned}
$$

$$
\begin{aligned}
\Psi = {} & \cos(\alpha\tfrac{\pi}{2})(X_{11}A^{\mathrm{T}} - AX_{11}) + \sin(\alpha\tfrac{\pi}{2})(AX_{12} - X_{12}^{T}A^{\mathrm{T}}) \\
& + \cos(\alpha\tfrac{\pi}{2})(AX_{21} - X_{21}A^{\mathrm{T}}) + \sin(\alpha\tfrac{\pi}{2})(AX_{22} - X_{22}^{\mathrm{T}}A^{\mathrm{T}}).
\end{aligned}
$$

So, it is easy to see that the matrix Ξ is the same as

$$
\Xi = \Xi_1 + \Xi_2 + \Xi_3 + \Xi_4,
$$

where

$$
\Xi_1 = \begin{bmatrix} \operatorname{sym}\left\{\sin(\alpha\frac{\pi}{2})AX_{11}\right\} & \cos(\alpha\frac{\pi}{2})(X_{11}A^{\mathrm{T}} - AX_{11}) \\ \cos(\alpha\frac{\pi}{2})(AX_{11} - X_{11}A^{\mathrm{T}}) & \operatorname{sym}\left\{\sin(\alpha\frac{\pi}{2})AX_{11}\right\} \end{bmatrix},
$$

$$
\Xi_2 = \begin{bmatrix} \operatorname{sym}\left\{\cos(\alpha\frac{\pi}{2})AX_{12}\right\} & \sin(\alpha\frac{\pi}{2})(AX_{12} - X_{12}^{T}A^{\mathrm{T}}) \\ \sin(\alpha\frac{\pi}{2})(X_{12}^{T}A^{\mathrm{T}} - AX_{12}) & \operatorname{sym}\left\{\cos(\alpha\frac{\pi}{2})AX_{12}\right\} \end{bmatrix},
$$

$$
\Xi_3 = \begin{bmatrix} \operatorname{sym}\left\{\sin(\alpha\frac{\pi}{2})AX_{21}\right\} & \cos(\alpha\frac{\pi}{2})(AX_{21} - X_{21}A^{\mathrm{T}}) \\ \cos(\alpha\frac{\pi}{2})(X_{21}A^{\mathrm{T}} - AX_{21}) & \operatorname{sym}\left\{\sin(\alpha\frac{\pi}{2})AX_{21}\right\} \end{bmatrix},
$$

$$
\Xi_4 = \begin{bmatrix} \operatorname{sym}\left\{-\cos(\alpha\frac{\pi}{2})AX_{22}\right\} & \sin(\alpha\frac{\pi}{2})(AX_{22} - X_{22}^{\mathrm{T}}A^{\mathrm{T}}) \\ \sin(\alpha\frac{\pi}{2})(X_{22}^{\mathrm{T}}A^{\mathrm{T}} - AX_{22}) & \operatorname{sym}\left\{-\cos(\alpha\frac{\pi}{2})AX_{22}\right\} \end{bmatrix}.
$$

With the help of the Kronecker product, the matrices $\Xi_i, i = 1, 2, 3, 4$ can be written as

$$\Xi_1 = \text{sym}\left\{\begin{bmatrix} \sin(\alpha\frac{\pi}{2}) & -\cos(\alpha\frac{\pi}{2}) \\ \cos(\alpha\frac{\pi}{2}) & \sin(\alpha\frac{\pi}{2}) \end{bmatrix} \otimes (AX_{11})\right\},$$

$$\Xi_2 = \text{sym}\left\{\begin{bmatrix} \cos(\alpha\frac{\pi}{2}) & \sin(\alpha\frac{\pi}{2}) \\ -\sin(\alpha\frac{\pi}{2}) & \cos(\alpha\frac{\pi}{2}) \end{bmatrix} \otimes (AX_{12})\right\},$$

$$\Xi_3 = \text{sym}\left\{\begin{bmatrix} \sin(\alpha\frac{\pi}{2}) & \cos(\alpha\frac{\pi}{2}) \\ -\cos(\alpha\frac{\pi}{2}) & \sin(\alpha\frac{\pi}{2}) \end{bmatrix} \otimes (AX_{21})\right\},$$

$$\Xi_4 = \text{sym}\left\{\begin{bmatrix} -\cos(\alpha\frac{\pi}{2}) & \sin(\alpha\frac{\pi}{2}) \\ -\sin(\alpha\frac{\pi}{2}) & cos(\alpha\frac{\pi}{2}) \end{bmatrix} \otimes (AX_{22})\right\}.$$

By defining

$$\Theta_{11} = \begin{bmatrix} \sin(\alpha\frac{\pi}{2}) & -\cos(\alpha\frac{\pi}{2}) \\ \cos(\alpha\frac{\pi}{2}) & \sin(\alpha\frac{\pi}{2}) \end{bmatrix}, \Theta_{12} = \begin{bmatrix} \cos(\alpha\frac{\pi}{2}) & \sin(\alpha\frac{\pi}{2}) \\ -\sin(\alpha\frac{\pi}{2}) & \cos(\alpha\frac{\pi}{2}) \end{bmatrix},$$
$$\Theta_{21} = \begin{bmatrix} \sin(\alpha\frac{\pi}{2}) & \cos(\alpha\frac{\pi}{2}) \\ -\cos(\alpha\frac{\pi}{2}) & \sin(\alpha\frac{\pi}{2}) \end{bmatrix}, \Theta_{22} = \begin{bmatrix} -\cos(\alpha\frac{\pi}{2}) & \sin(\alpha\frac{\pi}{2}) \\ -\sin(\alpha\frac{\pi}{2}) & cos(\alpha\frac{\pi}{2}) \end{bmatrix}.$$

It is clear that to check the inequality (2.23), it suffices to substitute $\Theta_{ij}, i, j = 1, 2$ in the matrix Ξ. This ends the proof. □

An other interesting result published in [10] on the stability of fractional linear systems based on $\mathcal{LMI}$ formulation This $\mathcal{LMI}$ is a rewrite of the $\mathcal{LMI}$ (2.20) in Lemma 2.4.1 and is given by the following theorem.

Theorem 2.4.1 *[10] Let $A \in \mathbb{R}^{n\times n}$, $0 < \alpha < 1$ and $\theta = (1-\alpha)\frac{\pi}{2}$. The fractional-order system $D^\alpha x(t) = Ax(t)$ is asymptotically stable if and only if there exist a positive definite Hermitian matrix $X = X^* \in \mathbb{C}^{n\times n}$ such that*

$$(rX + \bar{r}\overline{X})^T A^T + A(rX + \bar{r}\overline{X}) < 0, \tag{2.33}$$

where $r = e^{j(1-\alpha)\frac{\pi}{2}}$.

Proof We have to show that both $\mathcal{LMI}$ problem in Lemma 2.4.1 and Theorem 2.4.1 are equivalent. According to Lemma 2.4.1, the fractional-order system

$$D^\alpha x(t) = Ax(t)$$

is asymptotically stable if and only if there exist two complex matrices

$$X_1 = X_1^* > 0, X_2 = X_2^* > 0$$

such that the $\mathcal{LMI}$ (2.20) holds.

Define $L_1(X_1) = rX_1A^{\mathrm{T}} + \bar{r}AX_1$ and $L_2(X_2) = \bar{r}X_2A^{\mathrm{T}} + rAX_2$. It follows that $\mathcal{LMI}$ problem in Lemma 2.4.1 can be rewritten as

$$L_1(X_1) + L_2(X_2) < 0 \tag{2.34}$$

for some complex matrices $X_1 = X_1^* > 0$, $X_2 = X_2^* > 0$. Condition (2.33) is equivalent to

$$(rX^{\mathrm{T}} + \bar{r}\overline{X}^{\mathrm{T}})A^{\mathrm{T}} + A(rX + \bar{r}\overline{X}) < 0.$$

Since $X = X^*$, then $X^{\mathrm{T}} = \overline{X}$. In this case the last inequality can be rewritten as

$$(r\overline{X}A^{\mathrm{T}} + \bar{r}A\overline{X}) + (\bar{r}XA^{\mathrm{T}} + rAX) < 0$$

which means that the $\mathcal{LMI}$ problem of Theorem 2.4.1 is equivalent to

$$L_1(\overline{X}) + L_2(X) < 0 \tag{2.35}$$

for some complex matrix $X = X^* > 0$. Consequently, we have to prove that (2.34) and (2.35) are equivalent.

- If (2.35) holds, it suffices to take $X_1 = \overline{X}$ and $X_2 = X$ to see that (2.34) also holds.
- Conversely, if there exist complex matrices $X_1 = X_1^* > 0$, $X_2 = X_2^* > 0$ such that (2.34) holds then $\overline{L_1(X_1) + L_2(X_2)} < 0$ and consequently

$$L_1(X_1) + L_2(X_2) + \overline{L_1(X_1) + L_2(X_2)} < 0. \tag{2.36}$$

Few computations show that (2.36) is equivalent to

$$rX_1A^{\mathrm{T}} + \bar{r}AX_1 + \bar{r}X_2A^{\mathrm{T}} + rAX_2 + r\overline{X_1}A^{\mathrm{T}} + \bar{r}A\overline{X_1} + \bar{r}\overline{X_2}A^{\mathrm{T}} + rA\overline{X_2} < 0, \tag{2.37}$$

i.e.,

$$r(X_1 + \overline{X_2})A^{\mathrm{T}} + \bar{r}A(X_1 + \overline{X_2}) + \bar{r}(\overline{X_1} + X_2)A^{\mathrm{T}} + rA(\overline{X_1} + X_2) < 0.$$

It is easy to see that this last inequality is the same as

$$L_1(\overline{\overline{X_1} + X_2}) + L_2(\overline{X_1} + X_2) < 0. \tag{2.38}$$

From $X_1 > 0$ and $X_2 > 0$, we have $\overline{X_1} > 0$ and $\overline{X_1} + X_2 > 0$. So we get (2.35) for $X = \overline{X_1} + X_2$.

This ends the proof. □

An equivalent result is then deduced.

Lemma 2.4.4 *[12] Let $A \in \mathbb{R}^{n \times n}$ and $0 < \alpha < 1$. The fractional-order system $D^\alpha x(t) = Ax(t)$ is asymptotically stable if and only if there exist two real matrices $X_1, X_2 \in \mathbb{R}^{n \times n}$ such that*

$$\begin{bmatrix} X_1 & X_2 \\ -X_2 & X_1 \end{bmatrix} > 0, \tag{2.39}$$

$$\text{sym}\left\{A(\sin(\alpha\frac{\pi}{2})X_1 - \cos(\alpha\frac{\pi}{2})X_2)\right\} < 0. \tag{2.40}$$

Proof By setting in Theorem 2.4.1, $X = Z_1 + jZ_2$ and after a few computations, it is easy to see that the $\mathcal{LMI}$ (2.33) is equivalent to that(2.40) with

$$X_1 = 2Z_1, \ X_2 = 2Z_2.$$

Moreover, according to Lemma 2.4.3, $X > 0$ is equivalent to $\begin{bmatrix} X_1 & X_2 \\ -X_2 & X_1 \end{bmatrix} > 0$. This completes the proof. □

To illustrate the results in Theorems 2.3.1 and 2.4.1, numerical examples are provided.

Example 2.4.1 (1) Consider the fractional order system described by

$$D^\alpha x(t) = Ax(t),$$

with

$$A = \begin{bmatrix} -1 & 2 & 1 \\ -1 & -2 & 0.2 \\ 0.5 & 0 & -2 \end{bmatrix}, \ \alpha = 1.5.$$

To verify the stability of this system, we will show two different ways.

The stability of the system can be studied by the localization of the eigenvalues of the matrix A.

$$\text{spec}(A) = \{-1.5727 + 1.0923i, -1.5727 - 1.0923i, -1.8546\}.$$

Each eigenvalue λ of A satisfy the criteria $|\arg(\lambda)| > \alpha\frac{\pi}{2}$, which means that the system is asymptotically stable.

According to Theorem 2.3.1, a feasible solution of the linear matrix inequality (2.5) is as follows

$$P = \begin{bmatrix} 0.3112 & 0.1669 & 0.1384 \\ 0.1669 & 0.4876 & 0.1602 \\ 0.1384 & 0.1602 & 0.5463 \end{bmatrix}.$$

(2) Consider the fractional order system described by

$$D^\alpha x(t) = Ax(t),$$

with

$$A = \begin{bmatrix} -1 & 1 & 1 \\ 0 & -6 & 0 \\ 0 & -1 & -4 \end{bmatrix}, \ \alpha = 0.5.$$

The eigenvalues of the matrix A are as follows

$$\mathrm{spec}(A) = \{-1, -4, -6\}.$$

It is easy to verify that they lie in the stable region.

The stability of the system can also be verified through the resolution of the $\mathcal{LMI}$ (2.33) in Theorem 2.4.1. A feasible solution is

$$X = \begin{bmatrix} 625.6851 & 170.3633 & 214.7760 \\ 170.3633 & 313.7579 & 83.7615 \\ 214.7760 & 83.7615 & 359.0291 \end{bmatrix}.$$

2.5 An Equivalent LTI System for an FO-LTI System in the Sense of Stability

The stability of fractional order system was studied in [13] where an other interesting result on this topic was derived. It's about an equivalent LTI system for an FO-LTI system in the sense of stability. To this end, we need the following theorem which states a necessary and sufficient condition to place the eigenvalues of a real matrix in a specified sector.

Theorem 2.5.1 *[14, 15] Eigenvalues of an $n \times n$ matrix A lie within the region Ω in Fig. 2.3 $\Omega = \left\{\lambda \in \mathbb{C}/\, \mathrm{Re}(\lambda)\cos(\delta) \pm \mathrm{Im}(\lambda)\sin(\delta) \leq 0;\ 0 \leq \delta < \frac{\pi}{2}\right\}$ if and only if the eigenvalues of the $2n \times 2n$ matrix*

$$\tilde{A} = \begin{bmatrix} A\cos(\delta) & -A\sin(\delta) \\ A\sin(\delta) & A\cos(\delta) \end{bmatrix}$$

have negative real parts.

Theorem 2.5.1 has been proved using two different approaches in [14, 15]. If the characteristic equation of matrix A is available, then one can state Theorem 2.5.1 with a more useful expression as in Theorem 2.5.2.

Theorem 2.5.2 *[16] A necessary and sufficient condition that the zeros of an nth-order polynomial*

$$P(s) = s^n + a_{n-1}s^{n-1} + \ldots + a_1 s + a_0,$$

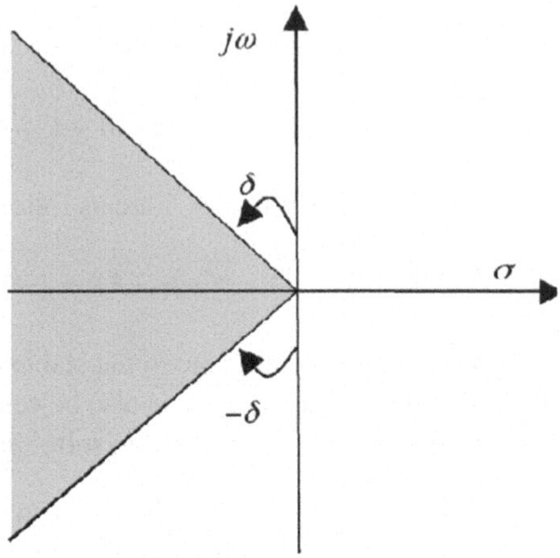

Fig. 2.3 Region Ω in Theorem 2.5.1 is shown in gray

lie within the region Ω, is that the 2nth-order polynomial

$$F(s) = P(se^{j\delta})P(se^{-j\delta}),$$

has zeros in the left half of the complex plane.

If the coefficients of $P(s)$ $(a_i, 0 \leq i < n)$ are real, then its roots are symmetric about the real axis of complex plane. Hence, the roots of $F(s)$ are also symmetric about the real axis and then polynomial $F(s)$ has real coefficients.

In fact, the region Ω can be interpreted as stable region of FO-LTI system $D^{\alpha}x(t) = Ax(t)$ where $\delta = (\alpha - 1)\frac{\pi}{2}$. Now, the following two theorems are proposed.

Theorem 2.5.3 *[13] The FO-LTI system $D^{\alpha}x(t) = Ax(t)$ with $1 \leq \alpha < 2$ is asymptotically stable if and only if the LTI system*

$$\dot{\tilde{x}}(t) = \begin{bmatrix} A\sin(\alpha\frac{\pi}{2}) & A\cos(\alpha\frac{\pi}{2}) \\ -A\cos(\alpha\frac{\pi}{2}) & A\sin(\alpha\frac{\pi}{2}) \end{bmatrix} \tilde{x}(t) \tag{2.41}$$

is asymptotically stable.

Proof According to Theorem 2.5.1 and the stability condition (2.4), the proof is obvious. □

Theorem 2.5.4 *[13] all eigenvalues FO-LTI system $D^{\alpha}x(t) = Ax(t)$ with $0 < \alpha \leq 1$ settle in the unstable region (Fig. 2.1) if and only if the LTI system*

$$\dot{\tilde{x}}(t) = -\begin{bmatrix} A\sin(\alpha\frac{\pi}{2}) & -A\cos(\alpha\frac{\pi}{2}) \\ A\cos(\alpha\frac{\pi}{2}) & A\sin(\alpha\frac{\pi}{2}) \end{bmatrix} \tilde{x}(t) \tag{2.42}$$

is asymptotically stable.

Proof (Necessity) Suppose that all the eigenvalues of the FO-LTI system $D^{\alpha}x(t) = Ax(t)$ with $0 < \alpha \leq 1$ are located in the unstable region ,shown in Fig. 2.1,

$$\Omega_1 = \left\{\lambda/\lambda \in \mathbb{C}, |\arg(\lambda)| < \alpha\frac{\pi}{2}\right\}$$

We know that if λ is an eigenvalue of the matrix A, $-\lambda$ is an eigenvalue of the matrix $-A$ and then it lies in the symmetrical region with respect to the imaginary axis

$$\Omega_2 = \left\{\lambda/\lambda \in \mathbb{C}, |\arg(\lambda)| > (2-\alpha)\frac{\pi}{2}\right\}$$

It is clear that the region Ω_2 is the stability region of the fractional order system

$$D^{(2-\alpha)}x(t) = -Ax(t), 1 \leq 2-\alpha < 2$$

Thus, according to Theorem 2.5.3, the system

$$\dot{\tilde{x}}(t) = \begin{bmatrix} -A\sin((2-\alpha)\frac{\pi}{2}) & -A\cos((2-\alpha)\frac{\pi}{2}) \\ A\cos((2-\alpha)\frac{\pi}{2}) & -A\sin((2-\alpha)\frac{\pi}{2}) \end{bmatrix} \tilde{x}(t)$$

is asymptotically stable. After some computations, one can see that this last system is exactly the system (2.42), so we get our goal.
(Sufficiency) Just follow similar reasoning. □

Theorems 2.5.3 and 2.5.4 relate stability of a fractional order system to stability of its equivalent ordinary system. Based on these two theorems, most of the stability related analysis in the ordinary systems is applicable to the fractional systems with commensurate order as well. As an illustration, in Example 2.5.1 the Routh–Hurwitz criterion is applied without any change to investigate stability of a fractional system.

Example 2.5.1 [13] In this example, we want to find the parameters which make the following FO-LTI system stable.

$$D^{\alpha}x(t) = Ax(t),$$

where $A = \begin{bmatrix} 0 & 1 \\ b & a \end{bmatrix}$ and $1 \leq \alpha < 2$ is given.

From (2.41), the equivalent LTI system in the sense of stability for the above system has the following characteristic polynomial

$$P(s) = s^4 + a_3 s^3 + a_2 s^2 + a_1 s + a_0,$$

where $a_0 = b^2,\ a_1 = 2ab\sin(\alpha\frac{\pi}{2}),\ a_2 = a^2 + 2ab\cos(\alpha\pi),\ a_3 = -2a\sin(\alpha\frac{\pi}{2})$.

By the Routh–Hurwitz criterion, the real parts of the roots of $P(s)$ are negative if and only if $a < 0,\ b < 0$ and $a^2 + 4b\cos^2(\alpha\frac{\pi}{2}) > 0$. Figure 2.4 shows the stable

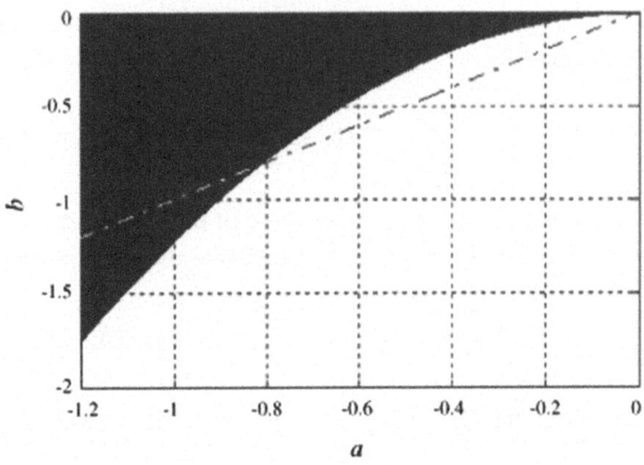

Fig. 2.4 Stable (white) and unstable (black) regions in a–b plane when $\alpha = 1.3$ (Example 2.5.1)

and unstable regions in a–b plane for $\alpha = 1.3$. If $a = b$, the above conditions reduce to

$$a < 0,\ b < 0,\ a < -4\cos^2(\alpha\frac{\pi}{2}).$$

Time response for the initial condition $(x(0), \dot{x}(0), y(0), \dot{y}(0)) = (1, 0, 1, 0)$ and fractional order $\alpha = 1.3$ is shown in Figs. 2.5 and 2.6.

In Fig. 2.5, system parameters are $a = b = -0.9$, hence system is stable but in Fig. 2.6, system parameters are chosen as $a = b = -0.7$, and consequently the system is unstable.

2.6 Controllability and Observability of FO-LTI Systems

Controllability and observability represent two major concepts of modern control system theory. These concepts were introduced by R. Kalman in 1960. They can be roughly defined as follows.

Controllability: In order to be able to do whatever we want with the given dynamic system under control input, the system must be controllable.

Observability: In order to see what is going on inside the system under observation, the system must be observable.

The controllability and observability conditions for commensurate-order systems can be seen in [4, 17–20] in which tests for these concepts are connected to the rank tests of certain matrices: the controllability and observability matrices as in the integer case due to Kalman [21] and the proofs are given following a method similar to that used for integer-order systems [22, 23].

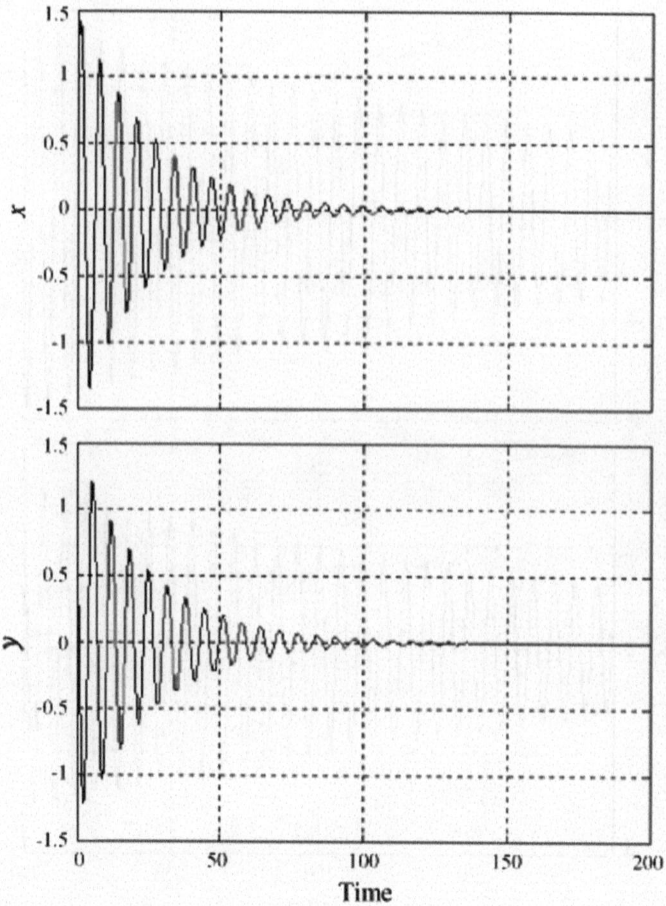

Fig. 2.5 Time response when $\alpha = 1.3$ and $a = b = -0.9$ (Example 2.5.1)

Consider the linear fractional-order invariant-time system described by

$$\begin{cases} D^{\alpha}x(t) = Ax(t) + Bu(t) \\ y(t) \quad = Cx(t) \\ x(0) \quad = x_0 \end{cases} \qquad 0 < \alpha < 2 \qquad (2.43)$$

where α is the fractional commensurate order of the system, $x(t) \in \mathbb{R}^n$ is the state vector, $u(t) \in \mathbb{R}^m$ is the input vector $y(t) \in \mathbb{R}^p$ is the output vector and A, B, C are constant matrices with appropriate dimensions.

The definition of both the controllability and the observability given below are the same as for integer order case [24]. In what follows, the initial time can be taken zero.

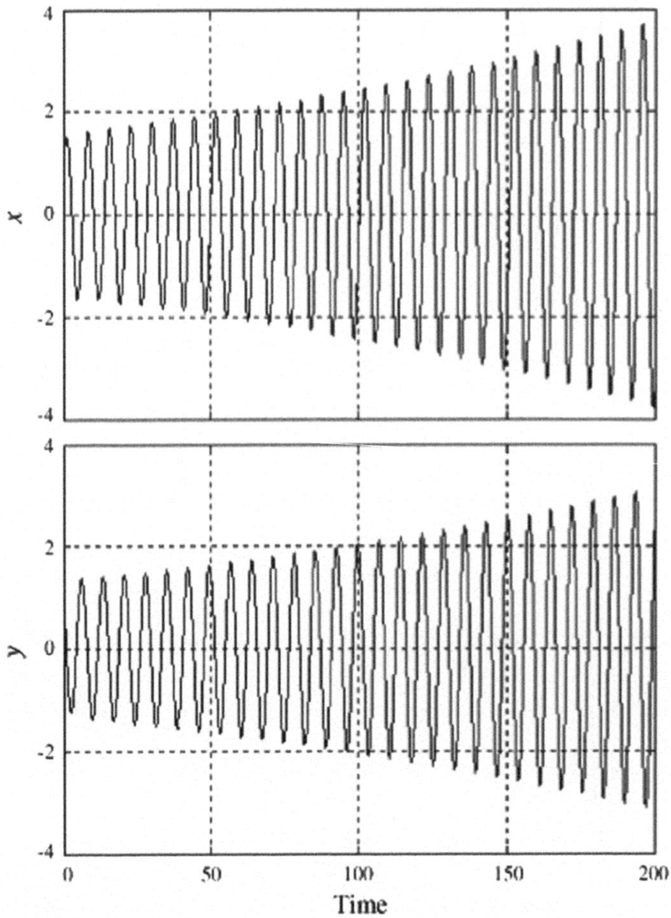

Fig. 2.6 Time response when $\alpha = 1.3$ and $a = b = -0.7$ (Example 2.5.1)

Definition 2.6.1 [18, 19] The system (2.43) is controllable if, for a given time t_0, there exists a finite time $t_f > t_0$ such that from any initial condition $x(t_0) = x_0$ and any $x(t_f) = x_f$ in the state-space, there exists an input $u(t), t \in [t_0, t_1]$ which can lead the system from x_0 to x_f at time t_f.

Controllability is an important concept since it establishes the fact that we can control the system to modify its behavior (stabilizing a system unstable, change specific dynamics). This concept plays a very important role in the synthesis theory of control systems in state-space.

A trivially uncontrollable system is the one whose input matrix is zero, $B = 0$.

Theorem 2.6.1 (Controllability Criterion) *[18, 19] The following statements are equivalent.*

- *The system given by* (2.43) *(or the pair* (A, B) *is controllable.*
- *The matrix* $\mathcal{C}$ *defined by*

$$\mathcal{C} = \left[B \; AB \; A^2B \; ... \; A^{n-1}B \right] \tag{2.44}$$

is full-rank .

- *For all* $\lambda \in \mathbb{C}$,

$$\text{rank}\left(\left[\lambda I_n - A \;\; B \right] \right) = n. \tag{2.45}$$

Definition 2.6.2 [18, 19] The system (2.43) is observable on $[t_0, t_1]$, $t_1 > 0$, if $x(t_0)$ can be deduced from the observation of the output $y(t)$ and the knowledge of the input $u(t)$ for $t \in [t_0, t_1]$.

Clearly, the notion of observability is crucial for systems where the state vector is not completely accessible to the measure but must be rebuilt, estimated or filtered from data supplied by the output.

Theorem 2.6.2 (Observability Criterion) *[18, 19] The following statements are equivalent.*

- *The system given by* (2.43) *(or the pair* (C, A) *is observable if and only if the matrix* $\mathcal{O}$ *defined by*

$$\mathcal{O} = \begin{bmatrix} C \\ CA \\ CA^2 \\ \cdots \\ CA^{n-1} \end{bmatrix} \tag{2.46}$$

is full-rank .

- *For all* $\lambda \in \mathbb{C}$

$$\text{rank}\left(\begin{bmatrix} \lambda I_n - A \\ C \end{bmatrix} \right) = n. \tag{2.47}$$

Testing the rank of the matrix $\mathcal{O}$ may pose numerical difficulties, since it requires the computation of powers of A up to A^{n-1}, an operation which may be ill-conditioned. Testing the rank of the matrix $\begin{bmatrix} \lambda I_n - A \\ C \end{bmatrix}$ for all $\lambda \in \text{spec}(A)$ is generally more robust numerically.

2.7 Summary

In this chapter, the reader will find an overview of the results dedicated to stability analysis of linear fractional differential systems in continuous time. It was found that the stability is equivalent to the repartition of the system poles in a restricted area of the complex plane. Based on this key formulation of stability and the use of convex optimization algorithms, the results are stated as linear matrix inequality conditions. Some properties of the controllability and observability for such class of systems have been analyzed which can be considered as a generalization of the known corresponding results of integer order case to the fractional order one.

References

1. Matignon, D.: Stability properties for generalized fractional differential systems. In: ESAIM Proceedings Fractional Differential Systems, Models Methods and Applications, vol. 5, pp. 145–158 (1998). http://www.emath.fr/proc/
2. Bonnet, C., Partington, J.R.: Analysis of fractional delay systems of retarded and neutral type. Automatica **38**, 1133–1138 (2002)
3. Moze, M., Sabatier, J., Oustaloup, A.: Lmi tools for stability analysis of fractional systems. In: Proceedings of the ASME IDET/CIE Conferences, Long-Beach, USA (2005)
4. Matignon, D.: Stability results for fractional differential equations with applications to control. In: Processing, Computational Engineering in Systems and Application Multi-Conference, IMACS, IEEE-SMC, vol. 2, pp. 963–968, Lille, France, July 1996
5. Sabatier, J., Moze, M., Farges, C.: On stability of fractional order systems. In: Proceedings of the Third IFAC Workshop on Fractional Differentiation and its Applications (FDA'08), Ankara, Turkey (2008)
6. Chilali, M., Gahinet, P.: Member IEEE. $\mathcal{H}_\infty$ design with pole placement constraints: an $\mathcal{LMI}$ approach. IEEE Trans. Autom. Control **41** (1996)
7. Sabatier, J., Moze, M., Farges, C.: $\mathcal{LMI}$ conditions for fractional order systems. Comput. Math. Appl. **59**, 1594–1609 (2010)
8. Chilali, M., Gahinet, P.: h_∞ design with pole placement constraints: an $\mathcal{LMI}$ approach. IEEE Trans. Automat. Contr. **41**, 358–367 (1996)
9. Bachelier, O.: Commande des systèmes linéaires incertains: placement de pôles robuste en d-stabilité. Ph.D. Thesis, INSA, Toulouse, France (1998)
10. Farges, C., Moze, M., Sabatier, J.: Pseudo-state feedback stabilization of commensurate fractional-order systems. Automatica **46**, 1730–1734 (2010)
11. Lu, J.G., Chen, Y.Q.: Robust stability and stabilization of fractional-order interval systems with the fractional order α: the $0<\alpha<1$ case. IEEE Trans. Autom. Control **55**, 152–158 (2010)
12. Zhang, X.F., Chen, Y.D.: D-stability based lmi criteria of stability and stabilization for fractional-order systems. In: Proceedings of the ASME 2015 International Design Engineering Technical Conference and Computers and Information in Engineering Conference, DETC2015-46692, Boston, Massachusetts, USA, pp. 1–6, Aug 2015
13. Tavazoei, M.S., Haeri, M.: A note on the stability of fractional order systems. Math. Comput. Simul. J. **79**, 1566–1576 (2009)
14. Anderson, B.D.O., Jury, E.I., Bose, N.K.: A simple test for zeros of a complex polynomial in a sector. IEEE Trans. Autom. Control **19**, 437–438 (1974)
15. Davison, E.J., Ramesh, N.: A note on the eigenvalues of a real matrix. IEEE Trans. Autom. Control **15**, 252–253 (1970)

16. Hostetter, G.H.: An improved test for the zeros of a polynomial in a sector. IEEE Trans. Autom. Control **20**, 433–434 (1975)
17. Matignon, D., André, B.: A-novel some results on controllability and observability of finite-dimensional fractional differential systems. In: Proceedings of the Mathematical Theory of Networks and Systems Symposium, Lille, France (1996)
18. Monje, C.A., Chen, Y.Q., Vinagre, B.M., Xue, D.Y., Feliu-Batlle, V.: Fractional-Order Systems and Controls: Fundamentals and Applications, pp. 37–78. Springer, London (2010)
19. Bettayeb, M., Djenoune, S.: New results on the controllability and observability of fractional dynamical systems. J. Vib. Control **14**, 1531–1541 (2008)
20. Vinagre, B.M., Monje, C.A., Caldero, A.J.: Fractional order systems and fractional order actions, tutorial workshop. In: Proceedings of the IEEE Conference on Decision and Control, Las Vegas, USA (2002)
21. Kalman, R.E.: Mathematical description of linear dynamical systems. SIAM J. Contr. **1**, 152–192 (1963)
22. Ogata, K.: Modern Control Engineering, 4th edn. Prentice Hall (2001)
23. Zadeh, L., Desoer, C.A.: Linear Systems Theory. McGraw-Hill (1963)
24. Chen, C.T.: Linear System Theory and Design. Holt, Rinehart and Wiston, New York (1984)

Chapter 3
Stabilization of Fractional-Order Systems

3.1 Introduction

Control is an interdisciplinary branch of engineering and mathematics that deals with the modification of dynamic systems to obtain a desired behaviour given in terms of a set of specifications or a reference model. Several control methods have been executed to reach system performances [1, 2]. Researches keep making immense efforts in developing various stabilizing control design schemes such as linear matrix inequality $\mathcal{LMI}$ based method which receives much attention in the last decades, see [3–5] and the references therein. Fractional-order systems represent a wider class of complex dynamical systems. This class of systems attracts much attention in both theory and applications in the latest four decades.

As for systems with integer order derivative, the stability of a fractional linear system depends on the location of the eigenvalues of the dynamic matrix in the complex plane [6, 7]. Recently, a considerable interest has been also devoted to the analysis of the stability and stabilization of linear systems derived from fractional order [5, 8–10], where some results have been fulfilled through $\mathcal{LMI}$ methods for mainly two types of fractional orders, $0 < \alpha \leq 1$ and $1 \leq \alpha < 2$. It is notable that the problem of the synthesis for FOS is difficult, however it was treated in recent papers. The most obtained results were in terms of $\mathcal{LMI}$ approach. In fact, Based on the generalization of Gronwall-Bellman Lemma, the analytical stability conditions and state feedback stabilization problem of non linear fractional-order differentiation equation have been studied in [11, 12]. the problem of stabilization and robust stabilization by pseudo-state feedback of commensurate polytopic systems, particularly on the case $0 < \alpha < 1$ was investigated in [13]. stabilization of fractional order systems with uncertainties was also highlighted in [14], where the method of observer-based control and static output feedback control in both cases $0 < \alpha < 1$ and $1 \leq \alpha < 2$ was treated and the formalism via $\mathcal{LMI}$ was derived. The robust asymptotical stability and stabilization problem for a class of uncertain FO-LTI interval systems with the fractional order α belonging to $0 < \alpha < 1$ and linear coupling relationships between

S. Marir and M. Chadli, *Control of Singular Fractional Order Systems: LMI Approach*, Studies in Systems, Decision and Control 292,
https://doi.org/10.1007/978-3-031-87382-9_3

the fractional order α and the system parameters were investigated in [15]. Accordingly, study on fractional-order systems has been extended to T-S fuzzy models. In [16, 17], the stability, robust stability and stabilization were investigated for such class of systems. Recently, the static output feedback stabilization for fractional-order systems in T-S fuzzy models has been developed in both papers [18, 19].

In this section, we present some existing results concerning the problem of stabilization of linear fractional-order systems. In first time, the problem is processed via the pole placement approach with $0 < \alpha < 2$ and secondly, by an $\mathcal{LMI}$ approach where $1 \leq \alpha < 2$ and a $\mathcal{GLMI}$ method in the case where $0 < \alpha \leq 1$. Parallel to integer order systems, the concerning basic definitions and relevants facts for FOS are given as follows.

3.2 Stabilization by State-Feedback

3.2.1 Eigenvalue Assignment

Consider the linear fractional-order invariant-time system described by (2.43).

To begin, it is assumed that all components of the state vector are measured which means that $C = I_n$. The problem we wish to consider now is to determine a state feedback control law $u(t) = Kx(t)$ having the property that the resulting closed-loop system is asymptotically stable.

Definition 3.2.1 [20] Le fractional system (2.43) of order $0 < \alpha \leq 1$ or $1 \leq \alpha < 2$ is stabilizable (or the pair (A, B) is stabilizable) if there exists a state feedback control $u = Kx$ such that the closed-loop system is asymptotically stable.

Theorem 3.2.1 *[20] The fractional-order system* (2.43) *is said to be stabilizable if there exists a state feedback control law defined by $u(t) = Kx(t)$, $K \in \mathbb{R}^{m\times n}$ such that*

$$|\arg(\text{spec}(A + BK)| > \alpha\frac{\pi}{2}.$$

Proof The proof of this theorem is obtained directly by replacing matrix A with $A = A + BK$ in Theorem 2.2.2. □

The matrix K is known as the state feedback gain matrix, and it affects the closed-loop system behavior. The main influence of K is through the matrix A, resulting in the matrix $A + BK$ of the closed-loop system. Clearly, K affects the eigenvalues of $A + BK$, and therefore, the modes of the closed-loop system. In other words, "the eigenvalues of the original system can arbitrarily be changed in this case." This last statement, commonly used in the literature, is rather confusing: The eigenvalues of a given system are not physically changed by the use of feedback. They are the same as they used to be before the introduction of feedback. Instead, the feedback law $u(t) = Kx(t)$, generates an input $u(t)$ that, when fed back to the system, makes it behave as if the eigenvalues of the system were at different locations [i.e., the input

$u(t)$ makes it behave as a different system, the behavior of which is, we hope, more desirable than the behavior of the original system].

It is quite clear that a given system (2.43) can be made asymptotically stable via the state feedback control law $u(t) = Kx(t)$ only when all the uncontrollable eigenvalues of A are already in the stable region. This is so because state feedback can alter only the controllable eigenvalues.

Definition 3.2.2 [20] The pair (A, B) is called stabilizable if and only if all its uncontrollable eigenvalues verify $|\arg(\mathrm{spec}(A))| > \alpha\frac{\pi}{2}$.

Example 3.2.1 Consider the stabilization of the following fractional order system described in (2.43) for the fractional order $0 < \alpha < 1$ and with parameters as follows

$$A = \begin{bmatrix} -1 & 0 & 1 \\ 0 & 1 & 2 \\ 5 & 2 & 1 \end{bmatrix}, \; B = \begin{bmatrix} -1 & 0 \\ 0 & 1 \\ 0 & 0 \end{bmatrix}, \; \alpha = 0.5.$$

The eigenvalues of the matrix A are $\lambda_1 = 3.6139$, $\lambda_2 = -2.8108$, $\lambda_3 = 0.1969$. According to Theorem 2.2.2, the system is unstable as can be seen in Fig. 3.1.

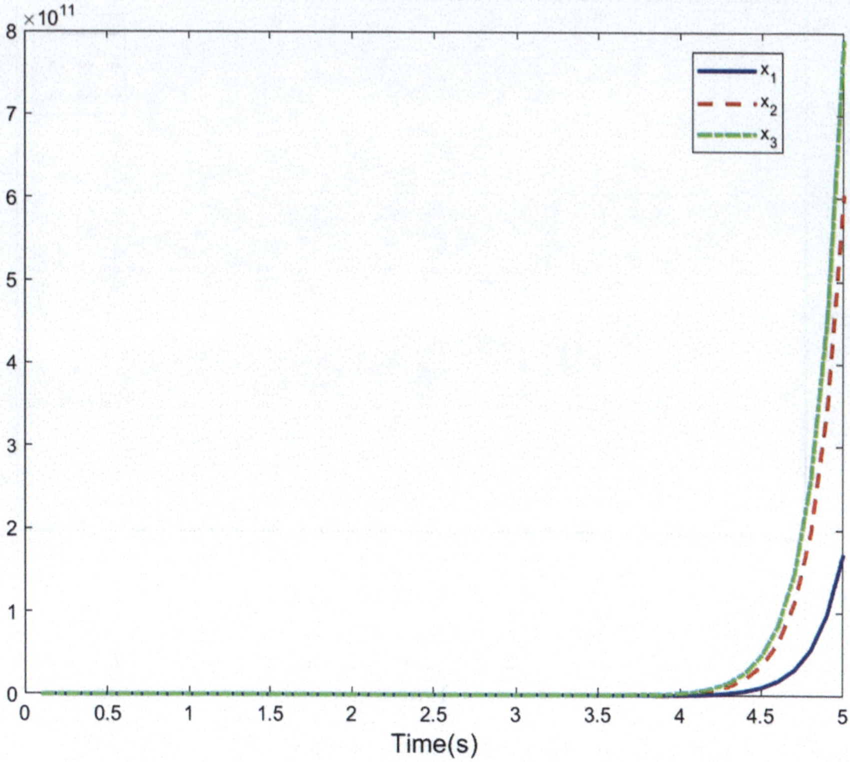

Fig. 3.1 State responses of the open-loop system with $\alpha = 0.5$

The controllability matrix is

$$\mathcal{C} = \begin{bmatrix} 1 & 0 & -1 & 0 & 6 & 2 \\ 0 & 1 & 0 & 1 & 10 & 5 \\ 0 & 0 & 5 & 2 & 0 & 4 \end{bmatrix},$$

and its rank is

$$\text{rank}(\mathcal{C}) = 3.$$

The system is then stabilizable. We wish to determine K so that the eigenvalues of $A + BK$ are at -1, -2.8108, -2. We obtain

$$K = \begin{bmatrix} 3.4337 & 0.9728 & 2.3199 \\ -0.9433 & -3.3771 & -2.5122 \end{bmatrix}$$

as the appropriate state feedback matrix. The resulting closed-loop system is then asymptotically stable as shown in Fig. 3.2.

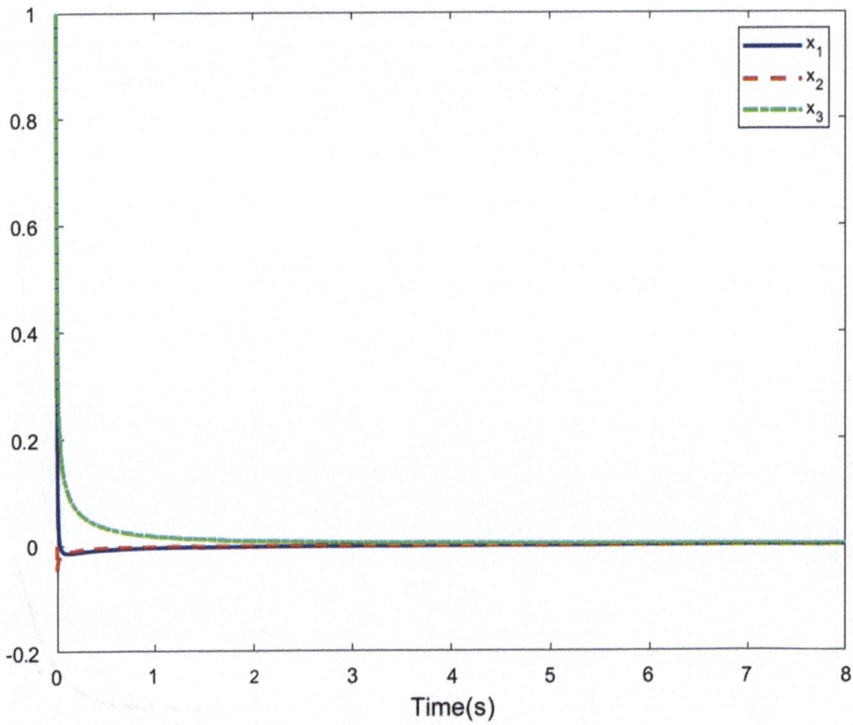

Fig. 3.2 State responses of the closed-loop system with $\alpha = 0.5$

3.2.2 LMI Conditions

According to Sect. 2.5 of Chap. 2, we have seen that there are two $\mathcal{LMI}$ conditions for stability of fractional order linear system $0 < \alpha < 2$. Which leads us to study stabilization of the system (2.43) in both cases $1 \leq \alpha < 2$ and $0 < \alpha < 1$. The aim is to compute a static state feedback control law of the form $u(t) = Kx(t)$ where $K \in \mathbb{R}^{m \times n}$ is a constant matrix gain such that the stability of the closed-loop system

$$D^{\alpha}x(t) = (A + BK)x(t) \tag{3.1}$$

is ensured, which means that

$$|\arg(\text{spec}(A + BK)| > \alpha\frac{\pi}{2}.$$

1. **Case** $1 \leq \alpha < 2$
   The problem, when $1 < \alpha < 2$, is a well known problem in $\mathcal{LMI}$ control theory. A solution of this problem is provided by the $\mathcal{LMI}$ region framework [21]. In order to obtain an $\mathcal{LMI}$ condition, a linearising change is applied and leads to Theorem 3.2.2.

Theorem 3.2.2 *[13] The fractional-order system* (2.43) *with* $1 < \alpha < 2$ *is stabilizable by state feedback control* $u(t) = Kx(t)$ *if and only if there exist matrices* $X \in \mathbb{R}^{n \times n} > 0$, $Y \in \mathbb{R}^{m \times n}$ *such that*

$$\begin{bmatrix} \Sigma_{11} & \Sigma_{12} \\ \Sigma_{12}^T & \Sigma_{22} \end{bmatrix} < 0 \tag{3.2}$$

with

$\Sigma_{11} = \Sigma_{22} = \left(AX + XA^T + BY + Y^TB^T\right)\sin(\alpha\frac{\pi}{2})$

$\Sigma_{12} = \left(AX - XA^T + BY - Y^TB^T\right)\cos(\alpha\frac{\pi}{2})$

In this case, a stabilizing controller is given by:

$$K = YX^{-1}. \tag{3.3}$$

Proof Necessity. Suppose there exist a gain matrix $K \in \mathbb{R}^{m \times n}$ such that the closed-loop system (3.1) is asymptotically stable. So there exists $X > 0$ such that the condition (2.12) holds since it is equivalent to that in Theorem 2.3.1 taking $A + BK$ instead of A which leads to a nonlinear matrix inequality. Using the linearising change $Y = KX$, the relation (3.2) is obtained.

Sufficiency. By substituting $Y = KX$ in (3.2), the asymptotic stability of the closed-loop system (3.1) is easy to check. □

Example 3.2.2 Consider the fractional order system (2.43) with the following coefficients: $\alpha = 1.5$, and

$$A = \begin{bmatrix} 1 & 0 & 0 & 0 \\ 0 & 2 & 0 & 0 \\ 0 & 0 & -3 & 0 \\ 0 & 0 & 0 & -4 \end{bmatrix}, \quad B = \begin{bmatrix} 1 & 0 & 0 \\ 0 & 1 & 0 \\ 0 & 0 & 1 \\ 1 & 1 & 1 \end{bmatrix}.$$

The eigenvalues of the matrix A are $\{1, 2, -3, -4\}$, thus they don't all lie in the stable region described by (2.4) which concludes the corresponding system is not asymptotically stable.

In this example, the objective is to design pseudo-state feedback controller $u(t) = Kx(t)$ such that the closed-loop system (3.1) is asymptotically stable. Solving $\mathcal{LMI}$ (3.2) in Theorem 3.2.2, it can be obtained that

$$X = 10^4 \begin{bmatrix} 4.9707 & 0.0643 & 0.9094 & -1.0784 \\ 0.0643 & 4.5115 & 1.2248 & -1.5149 \\ 0.9094 & 1.2248 & 4.3854 & 0.6673 \\ -1.0784 & -1.5149 & 0.6673 & 3.6336 \end{bmatrix},$$

$$Y = 10^5 \begin{bmatrix} -0.8244 & 0.0295 & -0.1614 & 0.1995 \\ -0.0094 & -1.2251 & -0.3042 & 0.3779 \\ 0.3321 & 0.4702 & 0.8586 & 0.3587 \end{bmatrix}.$$

The state feedback controller is given by

$$K = \begin{bmatrix} -1.6053 & 0.1730 & -0.1084 & 0.1647 \\ -0.0402 & -2.8028 & 0.1222 & -0.1627 \\ 0.7443 & 1.1880 & 1.2474 & 1.4743 \end{bmatrix}.$$

The eigenvalues of the matrix $A + BK$ are as follows

$$\{-3.6076, -0.6328, -0.7220, -0.7220\},$$

which confirms that the resulting closed-loop system is asymptotically stable, its state responses are illustrated in Fig. 3.3.

2. **Case** $0 < \alpha < 1$
   In order to apply Theorem 2.4.1 to synthesis, it is essential to discover a linearizing change of variable. The process of finding such a change of variable may seem more challenging compared to previous cases due to the involvement of variable X in $\mathcal{LMI}$ (2.33), which becomes complex when the controller gain K to be determined is real. Nevertheless, $\mathcal{LMI}$ (2.33) in Theorem 2.4.1 has been formulated to ensure such that $rX + \bar{r}\bar{X}$ is real. By substituting A with $A + BK$ and employing the change of variables $Y = K(rX + \bar{r}\bar{X})$ it becomes possible

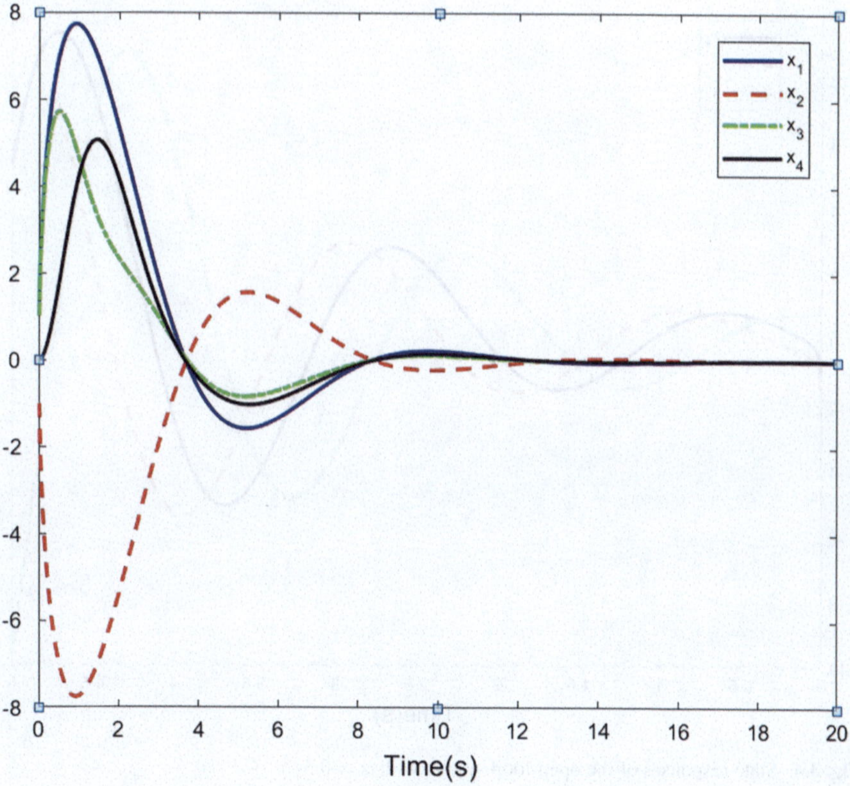

Fig. 3.3 State responses of the closed-loop system with $\alpha = 1.5$

to derive the following necessary and sufficient condition for the design of a pseudo-state feedback controller with real gain.

Theorem 3.2.3 *[13] Fractional system* (2.43) *of order* $0 < \alpha < 1$ *controlled by the static state feedback* $u(t) = Kx(t)$ *is asymptotically stable if and only if there exists matrices* $X \in \mathbb{C}^{n\times n} > 0$, $Y \in \mathbb{R}^{m\times n}$ *such that*

$$(rX + \bar{r}\bar{X})^T A^T + A(rX + \bar{r}\bar{X}) + Y^T B^T + BY < 0, \tag{3.4}$$

where $r = e^{j(1-\alpha)\frac{\pi}{2}}$. *A stabilizing controller gain is then:*

$$K = Y(rX + \bar{r}\bar{X})^{-1}. \tag{3.5}$$

Proof $\mathcal{LMI}$ (3.4) together with relation (3.5) follow directly from discussion above. The regularity of the matrix $rX + \bar{r}\bar{X}$ involved in (3.5) is now proved. For this purpose, let's demonstrate that the real matrix $rX + \bar{r}\bar{X}$ is antistable, meaning that all eigenvalues of $rX + \bar{r}\bar{X}$ reside within the right half of the complex plane. From

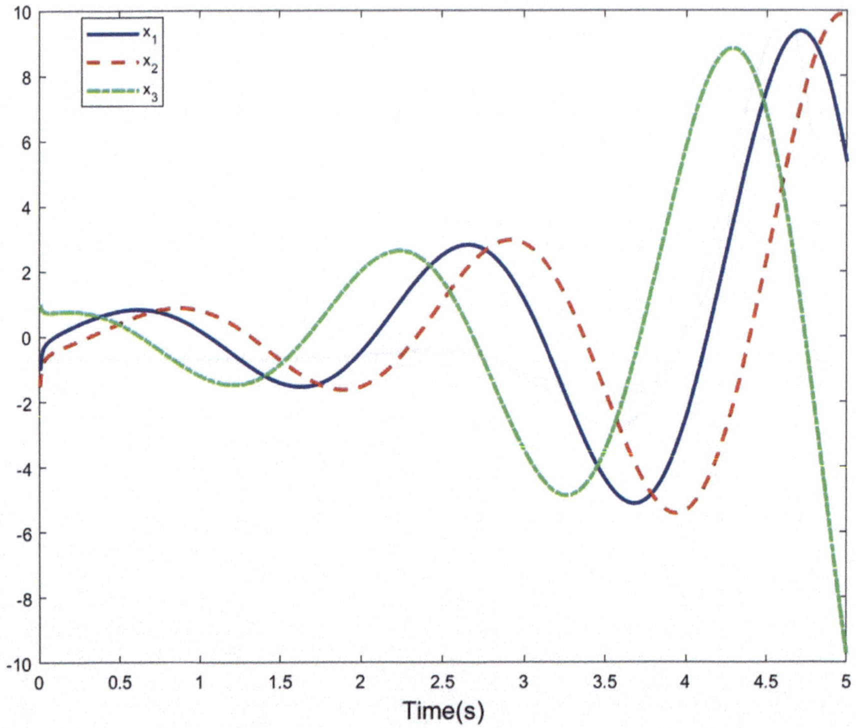

Fig. 3.4 State responses of the open-loop system with $\alpha = 0.5$

Lyapunov theory [22, 23], the matrix $rX + \bar{r}\bar{X}$ is antistable if and only if there exists a matrix $P = P^{\mathrm{T}} > 0$ such that

$$(rX + \bar{r}\bar{X})^{\mathrm{T}}P + P(rX + \bar{r}\bar{X}) > 0. \tag{3.6}$$

With $P = I > 0$, left term of inequality (3.6) becomes:

$$(r + \bar{r})(X + \bar{X}). \tag{3.7}$$

Since $r = \mathrm{e}^{(1-\alpha)\frac{\pi}{2}}$, with $0 < \alpha < 1$, $r + \bar{r}$ is a real positive scalar. So, positive definiteness of (3.7) is only subjected to positive definiteness of X. As in Theorem 3.2.3, $X > 0$ (and consequently $\bar{X} > 0$), (3.6) is verified with $P = I$ and $rX + \bar{r}\bar{X}$ is antistable. As matrix $rX + \bar{r}\bar{X}$ is antistable, it cannot have any null eigenvalue and is thus invertible. Consequently, relation (3.5) always holds. □

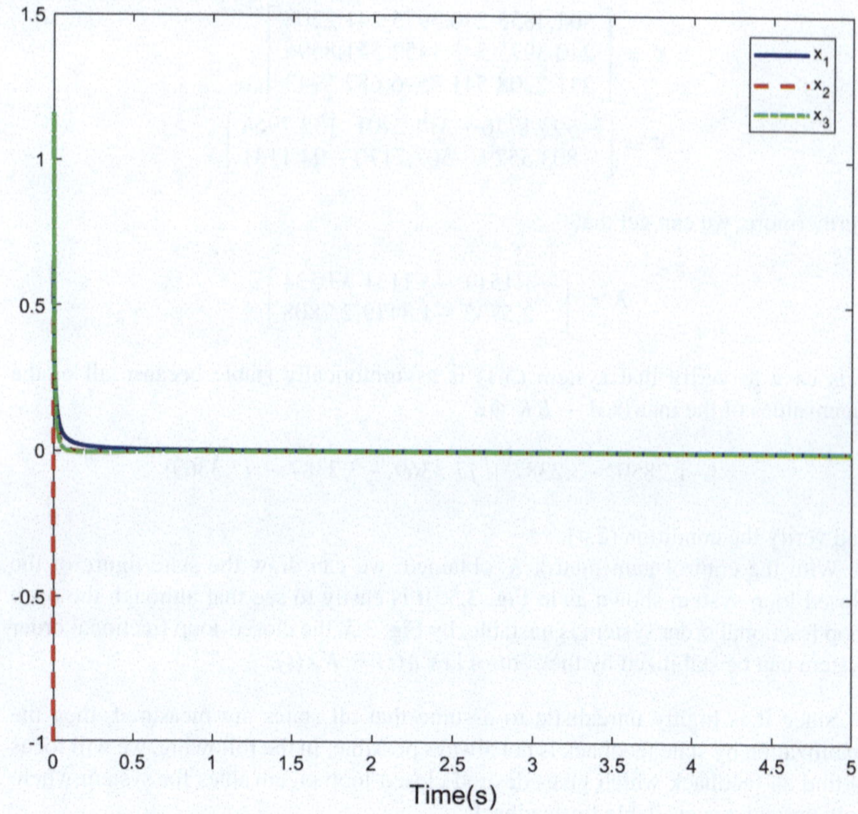

Fig. 3.5 State responses of the closed-loop system with $\alpha = 0.5$

Considering the continuous fractional-order system (2.43) with

$$A = \begin{bmatrix} 1 & 0 & 1 \\ 0 & 2 & 1 \\ 1 & -3 & 0 \end{bmatrix},\ B = \begin{bmatrix} 2 & 0 \\ 3.46 & 1.4 \\ 2.8 & 0.2 \end{bmatrix},\ \alpha = 0.5.$$

We can easily get that the open loop of the system (2.43) is unstable because all of the eigenvalues of the matrix A are

$$\{0.3177, 1.3412 + j1.1615, 1.3412 - j1.1615\},$$

and do not lie in the stable area described in Fig. 2.1. This can be confirmed in Fig. 3.4.

Using MATLAB LMI control Toolbox to solve (3.4) in Theorem 3.2.3, we can obtain the feasible solutions as follows

$$X = \begin{bmatrix} 504.4855 & 240.3975 & 241.2208 \\ 240.3975 & 543.1452 & 541.8596 \\ 241.2208 & 541.8596 & 687.5943 \end{bmatrix},$$

$$Y = \begin{bmatrix} -593.1776 & -314.2401 & 133.7964 \\ 803.5579 & -507.7170 & -94.1131 \end{bmatrix}.$$

Furthermore, we can get that

$$K = \begin{bmatrix} -1.1514 & -3.1151 & 3.0534 \\ 2.5733 & -4.8479 & 2.7808 \end{bmatrix}.$$

It is easy to verify that system (3.1) is asymptotically stable because all of the eigenvalues of the matrix $A + BK$ are

$$\{-1.2850, -3.2387 + j7.3369, -3.2387 - j7.3369\}$$

and verify the condition (2.4).

With the control gain matrix K obtained, we can draw the state figure of the closed-loop system shown as in Fig. 3.5. It is easily to see that although the open loop fractional order system is unstable, by Fig. 3.5, the closed-loop fractional order system can be stabilized by the control law $u(t) = Kx(t)$.

Since it is highly unrealistic to assume that all states are measured, then the stabilization by state feedback is not always possible. In the following, we will focus to find an feedback which gives desired closed loop eigenvalues for system where only outputs are available for feedback.

3.3 Stabilization by Static Output Feedback

3.3.1 Eigenvalues Assignment

The static output feedback problem concerns finding a static or constant feedback gain to achieve certain desired closed-loop characteristics. The output feedback control problem is much more difficult to solve when compared to state feedback control problem. Consider the system (2.43).

Definition 3.3.1 [20] The fractional-order system (2.43) is said to be detectable, or the pair (A, C) is detectable if there exists a gain $L \in \mathbb{R}^{n\times p}$ such that

$$|\arg(\text{spec}(A + LC)| > \alpha\frac{\pi}{2}.$$

Since, we can displace only the observable eigenvalues of the pair (A, C) from A to $A + LC$, we have the following. This concept can be formulated using the following two definitions.

Definition 3.3.2 [20] The pair (A, C) is called detectable if and only if all its unobservable eigenvalues verify $|\arg(\sec(A))| > \alpha\frac{\pi}{2}$.

The stabilization of (2.43) by static output feedback is now introduced.

Theorem 3.3.1 *[20] The system* (2.43) *is stabilizable by static output feedback if there exists a control law* $u(t) = Ly(t), L \in \mathbb{R}^{m\times p}$ *such that the closed-loop system*

$$D^{\alpha}x(t) = (A + BLC)x(t)$$

is asymptotically stable, or equivalently
i.e.,

$$|\arg(\text{spec}(A + BLC)| > \alpha\frac{\pi}{2}.$$

Proof La proof of this theorem is obtained directly by replacing matrix A with $A = A + BLC$ in Theorem 2.2.2. □

It is now quite clear that a given system (2.43) can be made asymptotically stable via the static output feedback control law $u(t) = Ly(t)$ only when all the unobservable and the uncontrollable eigenvalues of A are already in the stable region. This is so because static output feedback can alter only the observable and the controllable eigenvalues.

Example 3.3.1 Consider the fractional-order system (2.43) with the following parameters

$$A = \begin{bmatrix} 1\,4\,5 \\ 0\,2\,6 \\ 1\,0\,3 \end{bmatrix}, \quad B = \begin{bmatrix} 1\,1 \\ 1\,0 \\ 0\,0 \end{bmatrix},$$

$$C = \begin{bmatrix} 1\,0\,0 \\ 0\,1\,0 \end{bmatrix}, \quad \alpha = 1.5.$$

Obviously, when $u(t) = 0$, the system (2.43) is unstable because the eigenvalues of the matrix A are 5.5775+0.0000i, 0.2162+1.8829i and 0.2162-1.8829i, which are outside the stability area as can be shown in Fig. 3.6.

The observability matrix is

$$\mathcal{O} = \begin{bmatrix} 1 & 0 & 0 \\ 0 & 1 & 0 \\ 1 & 4 & 5 \\ 0 & 2 & 6 \\ 6 & 12 & 44 \\ 6 & 4 & 30 \end{bmatrix},$$

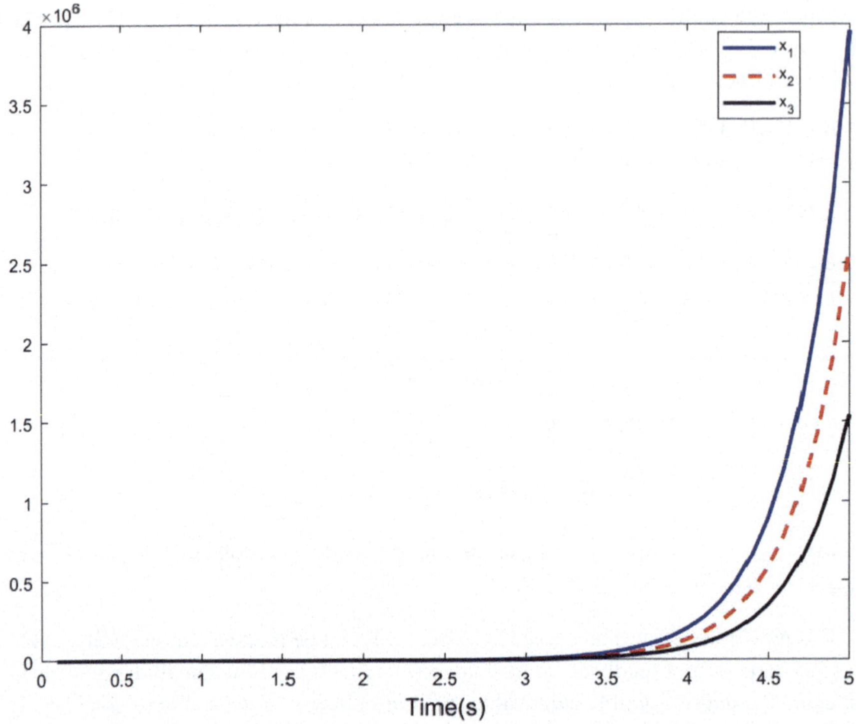

Fig. 3.6 State responses of the open-loop system with $\alpha = 1.5$

and the controllability matrix is

$$\mathcal{C} = \begin{bmatrix} 1 & 1 & 5 & 1 & 18 & 6 \\ 1 & 0 & 2 & 0 & 10 & 6 \\ 0 & 0 & 1 & 1 & 8 & 4 \end{bmatrix}.$$

Since $\text{rank}(\mathcal{O}) = 3$ and $\text{rank}(\mathcal{C}) = 3$, the system is observable and controllable, so all its eigenvalues can be shifted by static output feedback $u(t) = Ly(t)$. To get the eigenvalues of the closed-loop system at $\{-2.9847, -2.0251, -0.9902\}$, the static output feedback gain can be computed in several different ways. For simple problems it is convenient to introduce the elements of L as unknown parameters, determine the characteristic polynomial of the closed-loop system and identify it with the desired characteristic polynomial. the resulting static output feedback gain is

$$L = \begin{bmatrix} 147.1667 & -155 \\ -4.1667 & 1 \end{bmatrix}.$$

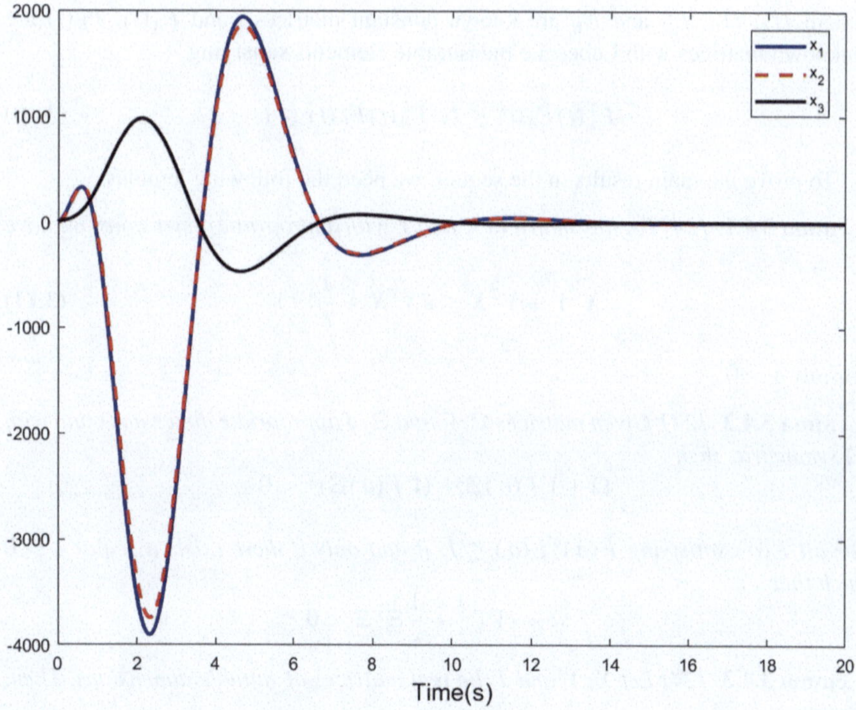

Fig. 3.7 State responses of the closed-loop system with $\alpha = 1.5$

The closed-loop system $D^{\alpha}x(t) = (A + BLC)x(t)$ is then asymptotically stable. This result is shown in Fig. 3.7.

3.4 Robust Control of Uncertain FO-LTI Systems

Consider the fractional-order linear system described by the following form:

$$\begin{cases} D^{\alpha}x(t) = (A + \Delta A(t))x(t) + (B + \Delta B(t))u(t), \\ \ \ y(t) \ \ \ = Cx(t), \end{cases} \tag{3.8}$$

where $0 < \alpha < 2$ is the fractional commensurate order, $x(t) \in \mathbb{R}^n$ is the state of the plant, $u(t) \in \mathbb{R}^m$ and $y(t) \in \mathbb{R}^p$ are the control input and output , respectively. A, B and C are constant matrices with appropriate dimensions. $\Delta A(t)$ and $\Delta B(t)$ represent the following admissible time-variant uncertainties:

$$\left[\, \Delta A(t) \ \ \Delta B(t) \,\right] = \left[\, D_A F_A(t) E_A \ \ D_B F_B(t) E_B \,\right], \tag{3.9}$$

where D_A, D_B, E_A and E_B are known constant matrices , and $F_A(t)$, $F_B(t)$ are unknown matrices with Lebesgue measurable elements satisfying

$$F_A^{\mathrm{T}}(t)F_A(t) \leq I,\ F_B^{\mathrm{T}}(t)F_B(t) \leq I. \tag{3.10}$$

To prove the main results in the sequel, we need the following lemmas.

Lemma 3.4.1 *[24] For any matrices X and Y with appropriate dimensions, we have*

$$X^TY + Y^TX \leq \delta X^TX + \frac{1}{\delta}Y^TY \tag{3.11}$$

for any $\delta > 0$

Lemma 3.4.2 *[24] Given matrices* Ω, Γ *and* Ξ *of appropriate dimensions and with* Ω *symmetric, then*

$$\Omega + \Gamma F(\sigma)\Xi + (\Gamma F(\sigma)\Xi)^T < 0$$

for all $F(\sigma)$ *satisfying* $F(\sigma)^TF(\sigma) \leq I$, *if and only if there exists a scalar* $\epsilon > 0$ *such that*

$$\Omega + \epsilon\Gamma\Gamma^T + \frac{1}{\epsilon}\Xi^T\Xi < 0.$$

Lemma 3.4.3 *[24] Let X, Y and F be real matrices of suitable dimensions. Then, for any* $x \in \mathbb{R}^n$

$$\max\left\{(x^TXFYx)^2/F^TF \leq I\right\} = (x^TXX^Tx)(x^TY^TYx).$$

Lemma 3.4.4 *[24] Let X, Y and Z be* $n \times n$ *symmetric matrices such as* $X \geq 0$, $Y < 0$ *and* $Z \geq 0$. *Furthermore, assume that* $(x^TYx)^2 - 4(x^TXx)(x^TZx) > 0$ *for all non-zero* $x \in \mathbb{R}^n$. *Then, there exists a constant* $\lambda > 0$ *such that*

$$\lambda^2X + \lambda Y + Z < 0.$$

Recall that for any matrix $\Pi \in \mathbb{R}^{m\times n}$ with full row rank, there exists a singular value decomposition (SVD) of Π as follows:

$$\Pi = U\begin{bmatrix} S & 0 \end{bmatrix}V^{\mathrm{T}}, \tag{3.12}$$

where $S = \mathrm{diag}\{\sigma_1, \sigma_2, ..., \sigma_m\}$, $\sigma_1 \geq \sigma_2 \geq ... \geq \sigma_p \geq 0$, $U \in \mathbb{R}^{m\times m}$, $V \in \mathbb{R}^{n\times n}$ are unitary matrices. In particular, the following lemma holds:

Lemma 3.4.5 *[25] Given matrix* $\Pi \in \mathbb{R}^{m\times n}$ *with* $\mathrm{rank}\,\Pi = p$, *assume that* $X \in \mathbb{R}^{n\times n}$ *is a symmetric matrix, then there exists a matrix* $\tilde{X} \in \mathbb{R}^{m\times m}$ *satisfying*

$$\Pi X = \tilde{X}\Pi$$

if and only if X can be expressed as

$$X = V \begin{bmatrix} X_{11} & 0 \\ 0 & X_{22} \end{bmatrix} V^T,$$

where $X_{11} \in \mathbb{R}^{m\times m}$, $X_{22} \in \mathbb{R}^{(n-m)\times(n-m)}$.

3.4.1 Observer-Based Robust Control of Uncertain FOS

1. **Case one: $0 < \alpha \leq 1$**
The purpose in this section is to study the problem of asymptotical stabilization of the fractional-order linear system (3.8) with the fractional-order derivative α satisfying $0 < \alpha \leq 1$ under a "Luenberger-type" fractional-order linear observer

$$\begin{cases} D^\alpha x(t) = (A + \Delta A(t))x(t) + (B + \Delta B(t))u(t) + L(y(t) - \hat{y}(t)), \\ \hat{y}(t) \quad = C\hat{x}(t), \end{cases} \tag{3.13}$$

coupled with a state estimated feedback controller

$$u(t) = K\hat{x}(t), \tag{3.14}$$

where $0 < \alpha < 1$.
Let $e(t) = x(t) - \hat{x}(t)$ denote the error signal, under the controller (3.14). The closed-loop system is given by

$$\begin{cases} D^\alpha \hat{x}(t) = (\tilde{A} + \tilde{B}K)\hat{x}(t) + LCe(t), \\ D^\alpha e(t) = (\tilde{A} - LC)e(t), \end{cases} \tag{3.15}$$

with

$$\tilde{A} = A + \Delta A(t), \ \tilde{B} = B + \Delta B(t). \tag{3.16}$$

The closed-loop system (3.15) can be rewritten by

$$D^\alpha z(t) = A_{cl} z(t), \tag{3.17}$$

with

$$z(t) = \begin{bmatrix} \hat{x}(t) \\ e(t) \end{bmatrix}, \ A_{cl} = \begin{bmatrix} \tilde{A} + \tilde{B}K & LC \\ 0 & \tilde{A} - LC \end{bmatrix} \tag{3.18}$$

The following results are obtained.

Theorem 3.4.1 *[14] The uncertain fractional-order linear system* (3.17) *is asymptotically stable if there exist two real symmetric positive definite matrices* $Q_k,\ k = 1, 2$ *and four real scalars* $\epsilon_i > 0,\ \varsigma_i > 0,\ i = 1, 2,$ *such that*

$$\begin{bmatrix} \Pi_{11} & \Pi_{12} \\ \Pi_{12}^T & \Pi_{22} \end{bmatrix} < 0, \tag{3.19}$$

where

$$\begin{aligned}
\Pi_{11} &= \sum_{i=1}^{2} \text{sym}\left(\Theta_{i1} \otimes \begin{bmatrix} \Phi_{11} & LCQ_2 \\ 0 & \Phi_{12} \end{bmatrix}\right) + \sum_{i=1}^{2} \epsilon_i \left(I_2 \otimes \begin{bmatrix} D_A D_A^T & 0 \\ 0 & D_A D_A^T \end{bmatrix}\right) \\
&\quad + \sum_{i=1}^{2} \varsigma_i \left(I_2 \otimes \begin{bmatrix} D_B D_B^T & 0 \\ 0 & 0 \end{bmatrix}\right), \\
\Phi_{11} &= (A + BK)Q_1,\ \Phi_{12} = (A - LC)Q_2, \\
\Pi_{12} &= \begin{bmatrix} \Xi_1 & \Xi_1 & \Xi_2 & \Xi_2 \end{bmatrix}, \\
\Xi_1 &= I_2 \otimes \begin{bmatrix} E_A Q_1 & 0 \\ 0 & E_A Q_2 \end{bmatrix}^T,\ \Xi_2 = I_2 \otimes \begin{bmatrix} E_B K Q_1 & 0 \\ 0 & 0 \end{bmatrix}^T, \\
\Pi_{22} &= \text{diag}\{-\epsilon_1, -\epsilon_2, -\varsigma_1, -\varsigma_2\} \otimes I_{4n},
\end{aligned}$$

and $\Theta_{i1},\ i = 1, 2$ *satisfy* (2.24)

Due to the existence of the nonlinear terms KQ_1 and LCQ_2, the matrix inequality (3.19) in Theorem 3.4.1 is not an $\mathcal{LMI}$. However, applying Lemma 3.4.5, it can be transformed into an $\mathcal{LMI}$, as shown below, which presents an $\mathcal{LMI}$- based design method of observer-based control for fractional-order uncertain system (3.8).

Theorem 3.4.2 *[14] For the uncertain fractional-order linear system* (3.8) *with the fractional-order derivative* α *satisfying* $0 < \alpha < 1$, *assume that the singular value decomposition (SVD) of the output matrix C is*

$$C = U \begin{bmatrix} S & 0 \end{bmatrix} V^T. \tag{3.20}$$

Then, under the control input (3.14), *the closed-loop system is robust asymptotically stable if there exist two matrices* $X_1 \in \mathbb{R}^{p\times n},\ X_2 \in \mathbb{R}^{n\times p}$ *and symmetric positive definite matrices* $Q_1 \in \mathbb{R}^{n\times n},\ Q_{11} \in \mathbb{R}^{p\times p},\ Q_{22} \in \mathbb{R}^{(n-p)\times(n-p)}$ *and four real scalars* $\epsilon_i > 0,\ \varsigma_i > 0,\ i = 1, 2,$ *such that*

$$\begin{bmatrix} \Sigma_{11} & \Sigma_{12} \\ \Sigma_{12}^T & \Sigma_{22} \end{bmatrix} < 0, \tag{3.21}$$

where

$$\Sigma_{11} = \sum_{i=1}^{2} \mathrm{sym}\left(\Theta_{i1} \otimes \begin{bmatrix} \Phi_{11} & X_2C \\ 0 & \Phi_{12} \end{bmatrix}\right) + \sum_{i=1}^{2} \epsilon_i \left(I_2 \otimes \begin{bmatrix} D_A D_A^T & 0 \\ 0 & D_A D_A^T \end{bmatrix}\right)$$
$$+ \sum_{i=1}^{2} \varsigma_i \left(I_2 \otimes \begin{bmatrix} D_B D_B^T & 0 \\ 0 & 0 \end{bmatrix}\right),$$
$$\Phi_{11} = AQ_1 + BX_1,\ \Phi_{12} = AQ_2 - X_2C,$$
$$\Sigma_{12} = \begin{bmatrix} \Xi_1 & \Xi_1 & \Xi_2 & \Xi_2 \end{bmatrix},$$
$$\Xi_1 = I_2 \otimes \begin{bmatrix} E_AQ_1 & 0 \\ 0 & E_AQ_2 \end{bmatrix}^T,\ \Xi_2 = I_2 \otimes \begin{bmatrix} E_BX_1 & 0 \\ 0 & 0 \end{bmatrix}^T,$$
$$\Sigma_{22} = \mathrm{diag}\{-\epsilon_1, -\epsilon_2, -\varsigma_1, -\varsigma_2\} \otimes I_{4n},$$
$$Q_2 = V \begin{bmatrix} Q_{11} & 0 \\ 0 & Q_{22} \end{bmatrix} V^T,$$

and $\Theta_{i1},\ i = 1, 2$ *satisfy* (2.24).

Moreover, the robustly asymptotically stabilizing gain matrices are given by

$$K = X_1Q_1^{-1},\ L = X_2USQ_{11}^{-1}S^{-1}U^{-1}. \tag{3.22}$$

Proof Since the matrix Q_2 has the form

$$Q_2 = V \begin{bmatrix} Q_{11} & 0 \\ 0 & Q_{22} \end{bmatrix} V^{\mathrm{T}},$$

then Lemma 3.4.5 ensures the existence a matrix

$$\tilde{Q}_2 = USQ_{11}S^{-1}U^{\mathrm{T}},$$

such that

$$CQ_2 = \tilde{Q}_2C.$$

Set $X_1 = KQ_1,\ X_2 = L\tilde{Q}_2$, so the inequality (3.19) is equivalent to (3.21).

The proof is complete. □

Example 3.4.1 [14] Consider the fractional-order uncertain linear system (3.8) with the fractional-order derivative $\alpha = 0.50$, $u(t) = 0$ and the following parameters

$$
\begin{aligned}
A &= \begin{bmatrix} 0.40 & 0.35 & 0.65 \\ -0.90 & -2.70 & 1.10 \\ -0.50 & -1.35 & -2.25 \end{bmatrix}, \; B = \begin{bmatrix} 1.45 \\ 0.75 \\ 0.75 \end{bmatrix}, \\
D_A &= \begin{bmatrix} -0.10 & 0.05 & 0.10 \\ -0.10 & -0.30 & 0.10 \\ -0.15 & -0.08 & -0.40 \end{bmatrix}, \; E_A = I_3, \\
D_B &= \begin{bmatrix} -0.01 & 0.05 & 0.01 \\ -0.01 & -0.03 & 0.01 \\ -0.15 & -0.08 & -0.40 \end{bmatrix}, \\
E_B &= \left[0.10 \; 0.15 \; 0.15 \right]^{\mathrm{T}}, \; C = \left[1.50 \; 2.00 \; 1.00 \right], \\
F_A(t) &= F_B(t) = \mathrm{diag}\left\{ \sin(0.1\pi t), \cos(0.1\pi t), \sin(0.1\pi t) \right\}.
\end{aligned}
\tag{3.23}
$$

Using the Matlab $\mathcal{LMI}$ tools, we find that the linear matrix inequality (3.21) in Theorem 3.4.2 is feasible, which concludes that the uncertain fractional-order system (3.8) with the parameters in (3.23) is asymptotically stabilizable. The stabilizing gain matrices are

$$K = \left[-0.3722 \; 0.2541 \; 0.1505 \right],$$

$$L = \left[1.2709 \; -0.1772 \; 0.4016 \right]^{\mathrm{T}}.$$

2. **Case two: $1 \leq \alpha < 2$**
   The study is concerned with a method of observer-based control for fractional-order uncertain linear systems (3.8) with the fractional commensurate order $\alpha (1 \leq \alpha < 2)$ based on linear matrix inequality ($\mathcal{LMI}$) approach. We present sufficient conditions under which an observer-based controller of the following form:

$$\begin{cases} D^{\alpha} \hat{x}(t) = \hat{A} x(t) + Bu(t) + L(y(t) - \hat{y}(t)), \\ \hat{y}(t) \;\; = C\hat{x}(t), \end{cases} \tag{3.24}$$

   coupled with a state estimated feedback controller

$$u(t) = K\hat{x}(t), \tag{3.25}$$

   where $1 \leq \alpha < 2$.

Remark 3.4.1 The advantage of the fractional-order linear observer (3.24), comparing it with (3.13), is the fact that it contains two unknown system matrices, $\hat{A}$ and L, which give an opportunity to better adjust the the dynamical properties of the observer-based fractional-order control system.

Let $e(t) = x(t) - \hat{x}(t)$ denote the error signal, under the controller (3.25), the closed-loop system is given by

$$\begin{cases} D^{\alpha}\hat{x}(t) = (\hat{A} + BK)\hat{x}(t) + LCe(t), \\ D^{\alpha}e(t) = (\tilde{A} + \Delta BK - \hat{A})\hat{x}(t) + (\tilde{A} - LC)e(t), \end{cases} \tag{3.26}$$

with

$$\tilde{A} = A + \Delta A(t), \; \tilde{B} = B + \Delta B(t) \tag{3.27}$$

The closed-loop system (3.26) can be rewritten by

$$D^{\alpha}X(t) = A_{cl}X(t), \tag{3.28}$$

with

$$X(t) = \begin{bmatrix} \hat{x}(t) \\ e(t) \end{bmatrix}, \; A_{cl} = \begin{bmatrix} \hat{A} + \tilde{B}K & LC \\ \tilde{A} + \Delta BK - \hat{A} & \tilde{A} - LC \end{bmatrix}, \tag{3.29}$$

Sufficient condition is first derived for robust asymptotic stability of uncertain fractional-order linear system (3.28). Next, based on the stability criterion, an $\mathcal{LMI}$ based approach is proposed for designing a state-estimated feedback controller (3.25) to asymptotically stabilise the uncertain fractional-order linear system (3.8).

Theorem 3.4.3 *[26] The uncertain fractional-order linear system* (3.28) *with* $1 \leq \alpha < 2$ *is asymptotically stable if there exist symmetrical matrices* $P_1 > 0, \; P_2 > 0$ *together with matrices* $\hat{A}, K, L$ *of appropriate dimensions and a real scalar* $\epsilon > 0$ *such that*

$$\begin{bmatrix} \Xi_1 & \Xi_2 \\ * & \Xi_3 \end{bmatrix} < 0 \tag{3.30}$$

holds, where

$$\Xi_1 = \begin{bmatrix} \Phi_{11} & \Phi_{12} & \Phi_{13} & \Phi_{14} \\ * & \Phi_{22} & \Phi_{23} & \Phi_{24} \\ * & * & \Phi_{33} & \Phi_{34} \\ * & * & * & \Phi_{44} \end{bmatrix},$$

$$\Phi_{11} = \Phi_{33} = (\hat{A}P_1 + P_1\hat{A}^T + BKP_1 + P_1K^TB^T)\sin(\theta)$$

$$\Phi_{12} = \Phi_{34} = [LCP_2 + P_1(A - \hat{A})^T]\sin(\theta),$$

$$\Phi_{13} = (\hat{A}P_1 - P_1\hat{A}^T + BKP_1 - P_1K^TB^T)\cos(\theta),$$

$$\Phi_{14} = [LCP_2 - P_1(A - \hat{A})^T]\cos(\theta),$$

$$\Phi_{22} = \Phi_{44} = [(A - LC)P_2 + P_2(A - LC)^T]\sin(\theta) + \epsilon(D_AD_A^T + D_BD_B^T),$$

$$\Phi_{23} = [(A - \hat{A})P_1 - P_2(LC)^T]\cos(\theta),$$

$$\Phi_{24} = [(A - LC)P_2 - P_2(A - LC)^T]\cos(\theta),$$

$$\Xi_2 = \begin{bmatrix} P_1 E_A^T & (E_B K P_1)^T & 0 & 0 \\ P_2 E_A^T & 0 & 0 & 0 \\ 0 & 0 & P_1 E_A^T & (E_B K P_1)^T \\ 0 & 0 & P_2 E_A^T & 0 \end{bmatrix},$$
$$\Xi_3 = \operatorname{diag}\{-\epsilon I, -\epsilon I, -\epsilon I, -\epsilon I\},$$
$$\theta = \pi - \alpha\frac{\pi}{2}.$$

The matrix inequality (3.30) in Theorem 3.4.3 is bilinear, due to the existence of the nonlinear terms such as $BKP_1,\ LCP_2,\ E_BKP_1$. However, applying Lemma 3.4.5, it can be transformed into an $\mathcal{LMI}$, as shown below, which presents an $\mathcal{LMI}$-based design method of observer-based robust control for fractional-order uncertain systems.

Theorem 3.4.4 *[26] For the uncertain fractional-order linear system* (3.8) *with the fractional-order derivative* $\alpha(1 \le \alpha < 2)$*, assume that the singular value decomposition (SVD) of the output matrix* C *is*

$$C = U\begin{bmatrix} S & 0 \end{bmatrix}V^T. \tag{3.31}$$

Then, the closed-loop control system under the control input (3.25) *is robust asymptotically stable if there exist symmetrical matrices* $P_1 > 0,\ P_{11} > 0,\ P_{22} > 0,$ *matrices* $M,\ X_1$ *and* X_2 *together with a real scalar* $\epsilon > 0$ *such that*

$$\begin{bmatrix} \Psi_1 & \Psi_2 \\ \Psi_2^T & \Psi_3 \end{bmatrix} < 0 \tag{3.32}$$

holds, where

$$\Psi_1 = \begin{bmatrix} \Omega_{11} & \Omega_{12} & \Omega_{13} & \Omega_{14} \\ * & \Omega_{22} & \Omega_{23} & \Omega_{24} \\ * & * & \Omega_{33} & \Omega_{34} \\ * & * & * & \Omega_{44} \end{bmatrix},$$
$$\Omega_{11} = \Omega_{33} = (M + M^T + BX_1 + X_1^T B^T)\sin(\theta),$$
$$\Omega_{12} = \Omega_{34} = (X_2 C + P_1 A^T - M^T)\sin(\theta),$$
$$\Omega_{13} = (M - M^T + BX_1 - X_1^T B^T)\cos(\theta),$$
$$\Omega_{14} = (X_2 C - P_1 A^T + M^T)\cos(\theta),$$
$$\Omega_{22} = \Omega_{44} = (AP_2 + P_2 A^T - X_2 C - C^T X_2^T)\sin(\theta) + \epsilon(D_A D_A^T + D_B D_B^T),$$
$$\Omega_{23} = (AP_1 - M - C^T X_2^T\cos(\theta),$$
$$\Omega_{24} = (AP_2 - P_2 A^T - X_2 C + C^T X_2^T)\cos(\theta),$$

$$\Omega_2 = \begin{bmatrix} P_1 E_A^T & X_1^T E_B^T & 0 & 0 \\ P_2 E_A^T & 0 & 0 & 0 \\ 0 & 0 & P_1 E_A^T & X_1^T E_B^T \\ 0 & 0 & P_2 E_A^T & 0 \end{bmatrix},$$

$$\Omega_3 = \mathrm{diag}\{-\epsilon I, -\epsilon I, -\epsilon I, -\epsilon I\},$$

$$P_2 = V \begin{bmatrix} P_{11} & 0 \\ 0 & P_{22} \end{bmatrix} V^T,\ \theta = \pi - \alpha\frac{\pi}{2}.$$

Moreover, the robustly asymptotically stabilising matrices, $\hat{A}$, K and L are given by

$$\hat{A} = M P_1^{-1},\ K = X_1 P_1^{-1}, \tag{3.33}$$

$$L = X_2 U S X_{11}^{-1} S^{-1} U^{-1}. \tag{3.34}$$

Proof As the matrix

$$P_2 = V \begin{bmatrix} P_{11} & 0 \\ 0 & P_{22} \end{bmatrix} V^T,$$

then according to Lemma 3.4.5, there exists a matrix $\bar{P}_2$ satisfying $C P_2 = \bar{P}_2 C$. It follows that $\bar{P}_2 = U S P_{11} S^{-1} U^{-1}$.

By setting

$$X_1 = K P_1,\ M = \hat{A} P_1,\ X_2 = L \bar{P}_2,$$

inequality (3.32) is equivalent (3.30) in Theorem 3.4.3. The proof is completed. □

In particular case where $\hat{A} = A$ in observer (3.24), we obtain the classical observer-based robust controller (3.13) for the system uncertain fractional-order linear system (3.8). The stabilising controller gain matrices K, L can be solved from the following corollary.

Corollary 3.4.1 *[26] For the uncertain fractional-order linear system* (3.8) *with the fractional-order derivative $\alpha(1 \leq \alpha < 2)$, assume that the singular value decomposition (SVD) of the output matrix C is*

$$C = U \left[S\ 0 \right] V^T. \tag{3.35}$$

Then, under the observer-based robust controller (3.13) *coupled with a state estimated feedback controller* (3.25)*, the uncertain fractional-order linear system* (3.8) *is robust asymptotically stabilizable if there exist symmetrical matrices $P_1 > 0$, $P_{11} > 0$, $P_{22} > 0$, matrices X_1 and X_2 together with a real scalar $\epsilon > 0$ such that*

$$\begin{bmatrix} \Lambda_1 & \Lambda_1 \\ & \Lambda_3 \end{bmatrix} \tag{3.36}$$

holds, where

$$\Lambda_1 = \begin{bmatrix} \Lambda_{11} & X_2C\sin(\theta) & \Lambda_{13} & X_2C\cos(\theta) \\ * & \Lambda_{22} & \Lambda_{23} & \Lambda_{24} \\ * & * & \Lambda_{33} & X_2C\sin(\theta) \\ * & * & * & \Lambda_{44} \end{bmatrix},$$

$$\Lambda_{11} = \Lambda_{33} = (AP_1 + P_1A^T + BX_1 + X_1^TB^T)\sin(\theta),$$
$$\Lambda_{13} = (AP_1 + P_1A^T + BX_1 + X_1^TB^T)\cos(\theta),$$
$$\Lambda_{22} = \Lambda_{44} = (AP_2 + P_2A^T - X_2C - C^TX_2^T)\sin(\theta) + \epsilon(D_AD_A^T + D_BD_B^T),$$
$$\Lambda_{23} = -C^TX_2^T\cos(\theta),$$
$$\Lambda_{24} = (AP_2 - P_2A^T - X_2C + C^TX_2^T)\cos(\theta),$$

$$\Lambda_2 = \begin{bmatrix} P_1E_A^T & X_1^TE_B^T & 0 & 0 \\ P_2E_A^T & 0 & 0 & 0 \\ 0 & 0 & P_1E_A^T & X_1^TE_B^T \\ 0 & 0 & P_2E_A^T & 0 \end{bmatrix},$$

$$\Lambda_3 = \mathrm{diag}\{-\epsilon I, -\epsilon I, -\epsilon I, -\epsilon I\},$$

$$P_2 = V\begin{bmatrix} P_{11} & 0 \\ 0 & P_{22} \end{bmatrix}V^T,\ \theta = \pi - \alpha\frac{\pi}{2}.$$

Moreover, the robustly asymptotically stabilising gain matrices, K and L are given by

$$K = X_1P_1^{-1},\ L = X_2USX_{11}^{-1}S^{-1}U^{-1}. \tag{3.37}$$

Example 3.4.2 Consider the fractional-order uncertain linear system (3.8) with the fractional order $\alpha = 1.50$, $u(t) = 0$ and

$$\begin{aligned} A &= \begin{bmatrix} -1.5 & 0.5 & 1.00 \\ -1.0 & -3.0 & 1.0 \\ -0.50 & -1.5 & -2.5 \end{bmatrix},\ B = \begin{bmatrix} 0.45 \\ 0.55 \\ 0.85 \end{bmatrix}, \\ D_A &= \begin{bmatrix} -0.10 & 0.05 & 0.10 \\ -0.10 & -0.230 & 0.10 \\ -0.15 & -0.2 & -0.140 \end{bmatrix},\ E_A = I_3, \\ D_B &= \begin{bmatrix} -0.1 & 0.5 & 0.1 \\ -0.1 & -0.3 & 0.1 \\ -0.15 & -0.28 & -0.40 \end{bmatrix}, \\ E_B &= \begin{bmatrix} 0.10\ 0.15\ 0.20 \end{bmatrix}^T,\ C = \begin{bmatrix} 1.50\ 2.00\ 1.00 \end{bmatrix}, \end{aligned} \tag{3.38}$$

$$F_A(t) = F_B(t) = \mathrm{diag}\{\sin(0.1\pi t), \cos(0.1\pi t), \sin(0.1\pi t)\}.$$

This example has been investigated in [27], which concluded that it was not asymptotically stable. (See Example 2 in [27].) Under the aforementioned conditions, the $\mathcal{LMI}$ (3.32) of Theorem 3.4.4 was found feasible, where

$$
\begin{aligned}
P_1 &= \begin{bmatrix} 0.1979 & -0.0450 & 0.0006 \\ -0.0450 & 0.0947 & -0.0347 \\ 0.0006 & -0.0347 & 0.0936 \end{bmatrix}, \\
\epsilon &= 1.3060, \ P_{11} = 0.2468, \\
P_{22} &= \begin{bmatrix} 0.5141 & 0.0626 \\ 0.0626 & 0.3218 \end{bmatrix}, \\
X_1 &= \begin{bmatrix} -0.0959 & -0.2662 & -0.6276 \end{bmatrix}, \\
X_2 &= \begin{bmatrix} 0.1451 & 0.0630 & -0.0496 \end{bmatrix}^{\mathrm{T}}, \\
\hat{A} &= \begin{bmatrix} -2.9920 & -0.1395 & 1.8769 \\ -1.6513 & -5.8810 & 0.4643 \\ -0.0107 & 0.0021 & -2.5483 \end{bmatrix}, \\
K &= \begin{bmatrix} -2.1063 & -7.2523 & -9.3876 \end{bmatrix}, \\
L &= \begin{bmatrix} 0.5879 & 0.2555 & -0.2009 \end{bmatrix}^{\mathrm{T}}.
\end{aligned}
$$

The state responses of the selected system is shown in Fig. 3.8, which shows that it asymptotically stable and its states converge to zero.

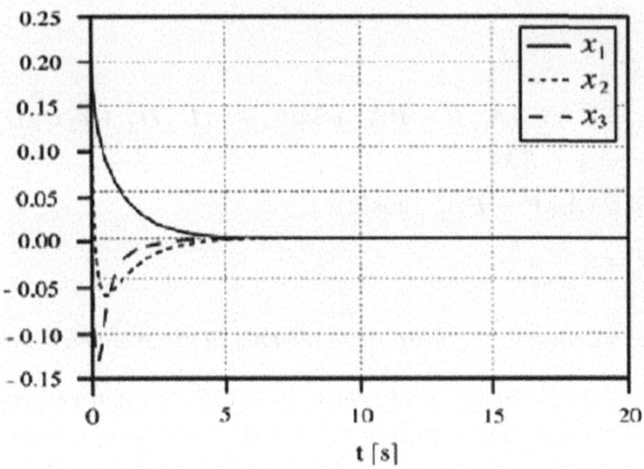

Fig. 3.8 State responses of the selected system in Example 3.4.2 with the observer-based control $u(t) = K\hat{x}(t)$ when $\alpha = 1.5$

3.4.2 Robust Static Output Feedback Control of Uncertain FOS: $1 \leq \alpha < 2$

For the uncertain fractional-order linear system (3.8) with the fractional-order derivative α such that $1 \leq \alpha < 2$ and under the controller

$$u(t) = Ky(t), \tag{3.39}$$

The closed-loop system is given by

$$D^{\alpha} x(t) = (\tilde{A} + \tilde{B}KC)x(t) = A_{lp}x(t). \tag{3.40}$$

$\tilde{A}$ and $\tilde{B}$ are defined in (3.16). We have the following results.

Theorem 3.4.5 *[14] The fractional-order uncertain linear system* (3.40) *with* $1 \leq \alpha < 2$ *is asymptotically stable if and only if there exist symmetrical matrix* $P > 0$ *together with gain matrix* K *and two scalars* $\epsilon_1 > 0$, $\epsilon_2 > 0$ *such that*

$$\begin{bmatrix} \Gamma_{11} & \Gamma_{12} & PE_A^T & 0 & PC^TK^TE_B^T & 0 \\ * & \Gamma_{22} & 0 & PE_A^T & 0 & PC^TK^TE_B^T \\ * & * & -\epsilon_1 I & 0 & 0 & 0 \\ * & * & * & -\epsilon_1 I & 0 & 0 \\ * & * & * & * & -\epsilon_2 I & 0 \\ * & * & * & * & * & -\epsilon_2 I \end{bmatrix} < 0 \tag{3.41}$$

holds, where

$$\begin{aligned} \Gamma_{11} &= \Gamma_{22} = (A_{oc}P + PA_{oc}^T)\sin(\theta) + \epsilon_1 D_A D_A^T + \epsilon_2 D_B D_B^T, \\ A_{oc} &= A + BKC, \\ \Gamma_{12} &= (A_{oc}P - PA_{oc}^T)\cos(\theta), \\ \theta &= \pi - \alpha\frac{\pi}{2}, \end{aligned}$$

and the symbol $*$ *denotes the transposed elements in the symmetric positions.*

Due to the existence of the nonlinear term KCP, the matrix inequality (3.41) in Theorem 3.4.5 is not linear. In order to get an $\mathcal{LMI}$, one apply Lemma 3.4.5, as shown below, which presents an $\mathcal{LMI}$-based design method of static output feedback control for fractional-order uncertain linear systems.

Theorem 3.4.6 *[14] For the uncertain fractional-order linear system* (3.8) *with the fractional-order derivative* $\alpha(1 \leq \alpha < 2)$*, assume that the singular value decomposition (SVD) of the output matrix* C *is*

$$C = U\begin{bmatrix} S & 0 \end{bmatrix}V^T. \tag{3.42}$$

Then, under the control input (3.39)*, the closed-loop system is robust asymptotically stable if and only if there exist symmetrical matrices* $P > 0,\ P_{11} > 0,\ P_{22} > 0,$ *matrix* X *together with* $\epsilon_1 > 0,\ \epsilon_2 > 0$ *such that*

$$\begin{bmatrix} \Phi_{11} & \Phi_{12} & PE_A^T & 0 & C^TX^TE_B^T & 0 \\ * & \Phi_{22} & 0 & PE_A^T & 0 & C^TX^TE_B^T \\ * & * & -\epsilon_1 I & 0 & 0 & 0 \\ * & * & * & -\epsilon_1 I & 0 & 0 \\ * & * & * & * & -\epsilon_2 I & 0 \\ * & * & * & * & * & -\epsilon_2 I \end{bmatrix} < 0 \tag{3.43}$$

holds, where

$$\Phi_{11} = \Phi_{22} = (AP + PA^T + BXC + C^TX^TB^T)\sin(\theta) + \epsilon_1 D_A D_A^T + \epsilon_2 D_B D_B^T,$$
$$\Phi_{12} = (AP - PA^T + BXC - C^TX^TB^T)\cos(\theta),$$
$$P = V\begin{bmatrix} P_{11} & 0 \\ 0 & P_{22} \end{bmatrix} V^T,\ \theta = (1-\alpha)\frac{\pi}{2},$$

and the symbol $*$ *denotes the transposed elements in the symmetric positions. Moreover, the robustly asymptotically stabilizing output feedback gain matrix* K *is given by*

$$K = XUSP_{11}^{-1}S^{-1}U^{-1}. \tag{3.44}$$

Proof Since

$$P = V\begin{bmatrix} P_{11} & 0 \\ 0 & P_{22} \end{bmatrix} V^T,$$

then according to Lemma 3.4.5, there exists $\tilde{P}$, such that $CP = \tilde{P}C$. It follows

$$\tilde{P}^{-1} = USP_{11}^{-1}S^{-1}U^{-1}.$$

Setting $X = K\tilde{P}$, inequality (3.41) is equivalent to inequality (3.43). The proof is completed. □

Example 3.4.3 [14] Consider the fractional-order uncertain linear system (3.8) with the fractional-order derivative $\alpha = 1.50$, $u(t) = 0$ and the following parameters

$$\begin{aligned} A &= \begin{bmatrix} -1.5 & 0.5 & 1.00 \\ -1.0 & -3.0 & 1.0 \\ -0.50 & -1.5 & -2.5 \end{bmatrix}, \; B = \begin{bmatrix} 0.45 \\ 0.55 \\ 0.85 \end{bmatrix}, \\ D_A &= \begin{bmatrix} -0.10 & 0.05 & 0.10 \\ -0.10 & -0.230 & 0.10 \\ -0.15 & -0.2 & -0.140 \end{bmatrix}, \; E_A = I_3, \\ D_B &= \begin{bmatrix} -0.1 & 0.5 & 0.1 \\ -0.1 & -0.3 & 0.1 \\ -0.15 & -0.28 & -0.40 \end{bmatrix}, \\ E_B &= \begin{bmatrix} 0.10 \; 0.15 \; 0.20 \end{bmatrix}^{\mathrm{T}}, \; C = \begin{bmatrix} 1.50 \; 2.00 \; 1.00 \end{bmatrix}, \\ F_A(t) &= F_B(t) = \mathrm{diag}\{\sin(0.1\pi t), \cos(0.1\pi t), \sin(0.1\pi t)\}. \end{aligned} \tag{3.45}$$

Using the Matlab $\mathcal{LMI}$ tools, we find that the linear matrix inequality (3.43) in Theorem 3.4.6 is feasible, which concludes that the uncertain fractional-order system (3.8) with the parameters in (3.38) is asymptotically stabilizable. The stabilizing gain matrix is

$$K = -1.0356$$

.

3.5 Summary

This chapter serves as an introduction to fractional-order control, providing foundational insights into this advanced field. It explores key results on control design techniques based on state-space representations, as well as the development of observers tailored for fractional-order systems. Furthermore, the chapter addresses the design of a robust stabilizing controller that accounts for system uncertainties, offering practical solutions to enhance stability and performance in fractional-order systems.

References

1. El-Khazali, R.: Fractional-order $pi^{\lambda}d^{\mu}$ controller design. Comput. Math. Appl. **66**, 639–646 (2013)
2. Liu, S., Wei, G., Liu, Y.: Error-constrained-reliable tracking control for discrete time-varying systems subject to quantization effects. Neurocomputing **174**, 897–905 (2016)
3. Lin, C., Chen, B., Zhuang, H., Zhang, L.: A direct method of static output feedback design for t-s fuzzy systems. In: Proceedings of the 11th IEEE International Conference on Control and Automation, pp. 445–448, Taichung, June 2014
4. Moreira, H.R., Oliveira, R.C.L.F., Peres, P.L.D.: Robust h_2 static output feedback design starting from a parameter dependent state feedback controller for time-invariant discrete-time polytopic systems. Optim. Control Appl. Methods **32**, 1–13 (2011)

5. Shu, Z., Lam, J., Xiong, J.: Static output-feedback stabilization of discrete-time Markovian jump linear systems: a system augmentation approach. Automatica **46**, 687–694 (2010)
6. Matignon, D.: Stability results for fractional differential equations with applications to control. In: Processing, Computational Engineering in Systems and Application Multi-Conference, IMACS, IEEE-SMC, vol. 2, pp. 963–968, Lille, France, July 1996
7. Matignon, D., André, B.: A-novel some results on controllability and observability of finite-dimensional fractional differential systems. In: Proceedings of the Mathematical Theory of Networks and Systems Symposium, Lille, France (1996)
8. Moze, M., Sabatier, J., Oustaloup, A.: Lmi tools for stability analysis of fractional systems. In: Proceedings of the ASME IDET/CIE Conferences, Long-Beach, USA (2005)
9. Sabatier, J., Moze, M., Farges, C.: On stability of fractional order systems. In: Proceedings of the Third IFAC Workshop on Fractional Differentiation and its Applications (FDA'08), Ankara, Turkey (2008)
10. Sabatier, J., Moze, M., Farges, C.: $\mathcal{LMI}$ conditions for fractional order systems. Comput. Math. Appl. **59**, 1594–1609 (2010)
11. N'Doye, I., Zasadzinski, M., Darouach, M., Radhy, N.: Observer-based control for fractional-order continuous-time systems. In: Joint 48th IEEE Conference on Decision and Control and 28th Chinese Control Conference, Shanghai, PR China, pp. 1932–1938, Dec 16–18 (2009)
12. N'Doye, I., Zasadzinski, M., Radhy, N.E., Bouaziz, A.: Stabilization of a class of nonlinear affine fractional-order systems using generalizations of Bellman-Gronwall lemma. In: 17th Mediterranean Conference on Control and Automation, Makedonia Palace, Thessaloniki, Greece, pp. 324–330, June 24–26 (2009)
13. Farges, C., Moze, M., Sabatier, J.: Pseudo-state feedback stabilization of commensurate fractional-order systems. Automatica **46**, 1730–1734 (2010)
14. Lan, Y.H., Zhou, Y.: Lmi-based robust control of fractional order uncertain linear systems. Comput. Math. Appl. **62**, 1460–1471 (2011)
15. Li, C., Wang, J.: Robust stability and stabilization of fractional order interval systems with coupling relationships: the $0<\alpha<1$ case. J. Franklin Inst. **349**, 2406–2419 (2012)
16. Li, Y.T., Li, J.M.: Stability analysis of fractional order systems based on t-s fuzzy model with the fractional order α:$0<\alpha<1$. Nonlinear Dyn. **78**, 2909–2919 (2014)
17. Li, J., Li, Y.: Robust stability and stabilization of fractional order systems based on uncertain Takagi-Sugeno fuzzy model with the fractional order $1 \leq \alpha<2$. J. Comput. Nonlinear Dyn. **8**, 1–7 (2013)
18. Ji, Y.D., Su, L.Q., Qiu, J.Q.: Design of fuzzy output feedback stabilization for uncertain fractional-order systems. Neurocomputing **173**, 1683–1693 (2016)
19. Lin, C., Chen, B., Wang, Q.G.: Static output feedback stabilization for fractional-order systems in t-s fuzzy models. http://dx.doi.org/10.1016/j.neucom.2016.08.085i
20. N'Doye, I.: Généralisation du lemme de Gronwall-Bellman pour la stabilisation des systèmes fractionnaires. Automatic. Université Henri Poincaré - Nancy I; Université Hassan II Ain Chock de Casablanca (2011). French. ¡tel-00584402¿
21. Chilali, M., Gahinet, P.: h_∞ design with pole placement constraints: an $\mathcal{LMI}$ approach. IEEE Trans. Automat. Contr. **41**, 358–367 (1996)
22. Boyd, S., Vandenberghe, L.: Convex Optimization. SIAM, Cambridge University Press, Philadelphia (2004)
23. Boyd, S., Elghaoui, L., Féron, E., Balakrishnan, V.: Linear Matrix Inequality in Systems and Control Theory. SIAM, Philadelphia (1994)
24. Khargonakar, P.P., Petersen, I.R., Zhou, K.: Robust stabilization of uncertain linear systems: quadratic stability and $\mathfrak{h}_\infty$ control theory. IEEE Trans. Automat. Contr. **35**, 356–361 (1990)
25. MacDuffee, C.C.: The Theory of Matrices. Dover Publications, New York (2004)
26. Lan, Y.H., Huang, H.X., Zhou, Y.: Observer-based robust control of $\alpha(1 \leq \alpha<2)$ fractional-order uncertain systems: a linear matrix inequality approach. IET Control Theory Appl. **6**, 229–234 (2011). https://doi.org/10.1049/iet-cta.2010.0484
27. Lu, J.G., Chen, G.R.: Robust stability and stabilization of fractional-order interval systems: an lmi approach. IEEE Trans. Autom. Control **54**, 1294–1300 (2009)

Chapter 4
H_∞ Control for Fractional-Order Systems

4.1 Introduction

Fractional-order systems have attracted extensive attention during the past few years. Furthermore, as a result of the great efforts invested by the researchers, many valuable results are derived about stability and controller synthesis of FOS. Nowadays, as the hot topics of researcher field in control theory, stability and H_∞ norm performance for this class of systems have been extensively studied [1–4]. The key idea of evaluating the H_∞-norm of FOS using $\mathcal{LMI}$-based approach is the bounded real lemma (BRL) [5, 6]. As a special case of the generalized Kalman-Yakubovic-Popov (KYP) Lemma [7, 8], versions of the bounded real lemmas for FOS have been presented in [5, 6, 9, 10] and the synthesis problems of H_∞ control were addressed in [4, 9, 11].

4.2 Preliminaries

Consider the following fractional order system admitting a pseudo state representation of the form

$$\begin{cases} D^\alpha x(t) = Ax(t) + Bw(t), \; 0 < \alpha < 2, \\ \quad y(t) = Cx(t) + Dw(t), \end{cases} \tag{4.1}$$

where α denotes the fractional order of the system; $x(t) \in \mathbb{R}^n$, $w(t) \in \mathbb{R}^m$ and $y(t) \in \mathbb{R}^p$ are the pseudo state, input vector and the output vector respectively. A, B, C, D are constant real matrices with appropriate sizes. The transfer function between $w(t)$ and $y(t)$ is given by

$$G_{wy}(s) = C(s^\alpha I - A)^{-1}B + D. \tag{4.2}$$

S. Marir and M. Chadli, *Control of Singular Fractional Order Systems: LMI Approach*, Studies in Systems, Decision and Control 292,
https://doi.org/10.1007/978-3-031-87382-9_4

The minimal realization is very useful for obtaining pseudo-state representations from fractional-order transfer functions.

Assumption 4.2.1 The fractional-order linear time invariant system (A, B, C, D) is the minimal realization of order α if, and only if, it is concurrently observable and controllable.

Definition 4.2.1 [12] The $\mathcal{H}_\infty$ norm of the transfer function $G_{wy(s)}$ is defined by

$$\left\|G_{wy}(s)\right\|_{\mathcal{H}_\infty} = \sup_{\mathrm{Re}(s)\geq 0} \overline{\sigma}(G_{wy}(s)). \tag{4.3}$$

Lemma 4.2.1 *[13] the transfer function $G_{wy(s)}$ is without poles in* $\mathrm{Re}(s) > 0$ *and* $\left\|G_{wy(s)}\right\|_{\mathcal{H}_\infty}$ *is bounded, then*

$$\left\|G_{wy(s)}\right\|_{\mathcal{H}_\infty} = \sup_{w\in\mathbb{R}} \bar{\sigma}(G_{wy}(jw)) = \sup_{w\geq 0} \bar{\sigma}(G_{wy}(jw)).$$

The frequency ranges can be characterized in the generalized Kalman-Yakubovic-Popov (KYP) lemma as the set of complex numbers that represent a certain class of curves.

Lemma 4.2.2 *[7] (Generalized KYP) Let matrices $A \in \mathbb{R}^{n\times n}$, $B \in \mathbb{R}^{n\times m}$, $\Theta \in \mathbb{H}_{n+m}$, $\Phi \in \mathbb{H}_2$ and $\Psi \in \mathbb{H}_2$. Set Λ is defined as*

$$\Lambda(\Theta, \Psi) = \left\{\lambda \in \mathbb{C}, \begin{bmatrix}\lambda \\ 1\end{bmatrix}^* \Phi \begin{bmatrix}\lambda \\ 1\end{bmatrix} = 0, \begin{bmatrix}\lambda \\ 1\end{bmatrix}^* \Psi \begin{bmatrix}\lambda \\ 1\end{bmatrix} \geq 0\right\}. \tag{4.4}$$

Consider the following two statements:

1. For $H(\lambda) = (\lambda I_n - A)^{-1}B$, there holds

$$\begin{bmatrix}H(\lambda) \\ I_m\end{bmatrix}^* \Theta \begin{bmatrix}H(\lambda) \\ I_m\end{bmatrix} < 0, \ \forall\lambda \in \Lambda. \tag{4.5}$$

2. There exist, $P, Q \in \mathbb{H}_n$, and $Q > 0$ such that

$$\begin{bmatrix}A & B \\ I_n & 0\end{bmatrix}^* (\Phi \otimes P + \Psi \otimes Q) \begin{bmatrix}A & B \\ I_n & 0\end{bmatrix} + \Theta < 0. \tag{4.6}$$

Then "2. $\Longrightarrow$ 1." always holds. Moreover, if Λ represents a curve in the complex plane, there holds "2. $\Longleftrightarrow$ 1.".

To extend the KYP lemma for linear fractional-order systems, the curves described by Λ can be relaxed to Υ as the following lemma.

Lemma 4.2.3 *(Generalized KYP) [9] Let set* $\Lambda(\Phi, \Psi)$ *in Lemma 4.2.2 be replaced by* $\Upsilon(\Phi, \Psi)$*, which is defined as*

$$\Upsilon(\Phi, \Psi) = \left\{\lambda \in \mathbb{C}, \begin{bmatrix} \lambda \\ 1 \end{bmatrix}^* \Phi \begin{bmatrix} \lambda \\ 1 \end{bmatrix} \geq 0, \begin{bmatrix} \lambda \\ 1 \end{bmatrix}^* \Psi \begin{bmatrix} \lambda \\ 1 \end{bmatrix} \geq 0\right\}. \tag{4.7}$$

Then, $\mathcal{LMI}$ *of* (4.5) *holds for* $\forall\lambda \in \Upsilon$ *if there exist matrices* $P > 0$ *and* $Q > 0$ *such that* $\mathcal{LMI}$ *of* (4.6) *holds.*

Proof See the Appendix D.

Lemma 4.2.4 (Schur Complement) *[14] Given any real matrices* Q, S *and* R *with* $Q = Q^{\mathrm{T}}$ *and* $R = R^{\mathrm{T}}$*. Then we have*

$$\begin{bmatrix} Q & S \\ S^{\mathrm{T}} & R \end{bmatrix} > 0 \tag{4.8}$$

if and only if

$$\begin{cases} R > 0, \\ Q - SR^{-1}S^{\mathrm{T}} > 0, \end{cases} \tag{4.9}$$

or equivalently

$$\begin{bmatrix} R & S^{\mathrm{T}} \\ S & Q \end{bmatrix} < 0, \tag{4.10}$$

or equivalently

$$\begin{cases} Q > 0, \\ R - S^{\mathrm{T}}R^{-1}S > 0. \end{cases} \tag{4.11}$$

4.3 Bounded Real Lemma for FOS

In this section, based on the generalized KYP, two bounded real lemmas of FOS are introduced in $\mathcal{LMI}$ formulations.

Theorem 4.3.1 *[9] Consider the fractional-order system* (4.1) *with its transfer function* $G_{wy}(s)$ *in* (4.2) *and* $\gamma > 0$*. Then* $\left\|G_{wy}(s)\right\|_{\mathcal{H}_\infty} < \gamma$ *holds if the following conditions are satisfied, depending on the value of* γ*.*

Case 1 $1 \leq \alpha < 2$: $\exists P \in \mathbb{H}_n$, $P > 0$ *such that*

$$\begin{bmatrix} \mathrm{Her}\left\{e^{j\theta}AP\right\} & * & * \\ Ce^{j\theta}P & -\gamma I_P & * \\ B^T & D^T & -\gamma I_m \end{bmatrix} < 0. \tag{4.12}$$

Case 2 $\mathbf{0 < \alpha < 1}$: $\exists P, Q \in \mathbb{H}_n,\ P > 0,\ Q > 0$ *such that*

$$\begin{bmatrix} \text{Her}\left\{A(e^{j\theta}P + e^{-j\theta}Q)\right\} & * & * \\ C(e^{j\theta}P + e^{-j\theta}Q) & -\gamma I_P & * \\ B^T & D^T & -\gamma I_m \end{bmatrix} < 0, \tag{4.13}$$

where $\theta = \dfrac{\pi}{2}(1-\alpha)$.

Moreover, in the case $1 \le \alpha < 2$, *the* $\mathcal{LMI}$ *condition* (4.12) *is also necessary.*

Proof Basically, the proof follows three steps:

1. Specifying a proper region corresponding to $\mathcal{H}_\infty$ norm.
2. Representing the region by $\mathcal{LMI}$ descriptions in (4.7).
3. Relaxing the norm condition into $\mathcal{LMI}$ by using Lemma 4.2.3.

Firstly, it is a fact that for any subregion Ω in the complex plane satisfying

$$\Omega \cup \bar{\Omega} = \{s^\alpha / s \in \mathbb{C}, \text{Re}(s) \ge 0\},$$

where $\bar{\Omega}$ is the symmetrical region of Ω with respect to the real axis, there must hold

$$\left\|G_{wy}(s)\right\|_{\mathcal{H}_\infty} = \sup_{\text{Re}(s)\ge 0} \overline{\sigma}(G_{wy}(s)) = \sup_{s\in\Omega} \overline{\sigma}(G_{wy}(s)).$$

This just follows from the maximum modulus principle and the complex conjugate symmetry of $G_{wy}(s)$.

Secondly, it is a easy to check that

$$\{s^\alpha / s \in \mathbb{C}, \text{Re}(s) \ge 0\} = \Upsilon \cup \bar{\Upsilon}, \tag{4.14}$$

where Υ is defined in (4.7) with its detailed data as

$$\Upsilon = \begin{cases} \Upsilon\left(\begin{bmatrix} 0 & e^{j\theta} \\ e^{-j\theta} & 0 \end{bmatrix}, \begin{bmatrix} 0 & e^{-j\theta} \\ e^{j\theta} & 0 \end{bmatrix}\right), & \text{if } 0 < \alpha < 1 \\ \Upsilon\left(\begin{bmatrix} 0 & e^{j\theta} \\ e^{-j\theta} & 0 \end{bmatrix}, 0\right), & \text{if } 1 \le \alpha < 2. \end{cases} \tag{4.15}$$

Suppose that the $\mathcal{LMI}s$ (4.12)and (4.13) are satisfied for some hermitian positive definite matrices P and Q. Let

$$X = \begin{cases} e^{j\theta}P + e^{-j\theta}Q, & \text{if } 0 < \alpha < 1, \\ e^{j\theta}P, & \text{if } 1 \le \alpha < 2. \end{cases}$$

So, we obtain the $\mathcal{LMI}$

$$\begin{bmatrix} \mathrm{Her}\{AX\} & * & * \\ CX & -\gamma I_P & * \\ B^T & D^T & -\gamma I_m \end{bmatrix} < 0, \tag{4.16}$$

which is the same as

$$\begin{bmatrix} \mathrm{Her}\{AX\} & * \\ CX & -\gamma^2 I_P \end{bmatrix} + \begin{bmatrix} B \\ D \end{bmatrix}\begin{bmatrix} B \\ D \end{bmatrix}^T < 0. \tag{4.17}$$

The last inequality is equivalent to

$$\begin{bmatrix} \mathrm{Her}\{AX\} & * \\ CX & 0 \end{bmatrix} + \begin{bmatrix} BB^T & BD^T \\ DB^T & DD^T - \gamma^2 I_P \end{bmatrix} < 0.$$

According to (4.15), the above $\mathcal{LMI}$ can be rewritten as

$$\begin{bmatrix} A^T & C^T \\ I_n & 0 \end{bmatrix}^T (\Phi \otimes P + \Psi \otimes Q) \begin{bmatrix} A^T & C^T \\ I_n & 0 \end{bmatrix} + \Theta < 0, \tag{4.18}$$

with

$$\Theta = \begin{bmatrix} BB^T & BD^T \\ DB^T & DD^T - \gamma^2 - I_n \end{bmatrix}. \tag{4.19}$$

Using Lemma 4.2.3, the inequality (4.18) implies that

$$\forall \lambda \in \Upsilon(\Phi, \Psi), \begin{bmatrix} H(\lambda) \\ I_n \end{bmatrix}^* \Theta \begin{bmatrix} H(\lambda) \\ I_n \end{bmatrix} < 0, \tag{4.20}$$

with $H(\lambda) = (\lambda I_n - A^T)^{-1} C^T$.

By some basic matrix calculations, the above $\mathcal{LMI}$ can be further simplified as

$$\forall \lambda \in \Upsilon(\Phi, \Psi), \left[C(\lambda I_n - A)^{-1} B + D\right]\left[C(\lambda I_n - A)^{-1} B + D\right]^* < \gamma^2 I_n. \tag{4.21}$$

Taking into account the identity (4.14), then

$$\forall \lambda \in \Upsilon(\Phi, \Psi), \lambda = s^\alpha, \text{ with } s \in \mathbb{C} \text{ and } \mathrm{Re}(s) \geq 0.$$

In this case, (4.21) is equivalent to

$$\forall s \in \mathbb{C}, \mathrm{Re}(s) \geq 0, \left[C(s^\alpha I_n - A)^{-1} B + D\right]\left[C(s^\alpha I_n - A)^{-1} B + D\right]^* < \gamma^2 I_n.$$

Then, it follows that

$$\left\|G_{wy}(s)\right\|_{\mathcal{H}_\infty} < \gamma.$$

The remaining proof is the necessity of $\mathcal{LMI}$ (4.12) for the case $1 \le \alpha < 2$.

Suppose that $1 \le \alpha < 2$ and $\left\|G_{wy}(s)\right\|_{\mathcal{H}_\infty} < \gamma$. Noting that the curve $\Lambda_1 = \Lambda(\Phi, \Psi)$ in (4.1) with $\Phi = 0$ and $\Psi = \begin{bmatrix} 0 & e^{j\theta} \\ e^{-j\theta} & 0 \end{bmatrix}$ belongs to $\{s^\alpha / s \in \mathbb{C}, \mathrm{Re}(s) \ge 0\}$, we have

$$\bar{\sigma}_{s^\alpha \in \Lambda_1}(G_{wy}(s)) \le \left\|G_{wy}(s)\right\|_{\mathcal{H}_\infty} < \gamma.$$

It follows that

$$\forall \lambda \in \Lambda_1, \begin{bmatrix} H(\lambda) \\ I_n \end{bmatrix}^* \Theta \begin{bmatrix} H(\lambda) \\ I_n \end{bmatrix} < 0,$$

where Θ is defined in (4.19). According to Lemma 4.2.2, it implies that there exists $P \in \mathbb{H}_n,\ P > 0$ satisfying

$$\begin{bmatrix} A^T & C^T \\ I_n & 0 \end{bmatrix}^T \left(\begin{bmatrix} 0 & e^{j\theta} \\ e^{-j\theta} & 0 \end{bmatrix} \otimes P \right) \begin{bmatrix} A^T & C^T \\ I_n & 0 \end{bmatrix} + \Theta < 0. \tag{4.22}$$

Moreover, applying the Shur complement Lemma 4.2.4, it follows that (4.22) is equivalent to (4.12). This ends the proof. □

In the following, an other result is devoted to $\mathcal{H}_\infty$ computation.

Theorem 4.3.2 *[4] Let γ be a positive real number. Then $\left\|G_{wy}(s)\right\|_{\mathcal{H}_\infty} < \gamma$ if the following conditions hold, depending on the value of γ.*

- ***Case 1* $1 \le \alpha < 2$:** *$\exists P \in \mathbb{H}_n > 0$ such that*

$$\begin{bmatrix} \mathrm{Her}\left\{e^{j\theta} A P\right\} & * & * \\ C e^{j\theta} P & -\gamma I_P & * \\ B^T & D^T & -\gamma I_m \end{bmatrix} < 0, \tag{4.23}$$

where $\theta = (1 - \alpha)\dfrac{\pi}{2}$.

- ***Case 2* $0 < \alpha < 1$:** *$\exists X \in \mathbb{H}_n > 0$ such that*

$$\begin{bmatrix} \mathrm{sym}\left\{A(e^{j\theta} X + e^{-j\theta} \bar{X})\right\} & * & * \\ C(e^{j\theta} X + e^{-j\theta} \bar{X}) & -\gamma I_P & * \\ B^T & D^T & -\gamma I_m \end{bmatrix} < 0, \tag{4.24}$$

where $\theta = (1 - \alpha)\dfrac{\pi}{2}$.

In case 1, $\mathcal{LMI}$ condition (4.23) is also necessary.

Proof Case 1 is the same as in Theorem 4.3.1. But for case 2, $\mathcal{LMI}$ (4.13) of Theorem 4.3.1 involved two variables P and Q when only one variable X is involved in $\mathcal{LMI}$ (4.24) of Theorem 4.3.2. The aim is to show the equivalence between these two conditions. To this purpose, the following notations are introduced:

$$L_1(P) = \begin{bmatrix} \text{Her}\left\{e^{j\theta} A P\right\} & * & * \\ e^{j\theta} C P & -\gamma I_P & * \\ B^T & D^T & -\gamma I_m \end{bmatrix}, \tag{4.25}$$

and

$$L_2(Q) = \begin{bmatrix} \text{Her}\left\{e^{-j\theta} A Q\right\} & * & * \\ e^{-j\theta} C Q & 0 & * \\ 0 & 0 & 0 \end{bmatrix}. \tag{4.26}$$

According to these notations, $\mathcal{LMI}$ (4.13) writes $L_1(P) + L_2(Q) < 0$ and $\mathcal{LMI}$ (4.24) writes $L_1(X) + L_2(\bar{X}) < 0$. We therefore want to prove that:

$$\exists P \in \mathbb{H}_n > 0, \exists Q \in \mathbb{H}_n > 0, \;L_1(P) + L_2(Q) < 0 \tag{4.27}$$

is equivalent to:

$$\exists X \in \mathbb{H}_n > 0, \;L_1(X) + L_2(\bar{X}) < 0. \tag{4.28}$$

(4.28) $\Longrightarrow$ (4.27) is obvious: it suffices to take $P = X$ and $Q = \bar{X}$.
(4.27) $\Longrightarrow$ (4.28) can be proved as follows.

Assume that there exist definite positive Hermitian matrices $P > 0$ and $Q > 0$ satisfying $L_1(P) + L_2(Q) < 0$, which is equivalent to $\overline{L_1(P) + L_2(Q)} < 0$. Thus, the matrices P and Q satisfy also $L_1(P) + L_2(Q) + \overline{L_1(P) + L_2(Q)} < 0$. This last inequality is the same as $L_1(P) + L_2(Q) + L_1(\bar{Q}) + L_2(\bar{P}) < 0$, which is equivalent to $L_1(P + \bar{Q}) + L_2(\bar{P} + Q) < 0$. So, the matrices P and Q verify the condition of case 2 in Theorem 4.3.2 with $X = P + \bar{Q}$. □

Remark 4.3.1 1. As indicated in [5, 6], it should be noted that in the case 1 $1 \leq \alpha < 2$ in both Theorems 4.3.1 and 4.3.2, unlike the classical bounded real lemma for integer-order systems, an issue worth mentioning is that stability of fractional system can be inferred by $P > 0$. Therefore, one more $\mathcal{LMI}$ constraint can be added.

2. Notice that in the case 1 $0 < \alpha < 1$, the number of variables involved for the $\mathcal{H}_\infty$ norm computation in Theorem 4.3.2, is divided by two (a single variable $P \in \mathbb{H}_n$) when compared with Theorem 4.3.1, where there are two variables $P \in \mathbb{H}_n$ and $Q \in \mathbb{H}_n$. Moreover, the proposed new formulation given by (4.24) is more suited to the design problems of the next chapter.

Example 4.3.1 [9] Consider the following fractional-order system

$$G(s) = \frac{s^{0.5} + 1}{s + 2s^{0.5} + 5}.$$

The pseudo state space realization is

$$\begin{cases} D^{0.5}x(t) = \begin{bmatrix} -1 & 2 \\ -2 & -1 \end{bmatrix} x(t) + \begin{bmatrix} 1 \\ 0 \end{bmatrix} w(t), \\ y(t) = \begin{bmatrix} 1 & 0 \end{bmatrix} x(t). \end{cases}$$

Using Theorem 4.3.1, we obtain

$$\|G(s)\|_{\mathcal{H}_\infty} < 0.2817.$$

The system is stable since its eigenvalues $\{-1 + 2j, -1 - 2j\}$ are located in the stability region. Therefore, we have

$$\|G(s)\|_{\mathcal{H}_\infty} = \sup_{w \geq 0} \bar{\sigma}(G(jw)).$$

So, we have to consider for $w \geq 0$,

$$G(jw) = \frac{(jw)^{0.5} + 1}{((jw)^{0.5} + 1)^2 + 4}.$$

Then, it follows from the bijection

$$\begin{array}{ccc} [0, +\infty[& \longrightarrow & [0, +\infty[\\ w & \longrightarrow & x = w^{\alpha} \end{array}$$

that $G(jw) = \tilde{G}(x)$, where

$$\tilde{G}(x) = \frac{e^{j\frac{\pi}{4}}x + 1}{(e^{j\frac{\pi}{4}}x + 1)^2 + 4}.$$

Then, it can be obtained that

$$\|G(s)\|_{\mathcal{H}_\infty} = \max_{x \geq 0} \left|\tilde{G}(x)\right| = 0.2774,$$

as shown in the image of $|G(jw)|$ in Fig. 4.1.

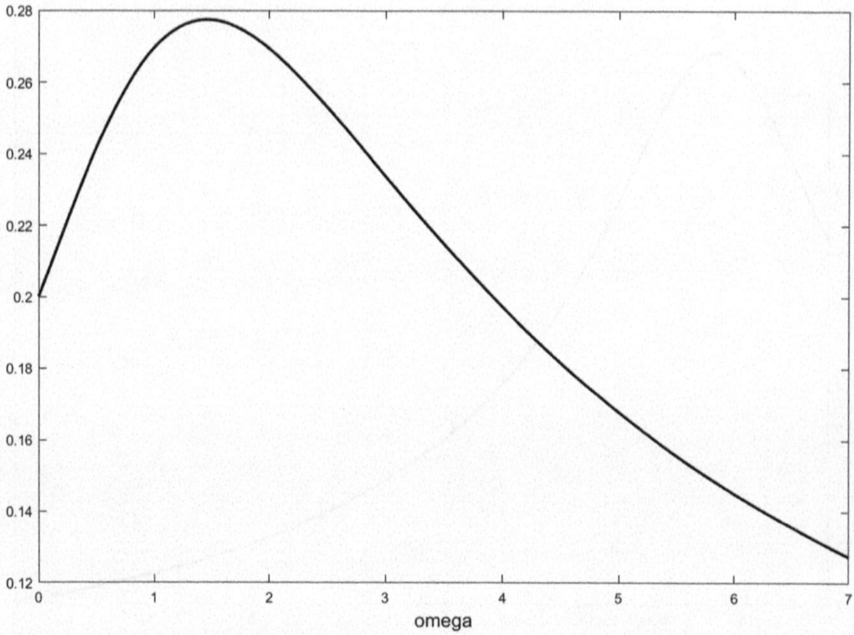

Fig. 4.1 Modulus of $G(jw)$ for $\alpha = 0.5$

Example 4.3.2 Consider system

$$\begin{cases} D^{\alpha}x(t) = Ax(t) + B_1u(t) + Bw(t), \\ y(t) = Cx(t), \end{cases} \tag{4.29}$$

with fractional-order $\alpha = 1.25$,

$$A = \begin{bmatrix} -1.5 & 0.5 \\ 0.5 & -1.5 \end{bmatrix},\ B_1 = \begin{bmatrix} 1 \\ 0.5 \end{bmatrix},\ B = \begin{bmatrix} 1 \\ 0 \end{bmatrix},\ C = \begin{bmatrix} 0.75 \\ 0.5 \end{bmatrix}^T \tag{4.30}$$

The open loop transfer function, i.e., $u(t) = 0$, is

$$G(s) = \frac{0.875s^{1.5} + 1.625}{s^3 + 3s^{1.5} + 2} \tag{4.31}$$

Result of Theorem 4.3.1 is applied. Solving the $\mathcal{LMI}$ (4.12) gives:

$$P = \begin{bmatrix} 2.5221 + 0.0000i & -1.5872 + 0.5829i \\ -1.5872 - 0.5829i & 4.1509 + 0.0000i \end{bmatrix},\ \gamma = 1.1530, \tag{4.32}$$

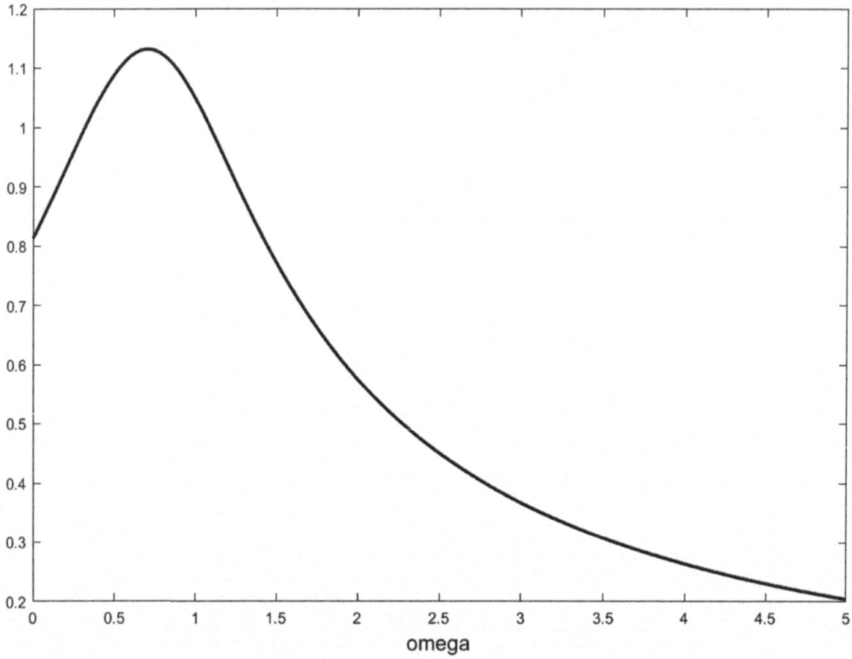

Fig. 4.2 Modulus of $G(jw)$ for $\alpha = 1.25$

which means that $\|G(s)\|_\infty < +\infty$, as can be shown in the image of $|G(j\omega)|$ for $\omega \geq 0$ in Fig. 4.2 and $\|G(s)\|_\infty = 1.1319$.

4.4 $\mathcal{H}_\infty$ Controller Synthesis for FOS

The bounded real lemma has played an important role in the synthesis problem of $\mathcal{H}_\infty$ control for integer-order systems. Considering this, the focus in this section will be on designing state feedback controllers to solve the $\mathcal{H}_\infty$ control problem for continuous-time, linear fractional-order systems by applying results of Theorems 4.3.1 and 4.3.2. Consider the following FOS

$$\begin{cases} D^\alpha x(t) = Ax(t) + B_1 u(t) + Bw(t), \\ \quad y(t) \;\; = Cx(t) + D_1 u(t) + Dw(t), \end{cases} \tag{4.33}$$

where $x(t) \in \mathbb{R}^n$ is the pseudo state vector, $u(t) \in \mathbb{R}^m$ is the control signal, $w(t) \in \mathbb{R}^q$ is the exogenous input, $y(t) \in \mathbb{R}^p$ is the regulated output. The system (4.33) is supposed to meet the Kalman controllability criterion.

The objective is to design a static state feedback control law of the form

$$u(t) = Kx(t), \tag{4.34}$$

where $K \in \mathbb{R}^{m\times n}$ is a constant matrix gain.

The state space representation of the closed loop system is

$$\begin{cases} D^\alpha x(t) = (A + B_1 K)x(t) + Bw(t), \\ y(t) \;\; = (C + D_1 K)x(t) + Dw(t), \end{cases} \tag{4.35}$$

and the transfer function from $w(t)$ to $y(t)$ has the form

$$G_{wy}(s) = (C + D_1 K)(s^\alpha I_n - (A + B_1 K))^{-1} B + D. \tag{4.36}$$

The following result is derived.

Theorem 4.4.1 *[9] Consider the FOS (4.33) with transfer function $G_{wy}(s)$ in (4.36), and the following $\mathcal{LMI}$*

$$\begin{bmatrix} \mathrm{Her}(AX + B_1 Y) & (CX + D_1 Y)^* & B \\ (CX + D_1 Y) & -\gamma I_p & D \\ B^T & D^T & -\gamma Iq \end{bmatrix} < 0. \tag{4.37}$$

Let $\theta = \frac{\pi}{2}(1-\alpha)$, then there holds $\left\|G_{wy}(s)\right\|_{H_\infty} < \gamma$ if

> ***Case 1*** $0 < \alpha < 1$*: There exist $P \in \mathbb{R}^{n\times n}$, $Q \in \mathbb{R}^{n\times n}$ and $Y \in \mathbb{R}^{m\times n}$ such that $P + jQ > 0$ and $X = \cos(\theta)P + \sin(\theta)Q$ satisfying $\mathcal{LMI}$ (4.37).*
> ***Case 2*** $1 \le \alpha < 2$*: There exist $P \in \mathbb{R}^{n\times n}$, $Q \in \mathbb{R}^{n\times n}$ such that $P > 0$, $X = e^{j\theta}P$, $Y = e^{j\theta}Q$ satisfying $\mathcal{LMI}$ (4.37).*

Moreover, the static state feedback gain matrix can be obtained by $K = YX^{-1}$.

Proof The case $1 \le \alpha < 2$ can be proved directly by utilizing Theorem 4.3.1 and is omitted.

Concerning the case $0 < \alpha < 1$, let $\hat{P} = \frac{1}{2}(P - jQ)$ and $\hat{Q} = \frac{1}{2}(P + jQ)$. Then we have $\hat{Q} > 0$ and $\hat{P} = \overline{\hat{Q}} > 0$ (the conjugate of a positive definite matrix is also positive definite).

After some basic computations, it follows that

$$X = \cos(\theta)P + \sin(\theta)Q = e^{j\theta}\hat{P} + e^{-j\theta}\hat{Q}.$$

According to Theorem 4.3.1, the feasibility of $\mathcal{LMI}$ (4.37) implies $\left\|G_{wy}(s)\right\|_{H_\infty} < \gamma$.

Finally, to show that the obtained gain matrix K is real and available, we first notice that the matrix X, involved in the $\mathcal{LMI}$ (4.37), is real since both matrices P and Q are real. The regularity of the matrix X is proved as follows.

From $\hat{Q} > 0$ and $\hat{P} > 0$, we have that for all non-zero vector $x \in \mathbb{C}^n$,

$$x^T\hat{Q}x > 0 \text{ and } x^T\hat{P}x > 0.$$

In particular, when $x \in \mathbb{R}^n$, $x^T \hat{Q} x > 0$ and $x^T \hat{P} x > 0$. Since, $x^T \hat{Q} x$ is a positive real, it follows

$$x^T \hat{Q} x = \overline{x^T \hat{Q} x} = x^T \overline{\hat{Q}} x.$$

This implies that for all nonzero vector $x \in \mathbb{R}^n$,

$$\begin{aligned} x^T X x &= x^T (e^{j\theta} \hat{P} + e^{-j\theta} \hat{Q}) x, \\ &= x^T (e^{j\theta} \overline{\hat{Q}} + e^{-j\theta} \hat{Q}) x, \\ &= e^{j\theta} (x^T \overline{\hat{Q}} x) + e^{-j\theta} (x^T \hat{Q} x), \\ &= e^{j\theta} (x^T \hat{Q} x) + e^{-j\theta} (x^T \hat{Q} x), \\ &= (e^{j\theta} + e^{-j\theta})(x^T \hat{Q} x). \end{aligned}$$

As $x^T \hat{Q} x > 0$ and $e^{j\theta} + e^{-j\theta}$ is a positive real scalar since $\theta = \frac{\pi}{2}(1 - \alpha)$ and $0 < \alpha < 1$, then $x^T X x > 0$ which means that the matrix X is positive definite and consequently X is non-singular.

Therefore, a real static state feedback gain matrix is available by $K = Y X^{-1}$. □

Remark 4.4.1 Compared to Theorem 4.3.1, the decision matrices in Theorem 4.4.1 are cast as real matrices. This will lead to some conservatism of the controller.

To solve the synthesis problem presented above, another result is derived by using Theorem 4.3.2.

Theorem 4.4.2 *[4] Fractional order system* (4.33) *is bounded input bounded output (BIBO) stabilizable by pseudo-state feedback control law* (4.34) *with* $\|G_{wy}(s)\|_{H_\infty} < \gamma$ *if the following conditions hold, depending on the value of* γ.

Case 1 **$0 < \alpha < 1$**: *There exist matrices* $X \in \mathbb{C}^{n \times n} > 0$ *and* $Y \in \mathbb{R}^{m \times n}$ *such that*

$$\begin{bmatrix} \mathrm{sym}\left\{A(e^{j\theta} X + e^{-j\theta} \bar{X}) + B_1 Y\right\} & (C(e^{j\theta} X + e^{-j\theta} \bar{X}) + D_1 Y)^T & B \\ C(e^{j\theta} X + e^{-j\theta} \bar{X}) + D_1 Y & -\gamma I_p & D \\ B^T & D^T & -\gamma I q \end{bmatrix} < 0. \tag{4.38}$$

A stabilizing controller gain is then $K = Y(e^{j\theta} X + e^{-j\theta} \bar{X})^{-1}$.

Case 2 **$1 \le \alpha < 21$**: *There exist matrices* $P \in \mathbb{R}^{n \times n} > 0$ *and* $Q \in \mathbb{R}^{m \times n}$ *such that*

$$\begin{bmatrix} \mathrm{Her}\left\{A e^{j\theta} P + B_1 e^{j\theta} Q\right\} & (C e^{j\theta} P + D_1 e^{j\theta} Q)^* & B \\ C e^{j\theta} P + D_1 e^{j\theta} Q & -\gamma I_p & D \\ B^T & D^T & -\gamma I q \end{bmatrix} < 0. \tag{4.39}$$

A stabilizing controller gain is then $K = Q P^{-1}$.

Proof Using the linearizing changes of variables $Q = KP$ for case 2 and $Y = K(e^{j\theta}X + e^{-j\theta}\bar{X})$ for case 1, then replacing A with $A + BK$ and C with $C + DK$ in $\mathcal{LMI}s$ (4.23) and (4.24) in Theorem 4.3.2, results of Theorem 4.4.2 are derived. □

Remark 4.4.2 • In order to generate a real controller K, matrix variable P for case 2 of Theorem 4.4.2 is imposed to be real, which leads to a certain conservatism in the $\mathcal{LMI}$ condition. This problem does not arise for case 1 since although the matrix X is complex but a real controller is generated because $e^{j\theta}X + e^{-j\theta}\bar{X}$ and Y are both real matrices.

- It is clear that results of both Theorems 4.4.1 and 4.4.2 for case 2 are the same. For case 1, results of both Theorems 4.4.1 and 4.4.2 are equivalent. To see that, it suffuses to take $X = \frac{P}{2} - j\frac{Q}{2}$ which is a positive definite matrix since $P + jQ > 0$ in (4.37) of Theorem 4.4.1 and then (4.38) in Theorem 4.4.2 is obtained. Conversely, by taking $X = \cos(\theta)P_1 + \sin(\theta)Q_1$ with $P_1 = 2P$ and $Q_1 = -2Q$, where $X = P + jQ > 0$ in (4.38) of Theorem 4.4.2, it is easy to see that (4.38) implies (4.37).

Example 4.4.1 Consider the following FOS

$$\begin{cases} D^\alpha x(t) = Ax(t) + B_1u(t) + Bw(t), \\ y(t) \quad = Cx(t), \end{cases} \tag{4.40}$$

with fractional order $\alpha = 1.5$,

$$A = \begin{bmatrix} -1 & 2 \\ -2 & -1 \end{bmatrix}, \quad B_1 = \begin{bmatrix} 1 \\ 1 \end{bmatrix}, \quad B = \begin{bmatrix} 1 \\ 0 \end{bmatrix}, \quad C = \begin{bmatrix} 1 \\ 0 \end{bmatrix}^T \tag{4.41}$$

The open loop transfer function, i.e., with $u(t) = 0$, is

$$G_{ol}(s) = \frac{s^{1.5} + 1}{s^3 + 2s^{1.5} + 5}. \tag{4.42}$$

$$\|G_{ol}(s\|_\infty = \sup_{\mathrm{Re}(s)\geq 0} \left| \frac{s^{1.5} + 1}{s^3 + 2s^{1.5} + 5} \right| \tag{4.43}$$

$$= \sup_{\mathrm{Re}(s)\geq 0} \frac{|s^{1.5} + 1|}{|s^{1.5} - (-1 - 2i)|\,|s^{1.5} - (-1 + 2i)|} \tag{4.44}$$

It is clear that the sup is reached for $s^{1.5} = -1 \pm 2i$, which satisfy $|\arg(-1 \pm 2i)| < 1.5 \times \frac{\pi}{2}$. Then, we obtain $\|G_{ol}(s\|_\infty$ tends to $+\infty$. This result can also be verified since the $\mathcal{LMI}$ condition (4.12) of the $\mathcal{H}_\infty$ is infeasible. In the other wise, the system is unstable because its eigenvalues $-1 \pm 2i$ don't belong to the stability region.

The closed loop transfer function is $G_{cl}(s) = C(s^\alpha I_n - A - B_1K)^{-1}B$. A control law K stabilizing the system and minimizing $\|G_{cl}(s)\|_\infty$ is designed using Theorem 4.4.1. By using Yalmip to solve $\mathcal{LMI}$ condition (4.37), the following feasible

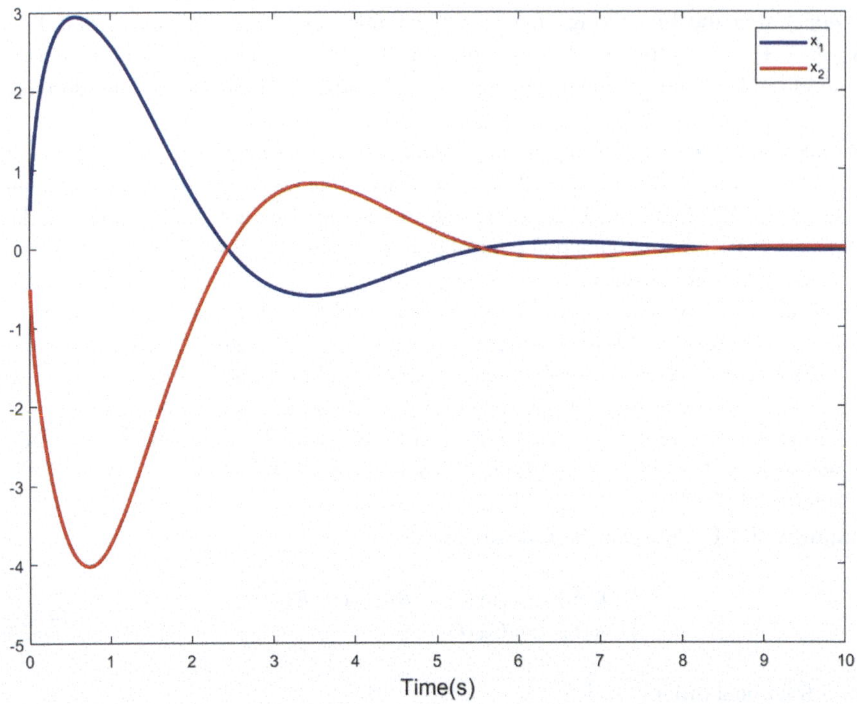

Fig. 4.3 Time response of the closed loop system with K

solutions are obtained,

$$P = \begin{bmatrix} 0.7858 & -0.7271 \\ -0.7271 & 1.2393 \end{bmatrix}, \; Q = \begin{bmatrix} 0.3897 & -2.7441 \end{bmatrix}, \tag{4.45}$$

with the minimum value of γ is 3.2053.

The gain matrix K of the controller (4.34) can be designed as

$$K = \begin{bmatrix} -3.3972 & -4.2073 \end{bmatrix}. \tag{4.46}$$

The eigenvalues λ of the matrix $A + B_1 K$ are $\{-1.3270, -8.2774\}$, which satisfy

$$|\arg(\lambda)| > 1.5 \times \frac{\pi}{2},$$

implying that the closed loop system is stable, as can be seen from Fig. 4.3 of the time response of the closed loop system of (4.40), where all states tend to 0.

The maximum singular values of $G_{cl}(jw)$ are plotted in Fig. 4.4, where the peak value is approximately 0.6409.

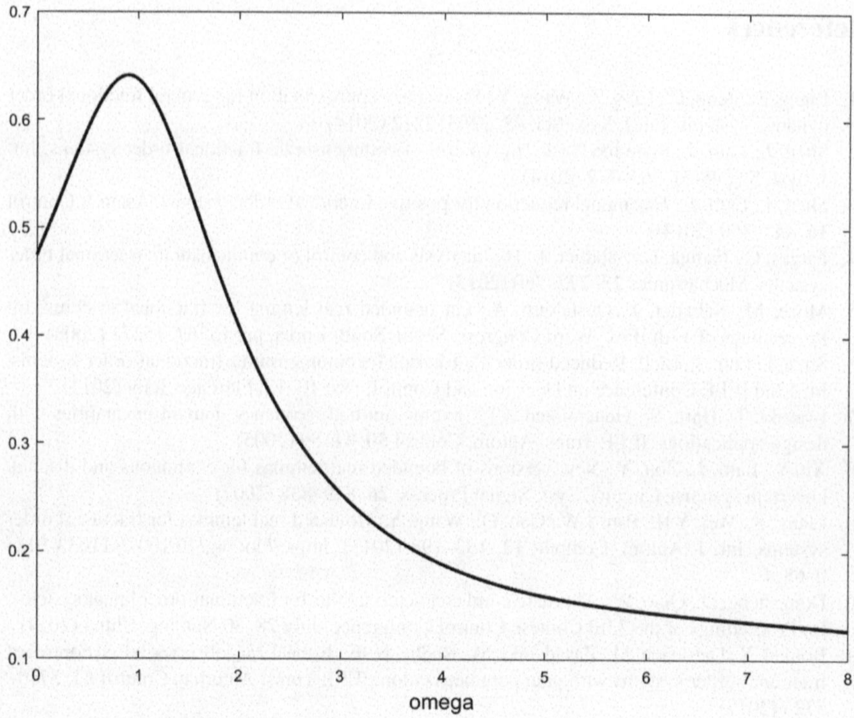

Fig. 4.4 Modulus of $G_{cl}(jw)$ for $\alpha = 1.5$

From the numerical simulations, the stability and the transfer function satisfying $\|G(s)\|_\infty < \gamma = 3.2053$ are both verified.

4.5 Summary

In this chapter, the problem of $\mathcal{H}_\infty$ control has been investigated for fractional order systems in both cases $0 < \alpha < 1$ and $1 \leq \alpha < 2$, with α is the fractional order derivative. Versions of bounded real lemmas have been presented for continuous time systems. Thus, the design methods of a desirable $\mathcal{H}_\infty$ controller for fractional order systems have been proposed. It should be pointed out that all the results are in the form of linear matrix inequalities and therefore can be solved easily by the existing soft wares.

References

1. Liang, S., Peng, C., Liao, Z., Wang, Y.: State space approximation for general fractional order dynamic systems. Int. J. Syst. Sci. **45**, 2203–2212 (2014)
2. Shen, J., Lam, J.: State feedback $\mathcal{H}_\infty$ control of commensurate fractional-order systems. Int. J. Syst. Sci. **45**(3), 363–372 (2014)
3. Shen, J., Lam, J.: $\mathcal{H}_\infty$ model reduction for positive fractional order systems. Asian J. Control **16**, 441–450 (2014)
4. Farges, C., Fadiga, L., Sabatier, J.: H_∞ analysis and control of commensurate fractional order systems. Mechatronics **23**, 772–780 (2013)
5. Moze, M., Sabatier, J., Oustaloup, A.: On bounded real lemma for fractional systems. In: Proceedings of 17th IFAC Word Congress, Seoul, South Korea, pp. 15267–15272 (2008)
6. Shen, J., Lam, J., Li, P.: Reduced-order $\mathfrak{h}_\infty$ filtering for commensurate fractional-order systems. In: 52nd IEEE Conference on Decision and Control, Dec 10–13. Florence, Italy (2013)
7. Iwasaki, T., Hara, S.: Generalized $\mathbb{KYP}$ lemma: unified frequency domain inequalities with design applications. IEEE Trans. Autom. Control **50**, 41–59 (2005)
8. Xu, S., Lam, J., Zou, Y.: New versions of bounded real lemmas for continuous and discrete uncertain systems. Circuits, Syst. Signal Process. **26**, 829–838 (2007)
9. Liang, S., Wei, Y.H., Pan, J.W., Gao, Q., Wang, Y.: Bounded real lemmas for fractional order systems. Int. J. Autom. Comput. **12**, 192–198 (2015). https://doi.org/10.1007/s11633-014-0868-4
10. Dongsheng, D., Qiao, W.: Alternative lmi characterizations for fractional-order linear systems. In: Proceedings of the 33rd Chinese Control Conference, July 28–30 Nanjing, China (2014)
11. Boukal, Y., Darouach, M., Zasadsinki, M., Radhy, N.-E.: Robust $\mathcal{H}_\infty$ observer-based control of fractional-order systems with gain parametrization. IEEE Trans. Automat. Control **62**, 5710–5723 (2017)
12. Green, M., Limebeer, D.J.N.: Linear Robust Control. Upper Saddle River, NJ, USA, Prentice-Hall, Inc (1995)
13. Francis, B.A.: A Course in $\mathcal{H}_\infty$ Control Theory. Springer, Heidelberg, Berlin, Germany (1987)
14. Boyd, S., Elghaoui, L., Féron, E., Balakrishnan, V.: Linear Matrix Inequality in Systems and Control Theory. SIAM, Philadelphia (1994)

Chapter 5
Admissibility Analysis of Singular Fractional-Order Systems

5.1 Introduction

Research on dynamic systems often necessitates the mathematical modeling of system behavior to understand and predict their performance. A significant class of physical systems can be represented using Differential-Algebraic Equations (DAEs), commonly referred to as singular systems. This chapter focuses on a novel category of singular linear systems characterized by the presence of non-integer order derivatives, offering a fresh perspective on the analysis and control of such systems.

These systems are described by the following fractional differential equation

$$ED^{\alpha}x(t) = Ax(t) \quad 0 < \alpha < 2 \tag{5.1}$$

The models of systems of the form (5.1), which are called fractional singular systems in [1–3], are of great interest for modelling many practical processes like their counterparts the singular linear systems of the form ($E\dot{x}(t) = Ax(t)$). To begin a motivation example of singular fractional electrical circuit is presented, then analysis of the singular fractional-order systems in the continuous context is initiated. As for integer-order systems systems, it was proved that the stability of linear fractional-order systems depends on the localization of the eigenvalues in the complex plan. Consequently, many works, specially those derived via linear matrix inequalities ($\mathcal{LMI}$s), have been proposed (see for example [2, 4–7]). Therefore, for a singular fractional-order system, it is important to develop conditions which guarantee that the given singular system is not only stable but also regular and impulse-free. In the literature, it is reported that such conditions can usually be obtained by decomposing singular systems into slow and fast sub-systems [8]. However, this may lead to some numerical problems. Furthermore, from the mathematical point of view, the decomposition approach is not elegant. Thus, attention in this chapter will be focused on the derivation of such conditions without decomposing the original singular system,

S. Marir and M. Chadli, *Control of Singular Fractional Order Systems: LMI Approach*, Studies in Systems, Decision and Control 292,
https://doi.org/10.1007/978-3-031-87382-9_5

and a linear matrix inequality ($\mathcal{LMI}$) approach will be developed. Necessary and sufficient conditions of admissibility for singular fractional-order continuous-time systems are proposed in strict $\mathcal{LMI}$ terms.

5.2 Illustrative Example

Consider electrical circuit [1] shown on Fig. 5.1 with given resistance R, capacitances C_1, C_2, C_3 and source voltages e_1 and e_2. Using the Kirchhoff's laws.

We can write for the electrical circuit the equations

$$e_1 = RC_1 + D^\alpha u_1 + u_1 + u_3, \tag{E1}$$

$$0 = C_1 D^\alpha u_1 + C_2 D^\alpha u_2 - C_3 D^\alpha u_3, \tag{E2}$$

$$e_2 = u_2 + u_3. \tag{E3}$$

The equations (E1)–(E2)–(E3) can be written as

$$\begin{bmatrix} RC_1 & 0 & 0 \\ C_1 & C_2 & -C_3 \\ 0 & 0 & 0 \end{bmatrix} D^\alpha \begin{bmatrix} u_1 \\ u_2 \\ u_3 \end{bmatrix} = \begin{bmatrix} -1 & 0 & -1 \\ 0 & 0 & 0 \\ 0 & -1 & -1 \end{bmatrix} \begin{bmatrix} u_1 \\ u_2 \\ u_3 \end{bmatrix} + \begin{bmatrix} 1 & 0 \\ 0 & 0 \\ 0 & 1 \end{bmatrix} \begin{bmatrix} e_1 \\ e_2 \end{bmatrix}, \tag{5.2}$$

which has the form

$$ED^\alpha u = Au + Be,$$

with

$$E = \begin{bmatrix} RC_1 & 0 & 0 \\ C_1 & C_2 & -C_3 \\ 0 & 0 & 0 \end{bmatrix}, \; A = \begin{bmatrix} -1 & 0 & -1 \\ 0 & 0 & 0 \\ 0 & -1 & -1 \end{bmatrix}, \; B = \begin{bmatrix} 1 & 0 \\ 0 & 0 \\ 0 & 1 \end{bmatrix}, \tag{5.3}$$

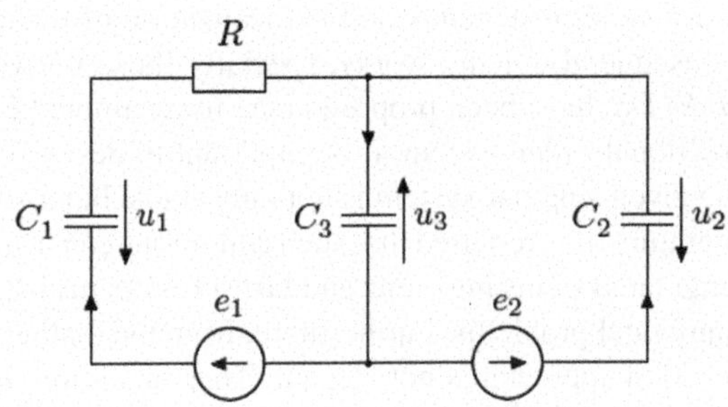

Fig. 5.1 Electrical circuit

and

$$u = \begin{bmatrix} u_1 \\ u_2 \\ u_3 \end{bmatrix}, \; e = \begin{bmatrix} e_1 \\ e_2 \end{bmatrix}. \tag{5.4}$$

Note that the matrix E is singular since $\det(E) = 0$ but the pencil

$$\begin{aligned} \det(s^\alpha E - A) &= \begin{bmatrix} RC_1 s^\alpha + 1 & 0 & 1 \\ C_1 s^\alpha & C_2 s^\alpha & -C_3 s^\alpha \\ 0 & 1 & 1 \end{bmatrix}, \\ &= (RC_1 s^\alpha + 1)(C_2 s^\alpha + C_3 s^\alpha) + C_1 s^\alpha \end{aligned} \tag{5.5}$$

is not identically zero, then it is regular. Therefore, the electrical circuit is a singular fractional linear system.

Remark 5.2.1 If the electrical circuit contains at least one mesh consisting of branches with only ideal supercondensators and voltage sources then its matrix E is singular since the row corresponding to this mesh is zero row. This follows from the fact that the equation written by the use of the voltage Kirchhoff's law is algebraic one.

5.3 Generalities About Singular Fractional-Order Systems

5.3.1 Preliminaries Results

Considering the following singular fractional-order (SFO) continuous time system described by

$$\begin{aligned} ED^\alpha x(t) &= Ax(t) + Bu(t), 0 < \alpha < 2, \\ y(t) &= Cx(t). \end{aligned} \tag{5.6}$$

where α is the time fractional derivative order, $x(t) \in \mathbb{R}^n$ is the pseudo-state, $u(t) \in \mathbb{R}^m$ is the control input and $y(t) \in \mathbb{R}^p$ is the output. $A \in \mathbb{R}^{n\times n}$, $B \in \mathbb{R}^{n\times m}$ and $C \in \mathbb{R}^{p\times n}$ with rank $C = p$ and matrix $E \in \mathbb{R}^{n\times n}$ is singular with rank $E = r < n$.

The unforced SFO system of (5.6) can be written as

$$ED^\alpha x(t) = Ax(t). \tag{5.7}$$

First, we present some preliminaries results on singular fractional-order systems

Definition 5.3.1 [3] The SFO system (5.7) or the pair (E, A) is said to be regular if there exists an unique solution $x(t)$ for a given initial conditions.

As in [9], the following lemmas which generalize well known results of singular "integer-order" systems to singular fractional-order systems were proposed in [3, 10, 11].

Lemma 5.3.1 *[3] The SFO system* (5.7) *or the pair* (E, A) *is said to be regular if and only if the pseudo-polynomial* $\det(s^{\alpha}E - A)$ *does not vanish identically, where* $s \in \mathbb{C}$.

Proof In order to give proof for the general case, We consider the singular fractional-order system

$$ED^{\alpha}x(t) = Ax(t), \quad m-1 < \alpha < m \tag{5.8}$$

$$x(0) = x_0, x'(0) = x_1, x^{(2)}(0) = x_2, ..., x^{(m-1)}(0) = x_{m-1}, \tag{5.9}$$

where m is the smallest non-zero integer checking $m - 1 < \alpha < m$.

Let $X(s) = \mathcal{L}(x(t))$ and applying Laplace transform on both sides of system (5.8) using the definition given in (1.29) and then it has

$$E(s^{\alpha}X(s) - \sum_{k=0}^{k=m-1} s^{\alpha-k-1}x^{(k)}(0)) = AX(s). \tag{5.10}$$

From this equation we have

$$(s^{\alpha}E - A)X(s) = \sum_{k=0}^{k=m-1} s^{\alpha-k-1}x^{(k)}(0). \tag{5.11}$$

It is easy to show that the system (5.11) has a unique solution if and only if the pencil $s^{\alpha}E - A$ is invertible i.e., $\det(s^{\alpha}E - A)$ is not identically zero. □

Lemma 5.3.2 *[3, 11] The pair* (E, A) *is regular if and only if there exist two invertible matrices M and N satisfying*

$$MEN = \begin{bmatrix} I_r & 0 \\ 0 & F \end{bmatrix}, \ MAN = \begin{bmatrix} A_1 & 0 \\ 0 & I_{n-r} \end{bmatrix}, \tag{5.12}$$

where $A_1 \in \mathbb{R}^{r \times r}$ *and* $F \in \mathbb{R}^{(n-r)\times(n-r)}$ *is a nilpotent matrix with the index* μ *(i.e.,* $F^{\mu} = 0$ *and* $F^{\mu-1} \neq 0$*).*

Proof Sufficiency. Let (5.12) be true. Then

$$\det(s^{\alpha}E - A) = \det(Q^{-1}\begin{bmatrix} s^{\alpha}I_r - A_1 & 0 \\ 0 & s^{\alpha}F - I \end{bmatrix}P^{-1}, \tag{5.13}$$

$$= \det(Q^{-1})\det(P^{-1})\det(s^{\alpha}I_r - A_1)\det(s^{\alpha}F - I). \tag{5.14}$$

Since the pencils $s^{\alpha} I_r - A_1$ and $s^{\alpha} F - I$ are invertible, so

$$\det(s^{\alpha} E - A) \neq 0. \tag{5.15}$$

Thus the pencil (E, A) is regular.

Necessity. If (E, A) is regular, then there exists $s_0 \in \mathbb{C}$ such that $\det(s_0^{\alpha} E - A) \neq 0$. Consider the pencil

$$\hat{E} = (s_0^{\alpha} E - A)^{-1} E, \quad \hat{A} = (s_0^{\alpha} E - A)^{-1} A. \tag{5.16}$$

It is easy to obtain

$$\hat{A} = s_0^{\alpha} \hat{E} - I. \tag{5.17}$$

On the other hand, it follows immediately from the Jordan canonical form decomposition in matrix theory that there exists a nonsingular matrix T such that

$$T \hat{E} T^{-1} = \text{diag}(\hat{E}_1, \hat{E}_2), \tag{5.18}$$

where $\hat{E}_1 \in \mathbb{R}^{r \times r}$ is nonsingular, $\hat{E}_2 \in \mathbb{R}^{(n-r)\times(n-r)}$ is a nilpotent matrix with the index μ ($\hat{E}_2^{\mu} = 0$ and $\hat{E}_2^{\mu-1} \neq 0$). This means that $I - s_0^{\alpha} \hat{E}_2$ is nonsingular.

Let $Q = \text{diag}\left(\hat{E}_1^{-1}, (s_0^{\alpha} \hat{E}_2 - I)^{-1}\right) T (s_0^{\alpha} E - A)^{-1}$ and $P = T^{-1}$.

From (5.16) and (5.17), we have

$$QEP = \text{diag}(I_r, F), \quad QAP = \text{diag}(A_1, I_{n-r}), \tag{5.19}$$

where $A_1 = \hat{E}_1^{-1}(s_0^{\alpha} \hat{E}_1 - I)$ and $F = (s_0^{\alpha} \hat{E}_2 - I)^{-1} \hat{E}_2$.

It is easy to check that

$$(s_0^{\alpha} \hat{E}_2 - I)^{-1} = - \sum_{k=0}^{k=\mu-1} s_0^{k\alpha} \hat{E}_2^k, \tag{5.20}$$

since

$$(s_0^{\alpha} \hat{E}_2 - I) \left(- \sum_{k=0}^{k=\mu-1} s_0^{k\alpha} \hat{E}_2^k \right) = I. \tag{5.21}$$

and $\hat{E}_2^k = 0, k = \mu, \mu + 1, \ldots$.. It is clear that the matrices $(s_0^{\alpha} \hat{E}_2 - I)^{-1}$ and $\hat{E}_2$ commute so the matrix F is nilpotent. □

Definition 5.3.2 [11] The SFO system (5.7) or the pair (E, A) is said to be impulse free if (5.7) is regular and has exactly r finite poles, i.e., $\deg(\det(\lambda E - A)) = \text{rank } E$, where $\lambda \in \mathbb{C}$.

Lemma 5.3.3 *[3, 11] The SFO system* (5.7) *or the pair* (E, A) *is said to be impulse free if and only if* (5.7) *is regular and the matrix* F *in* (5.12) *is zero i.e.,* $F = 0$

Proof Under the regularity assumption for (E, A), by Lemma 5.3.2 there exist nonsingular matrices Q and P satisfying (5.12). Then

$$\det(s^\alpha E - A) = \det(Q^{-1}\begin{bmatrix} s^\alpha I_r - A_1 & 0 \\ 0 & s^\alpha \digamma - I \end{bmatrix} P^{-1}, \tag{5.22}$$

$$= \det(Q^{-1})\det(P^{-1})\det(s^\alpha I_r - A_1)\det(s^\alpha \digamma - I). \tag{5.23}$$

If $\digamma = 0$, it's obvious that $\deg(\det(s^\alpha E - A)) = \deg(\det(s^\alpha I_r - A_1)) = r$.

If $\deg(\det(s^\alpha E - A)) = r$ and since $\deg(\det(s^\alpha I_r - A_1)) = r$ then $\det(s^\alpha \digamma - I)$ does not depend on s, and is different from zero so $\digamma = 0$. □

When the regularity of the pair (E, A) is unknown, then it is always possible to find non-singular matrices M and N such that

$$MEN = \begin{bmatrix} I_r & 0 \\ 0 & 0 \end{bmatrix}, \; MAN = \begin{bmatrix} A_{11} & A_{12} \\ A_{21} & A_{22} \end{bmatrix}. \tag{5.24}$$

This decomposition can be obtained via singular value decomposition of matrix E followed by scaling of the bases. Then we have the following lemma.

Lemma 5.3.4 *[10] The SFO system* (5.7) *or the pair* (E, A) *is impulse free if and only if* A_{22} *in* (5.24) *is invertible.*

Proof From the decomposition (5.24), we obtain

$$E = M^{-1}\begin{bmatrix} I_r & 0 \\ 0 & 0 \end{bmatrix} N^{-1}, \; A = M^{-1}\begin{bmatrix} A_{11} & A_{12} \\ A_{21} & A_{22} \end{bmatrix} N^{-1}. \tag{5.25}$$

Then,

$$\det(s^\alpha E - A) = \det(M^{-1})\det\begin{bmatrix} s^\alpha I_r - A_{11} & -A_{12} \\ -A_{21} & -A_{22} \end{bmatrix}\det(N^{-1}). \tag{5.26}$$

It is clear that

$$\deg(\det(s^\alpha E - A)) = \deg\left(\det\begin{bmatrix} s^\alpha I_r - A_{11} & -A_{12} \\ -A_{21} & -A_{22} \end{bmatrix}\right) \le \operatorname{rank}(E).$$

The equality holds if and only if $\det(-A_{22}) \neq 0$, i.e., $\det(A_{22}) \neq 0$. Therefore, the system (5.7) is impulse free if and only if the matrix A_{22} is invertible which implies the regularity of the system. □

5.4 Admissibility of Singular Fractional-Order Systems

Parallel to singular linear integer order systems, the admissibility definition for singular linear fractional-order system is given as follows

Definition 5.4.1 [3, 11] The singular linear fractional-order system (5.7) is admissible if

(i) the pair (E, A) is regular.
(ii) the pair (E, A) is impulse free.
(iii) the pair (E, A) is stable.

Lemma 5.4.1 *[3] The SFO system* (5.7) *is admissible if and only if*

(i) The pair (E, A) *is regular, that is* $\det(\lambda E - A)$ *is not identically zero, where* $\lambda \in \mathbb{C}$, *i.e., there exist two nonsingular matrices* P *and* Q *such that*

$$PEQ = \text{diag}(I_r, F),\ PAQ = \text{diag}(A_1, I_{n-r}), \tag{5.27}$$

where $A_1 \in \mathbb{R}^{r\times r}$, $F \in \mathbb{R}^{(n-r)\times(n-r)}$ *is nilpotent matrix.*

(ii) the pair (E, A) *is impulse free, that is* $\deg\det(\lambda E - A) = \text{rank}\, E$, *where* $\lambda \in \mathbb{C}$, *i.e.,* $F = 0$ *in* (5.27).
(iii) the finite modes of the pair (E, A), *which are the eigenvalues* λ_i *of matrix* A_1 *in system* (5.27), *satisfy* $|\arg(\lambda_i)| > \alpha\frac{\pi}{2}$ *for* $i = 1, \ldots, r$.

Proof Assume that rank $E = r < n$. The proofs of (i) and (ii) are given above (see Lemmas 5.3.1, 5.3.2 and 5.3.3).

When items (i) and (ii) hold, the system (5.7) is restricted system equivalent to

$$PEQQ^{-1}D^{\alpha}x(t) = PAQQ^{-1}x(t), \tag{5.28}$$

i.e.,

$$\begin{bmatrix} I_r & 0 \\ 0 & 0 \end{bmatrix}\begin{bmatrix} D^{\alpha}x_1(t) \\ D^{\alpha}x_2(t) \end{bmatrix} = \begin{bmatrix} A_1 & 0 \\ 0 & I_{n-r} \end{bmatrix}\begin{bmatrix} x_1(t) \\ x_2(t) \end{bmatrix}, \tag{5.29}$$

with the new state vector introduced

$$\bar{x}(t) = Q^{-1}x(t) = \begin{bmatrix} x_1(t) \\ x_2(t) \end{bmatrix},\ x_1(t) \in \mathbb{R}^r,\ x_2(t) \in \mathbb{R}^{n-r}. \tag{5.30}$$

The asymptotic stability of the singular linear fractional-order system (5.7) depends entirely on $x_1(t)$ irrespective of $x_2(t)$. Then, according to Theorem 2.2.2, the asymptotic stability is achieved if and only if item (iii) holds. this ends the proof. □

Lemma 5.4.2 *[10] The SFO system (5.7) or the pair (E, A) is admissible if and only if the matrix A_{22} in (5.24) is invertible and*

$$\left|\arg(\operatorname{spec}(A_{11} - A_{12}A_{22}^{-1}A_{21}))\right| > \alpha\frac{\pi}{2}. \tag{5.31}$$

Proof Without loss of generality, we suppose E and A have the following form

$$E = \begin{bmatrix} I_r & 0 \\ 0 & 0 \end{bmatrix}, \quad A = \begin{bmatrix} A_{11} & A_{12} \\ A_{21} & A_{22} \end{bmatrix}. \tag{5.32}$$

According to Lemma 5.3.4, the SFO system (5.7) is regular and impulse free if and only if the matrix A_{22} is invertible. Now, we will present the stability condition proposed.

Let

$$x(t) = \begin{bmatrix} x_1(t) \\ x_2(t) \end{bmatrix}, \tag{5.33}$$

then, the system (5.7) can be decomposed as

$$D^{\alpha}x_1(t) = A_{11}x_1(t) + A_{12}x_2(t), \tag{5.34}$$

$$0 = A_{21}x_1(t) + A_{22}x_2(t), \tag{5.35}$$

where $x_1(t) \in \mathbb{R}^r$ and $x_2(t) \in \mathbb{R}^{n-r}$.

From (5.35), we can derive

$$x_2(t) = -A_{22}^{-1}A_{21}x_1(t). \tag{5.36}$$

This, together with (5.34), we have

$$D^{\alpha}x_1(t) = (A_{11} - A_{12}A_{22}^{-1}A_{21})x_1(t). \tag{5.37}$$

It is easy to see that the stability condition of the (5.7) is equivalent to that of the system (5.37). Using Theorem 2.2.2, the FO LTI system (5.37) is asymptotically stable if and only $\left|\arg(\operatorname{spec}(A_{11} - A_{12}A_{22}^{-1}A_{21}))\right| > \alpha\frac{\pi}{2}$. This completes the proof. □

In the following, $\mathcal{LMI}$ admissibility conditions are given for open-loop system (5.7). Both cases $0 < \alpha \leq 1$ and $1 \leq \alpha < 2$ are treated separately.

5.4.1 Admissibility of Singular Fractional-Order Systems: Case $1 \leq \alpha < 2$

Using the decomposition (5.24), we have the following result.

Lemma 5.4.3 *[10] The system (5.7) with fractional-order derivative α satisfying $1 \leq \alpha < 2$ is admissible if and only if the following conditions are satisfied.*
(1) The matrix A_{22} in (5.24) is invertible.
(2) There exists a symmetric matrix $P > 0$ such that

$$\text{sym}\left\{\Theta \otimes \left(\bar{A}P\right)\right\} < 0, \tag{5.38}$$

where

$$\Theta = \begin{bmatrix} \sin(\alpha\frac{\pi}{2}) & \cos(\alpha\frac{\pi}{2}) \\ -\cos(\alpha\frac{\pi}{2}) & sin(\alpha\frac{\pi}{2}) \end{bmatrix}, \quad \bar{A} = A_{11} - A_{12}A_{22}^{-1}A_{21}. \tag{5.39}$$

Proof The result of this theorem is a direct consequence of Lemma 5.4.2 and Theorem 2.3.1. □

Remark 5.4.1 Lemmas 5.4.1, 5.4.2 and 5.4.3 present equivalent conditions on the admissibility of the system (5.7). It is noted that both conditions involve the decomposition of the matrices of the original singular system, which is sometimes numerically unreliable, especially when the order of the system is relatively large.

In view of Remark 5.4.1, next we will present another equivalent admissibility condition in terms of $\mathcal{LMI}$s, which involves no decomposition of the system matrices. Before giving such a condition, the following lemma, which will be useful in the following, are introduced.

Lemma 5.4.4 *[8] Let*

$$\mathcal{N} = \begin{bmatrix} \mathcal{P} & \mathcal{X} \\ \mathcal{Y} & \mathcal{Z} \end{bmatrix}, \tag{5.40}$$

where $\mathcal{P}$, $\mathcal{X}$, $\mathcal{Y}$ and $\mathcal{Z}$ are any real given matrices with appropriate dimensions such that

$$\mathcal{N} + \mathcal{N}^T < 0. \tag{5.41}$$

Then, $\mathcal{Z}$ is nonsingular and

$$\mathcal{P} + \mathcal{P}^T - \mathcal{X}\mathcal{Z}^{-1}\mathcal{Y} - \mathcal{Y}^T\mathcal{Z}^{-T}\mathcal{X}^T < 0.$$

Now, necessary and sufficient conditions for the admissibility of SFO system (5.7) in terms of $\mathcal{LMI}$s are presented.

Theorem 5.4.1 *[12] The SFO system* (5.7) *is admissible if and only if the following equivalent statements hold:*

(i) There exist matrices $X > 0$ *and* Y *satisfying*

$$\mathrm{sym}\{\Theta \otimes A^T(XE + E_0 Y^T)\} < 0, \tag{5.42}$$

where $E_0 \in \mathbb{R}^{n\times(n-r)}$ *is any matrix of full column rank and satisfies* $E^T E_0 = 0$.

(ii) There exist matrices $X > 0$ *and* Y *satisfying*

$$\mathrm{sym}\{\Theta \otimes A(XE^T + E_0 Y^T)\} < 0, \tag{5.43}$$

where $E_0 \in \mathbb{R}^{n\times(n-r)}$ *is any matrix of full column rank and satisfies* $EE_0 = 0$.

(iii) There exists a matrix P *satisfying*

$$E^T P = P^T E,\ E^T P \geq 0,\ \mathrm{sym}\{\Theta \otimes A^T P\} < 0. \tag{5.44}$$

with

$$\Theta = \begin{bmatrix} \sin(\pi - \alpha\frac{\pi}{2}) & \cos(\pi - \alpha\frac{\pi}{2}) \\ -\cos(\pi - \alpha\frac{\pi}{2}) & \sin(\pi - \alpha\frac{\pi}{2}) \end{bmatrix}, 1 \leq \alpha < 2,$$

Proof First, we prove that the admissibility and (i) are equivalent.

Sufficiency. Assume that the inequality (5.42) holds for some matrices $X > 0$ and Y. Since rank $E = r < n$, invertible matrices M and N can always be found by rank decomposition or by singular value decomposition such that

$$MEN = \begin{bmatrix} I_r & 0 \\ 0 & 0 \end{bmatrix}. \tag{5.45}$$

Let

$$MAN = \begin{bmatrix} A_{11} & A_{12} \\ A_{21} & A_{22} \end{bmatrix}, E_0 = M^{\mathrm{T}} \begin{bmatrix} 0 \\ I_{n-r} \end{bmatrix}. \tag{5.46}$$

From the invertibility of the matrix M we have that the matrix E_0 is of full column rank and satisfies $E^{\mathrm{T}} E_0 = 0$.

Let

$$X = M^{\mathrm{T}} \begin{bmatrix} X_{11} & X_{12} \\ X_{21} & X_{22} \end{bmatrix} M, Y = N^{-T} \begin{bmatrix} Y_1 \\ Y_2 \end{bmatrix}. \tag{5.47}$$

From (5.45), (5.46) and (5.47), we obtain

$$A^{\mathrm{T}}\left(XE + E_0 Y^{\mathrm{T}}\right) = N^{-T} \overline{A} N^{-1},$$

with

$$\overline{A} = \begin{bmatrix} A_{11}^{\mathrm{T}} X_{11} + A_{21}^{\mathrm{T}} Y_1^{\mathrm{T}} & A_{21}^{\mathrm{T}} Y_2 \\ A_{12}^{\mathrm{T}} X_{11} + A_{22}^{\mathrm{T}} X_{12}^{\mathrm{T}} + A_{22}^{\mathrm{T}} Y_1^{\mathrm{T}} & A_{22}^{\mathrm{T}} Y_2^{\mathrm{T}} \end{bmatrix}.$$

By using the property of the Kronecker product, we get

$$\begin{aligned}\Theta \otimes \left(A^{\mathrm{T}}\left(XE + E_0 Y^{\mathrm{T}}\right)\right) &= (I_2.\Theta) \otimes \left(N^{-T}.\left(\overline{A}N^{-1}\right)\right),\\ &= \begin{bmatrix} N^{-T} & 0 \\ 0 & N^{-T} \end{bmatrix} (\Theta \otimes \overline{A}) \begin{bmatrix} N^{-1} & 0 \\ 0 & N^{-1} \end{bmatrix},\end{aligned}$$

with

$$\Theta \otimes \overline{A} = \begin{bmatrix} \overline{A}_{11} & \overline{A}_{12} \\ -\overline{A}_{12} & \overline{A}_{11} \end{bmatrix},$$

and

$$\overline{A}_{11} = \begin{bmatrix} \left(A_{11}^{\mathrm{T}} X_{11} + A_{21}^{\mathrm{T}} Y_1^{\mathrm{T}}\right) \sin\theta & A_{21}^{\mathrm{T}} Y_2 \sin\theta \\ \left(A_{12}^{\mathrm{T}} X_{11} + A_{22}^{\mathrm{T}} X_{12}^{\mathrm{T}} + A_{22}^{\mathrm{T}} Y_1^{\mathrm{T}}\right) \sin\theta & A_{22}^{\mathrm{T}} Y_2^{\mathrm{T}} \sin\theta \end{bmatrix},$$

$$\overline{A}_{12} = \begin{bmatrix} A_{11}^{\mathrm{T}} X_{11} + A_{21}^{\mathrm{T}} Y_1^{\mathrm{T}}) \cos\theta & A_{21}^{\mathrm{T}} Y_2 \cos\theta \\ \left(A_{12}^{\mathrm{T}} X_{11} + A_{22}^{\mathrm{T}} X_{12}^{\mathrm{T}} + A_{22}^{\mathrm{T}} Y_1^{\mathrm{T}}\right) \cos\theta & A_{22}^{\mathrm{T}} Y_2^{\mathrm{T}} \cos\theta \end{bmatrix}.$$

Then

$$\mathrm{sym}\{\Theta \otimes (A^{\mathrm{T}}(XE + E_0 Y^{\mathrm{T}}))\} =$$

$$\begin{bmatrix} N^{-T} & 0 \\ 0 & N^{-T} \end{bmatrix} \begin{bmatrix} \overline{A}_{11} + \overline{A}_{11}^{\mathrm{T}} & \overline{A}_{12} - \overline{A}_{12}^{\mathrm{T}} \\ \overline{A}_{12}^{\mathrm{T}} - \overline{A}_{12} & \overline{A}_{11} + \overline{A}_{11}^{\mathrm{T}} \end{bmatrix} \begin{bmatrix} N^{-1} & 0 \\ 0 & N^{-1} \end{bmatrix}.$$

Consequently

$$\mathrm{sym}\{\Theta \otimes (A^{\mathrm{T}}(XE + E_0 Y^{\mathrm{T}}))\} < 0 \Longrightarrow \overline{A}_{11} + \overline{A}_{11}^{\mathrm{T}} < 0.$$

Knowing that

$$\overline{A}_{11} + \overline{A}_{11}^{\mathrm{T}} = \begin{bmatrix} * & * \\ * & \left(A_{22}^{\mathrm{T}} Y_2^{\mathrm{T}} + Y_2 A_{22}\right) \sin\theta \end{bmatrix},$$

we deduce that

$$A_{22}^{\mathrm{T}} Y_2^{\mathrm{T}} + Y_2 A_{22} < 0,$$

since $\sin\theta > 0$. Therefore A_{22} invertible, which means that the system (5.7) is regular and impulse free (see Lemma 5.3.4).

Since the system (5.7) is regular and impulse free, invertible matrices L and R can always be found such that

$$LER = \begin{bmatrix} I_r & 0 \\ 0 & 0 \end{bmatrix}, \quad LAR = \begin{bmatrix} A_1 & 0 \\ 0 & I_{n-r} \end{bmatrix}. \tag{5.48}$$

Let

$$X = L^{\mathrm{T}} \begin{bmatrix} X_{11} & X_{12} \\ X_{21} & X_{22} \end{bmatrix} L, \; Y = R^{-T} \begin{bmatrix} Y_1 \\ Y_2 \end{bmatrix}, \; E_0 = L^{\mathrm{T}} \begin{bmatrix} 0 \\ I_{n-r} \end{bmatrix}. \tag{5.49}$$

From (5.48) and (5.49), inequality (5.43) is equivalent to

$$\begin{bmatrix} R^{-T} & 0 \\ 0 & R^{-T} \end{bmatrix} \begin{bmatrix} \delta_{11} & \delta_{12} \\ \delta_{12}^{\mathrm{T}} & \delta_{11} \end{bmatrix} \begin{bmatrix} R^{-1} & 0 \\ 0 & R^{-1} \end{bmatrix} < 0, \tag{5.50}$$

with

$$\delta_{11} = \begin{bmatrix} \left(A_1^{\mathrm{T}} X_{11} + X_{11}^{\mathrm{T}} A_1\right) \sin\theta & (X_{12} + Y_1) \sin\theta \\ \left(X_{12}^{\mathrm{T}} + Y_1^{\mathrm{T}}\right) \sin\theta & \left(Y_2^{\mathrm{T}} + Y_2\right) \sin\theta \end{bmatrix},$$

$$\delta_{12} = \begin{bmatrix} \left(A_1^{\mathrm{T}} X_{11} - X_{11}^{\mathrm{T}} A_1\right) \cos\theta & -(X_{12} + Y_1) \cos\theta \\ \left(X_{12}^{\mathrm{T}} + Y_1^{\mathrm{T}}\right) \cos\theta & \left(Y_2^{\mathrm{T}} - Y_2\right) \cos\theta \end{bmatrix}.$$

The inequality (5.50) is equivalent to

$$\begin{bmatrix} \delta_{11} & \delta_{12} \\ \delta_{12}^{\mathrm{T}} & \delta_{11} \end{bmatrix} < 0, \tag{5.51}$$

which is equivalent to

$$\begin{bmatrix} \left(A_1^{\mathrm{T}} X_{11} + X_{11}^{\mathrm{T}} A_1\right) \sin\theta & \left(A_1^{\mathrm{T}} X_{11} - X_{11}^{\mathrm{T}} A_1\right) \sin\theta \\ \left(X_{11}^{\mathrm{T}} A_1 - A_1^{\mathrm{T}} X_{11}\right) \cos\theta & \left(A_1^{\mathrm{T}} X_{11} + X_{11}^{\mathrm{T}} A_1\right) \sin\theta \\ \left(X_{12}^{\mathrm{T}} + Y_1^{\mathrm{T}}\right) \sin\theta & \left(X_{12}^{\mathrm{T}} + Y_1^{\mathrm{T}}\right) \cos\theta \\ -\left(X_{12}^{\mathrm{T}} + Y_1^{\mathrm{T}}\right) \cos\theta & \left(X_{12}^{\mathrm{T}} + Y_1^{\mathrm{T}}\right) \sin\theta \end{bmatrix.$$
$$\left.\begin{matrix} (X_{12} + Y_1) \sin\theta & -(X_{12} + Y_1) \cos\theta \\ (X_{12} + Y_1) \cos\theta & (X_{12} + Y_1) \sin\theta \\ \left(Y_2^{\mathrm{T}} + Y_2\right) \sin\theta & \left(Y_2^{\mathrm{T}} - Y_2\right) \cos\theta \\ \left(Y_2 - Y_2^{\mathrm{T}}\right) \cos\theta & \left(Y_2^{\mathrm{T}} + Y_2\right) \sin\theta \end{matrix}\right] < 0.$$

Then, we deduce that

$$\begin{bmatrix} \left(A_1^{\mathrm{T}} X_{11} + X_{11}^{\mathrm{T}} A_1\right) \sin\theta & \left(A_1^{\mathrm{T}} X_{11} - X_{11}^{\mathrm{T}} A_1\right) \cos\theta \\ \left(X_{11}^{\mathrm{T}} A_1 - A_1^{\mathrm{T}} X_{11}\right) \cos\theta & \left(A_1^{\mathrm{T}} X_{11} + X_{11}^{\mathrm{T}} A_1\right) \sin\theta \end{bmatrix} < 0 \tag{5.52}$$

From Theorems 2.2.2 and 2.3.1 and since $X_{11} > 0$, it follows that the inequality (5.52) guarantees the asymptotic stability of the system $D^{\alpha} x_1(t) = A_1 x_1(t), \; x_1(t) \in$

$\mathbb{R}^r$. Then system (5.7) is stable. Consequently system (5.7) is admissible since it is regular, impulse free and stable.

Necessity. Assume that the system (5.7) is admissible, then it is always possible to find non-singular matrices L and R such that (5.48) holds. From Lemma 5.4.1, we have $\text{spec}(E, A) = \text{spec}(A_1)$. Then we get

$$|\arg(\text{spec}(A_1))| > \frac{\alpha\pi}{2}.$$

According to Theorem 2.3.1, there exists a matrix $X_1 \in \mathbb{R}^{r\times r} > 0$ such that

$$\text{sym}\{\Theta \otimes (A_1^{\mathrm{T}} X_1)\} < 0. \tag{5.53}$$

Let

$$E_0 = L^{\mathrm{T}} \begin{bmatrix} 0 \\ I_{n-r} \end{bmatrix}, \; Y = R^{-T} \begin{bmatrix} Y_1 \\ Y_2 \end{bmatrix}, \; X = L^{\mathrm{T}} \begin{bmatrix} X_1 & 0 \\ 0 & I_{n-r} \end{bmatrix} L, \tag{5.54}$$

where $E^{\mathrm{T}} E_0 = 0_n$.

Using (5.48) and (5.54) and some direct computation, we obtain

$$\text{sym}\{\Theta \otimes (A^{\mathrm{T}}(XE + E_0 Y^{\mathrm{T}}))\} = \begin{bmatrix} R^{-T} & 0 \\ 0 & R^{-T} \end{bmatrix} \left[\overline{A} + \overline{A}^{\mathrm{T}}\right] \begin{bmatrix} R^{-1} & 0 \\ 0 & R^{-1} \end{bmatrix},$$

where

$$\overline{A} = \begin{bmatrix} A_1^{\mathrm{T}} X_1 \sin\theta & 0 & A_1^{\mathrm{T}} X_1 \cos\theta & 0 \\ 0 & -I_{n-r}\sin\theta & 0 & -I_{n-r}\cos\theta \\ -A_1^{\mathrm{T}} X_1 \cos\theta & 0 & A_1^{\mathrm{T}} X_1 \sin\theta & 0 \\ 0 & I_{n-r}\cos\theta & 0 & -I_{n-r}\sin\theta \end{bmatrix},$$

and

$$\overline{A} + \overline{A}^{\mathrm{T}} = \begin{bmatrix} \chi_1 & \chi_2 \\ \chi_2^{\mathrm{T}} & \chi_1 \end{bmatrix},$$

with

$$\chi_1 = \begin{bmatrix} (A_1^{\mathrm{T}} X_1 + X_1 A_1)\sin\theta & 0 \\ 0 & -2I_{n-r}\sin\theta \end{bmatrix}, \; \chi_2 = \begin{bmatrix} (A_1^{\mathrm{T}} X_1 - X_1 A_1)\cos\theta & 0 \\ 0 & 0 \end{bmatrix}.$$

According to Shur complement Lemma 4.2.4, to prove

$$\text{sym}\{\Theta \otimes (A^{\mathrm{T}}(XE + E_0 Y^{\mathrm{T}}))\} < 0,$$

it suffices to prove

$$\chi_1 < 0, \; \chi_1 - \chi_2 \chi_1^{-1} \chi_2^{\mathrm{T}} < 0.$$

Indeed, from inequality (5.53), we have $\chi_1 < 0$. We have also

$$\chi_1 - \chi_2 \chi_1^{-1} \chi_2^{\mathrm{T}} = \begin{bmatrix} (A_1^{\mathrm{T}} X_1 + X_1 A_1)\sin\theta & 0 \\ 0 & -2I_{n-r}\sin\theta \end{bmatrix} - \begin{bmatrix} (A_1^{\mathrm{T}} X_1 - X_1 A_1)\cos\theta & 0 \\ 0 & 0 \end{bmatrix} \times$$
$$\begin{bmatrix} (A_1^{\mathrm{T}} X_1 + X_1 A_1)^{-1}\frac{1}{\sin\theta} & 0 \\ 0 & \frac{-1}{2\sin\theta} I_{n-r} \end{bmatrix} \begin{bmatrix} (X_1 A_1 - A_1^{\mathrm{T}} X_1)\cos\theta & 0 \\ 0 & 0 \end{bmatrix},$$
$$= \begin{bmatrix} \Sigma & 0 \\ 0 & -2I_{n-r}\sin\theta \end{bmatrix},$$

where

$$\Sigma = (A_1^{\mathrm{T}} X_1 + X_1 A_1)\sin\theta - (A_1^{\mathrm{T}} X_1 - X_1 A_1)(A_1^{\mathrm{T}} X_1 + X_1 A_1)^{-1}(X_1 A_1 - A_1^{\mathrm{T}} X_1)\frac{\cos^2\theta}{\sin\theta}.$$

From (5.53), we get $\Sigma < 0$ and then $\chi_1 - \chi_2 \chi_1^{-1} \chi_2^{\mathrm{T}} < 0$ which guarantees

$$\mathrm{sym}\{\Theta \otimes (A^{\mathrm{T}}(XE + E_0 Y^{\mathrm{T}}))\} < 0.$$

This completes the proof of the equivalence between the admissibility of the system (5.7) and (i).

The equivalence between items (i) and (ii) can be obtained from the fact that the pair (E, A) is admissible if and only if the pair $\left(E^{\mathrm{T}}, A^{\mathrm{T}}\right)$ is admissible.

Indeed, we have

$$\det(s^{\alpha}E - A) = \det(s^{\alpha}E - A)^{\mathrm{T}} = \det(s^{\alpha}E^{\mathrm{T}} - A^{\mathrm{T}}),$$

and

$$\deg\left(\det(\lambda E - A)\right) = \deg\left(\det(\lambda E - A)^{\mathrm{T}}\right) = \deg\left(\det(\lambda E^{\mathrm{T}} - A^{\mathrm{T}})\right),$$

which shows that the pair (E, A) is regular and impulse free if and only if the pair $\left(E^{\mathrm{T}}, A^{\mathrm{T}}\right)$ is regular and impulse free. On the other hand, from Lemma 5.3.2, if the pair (E, A) is regular and impulse free, there exist two non-singular matrices M and N such that

$$E = M\begin{bmatrix} I_r & 0 \\ 0 & 0 \end{bmatrix}N, \; A = M\begin{bmatrix} A_1 & 0 \\ 0 & I_{n-r} \end{bmatrix}N,$$

then we have

$$E^{\mathrm{T}} = N^{\mathrm{T}}\begin{bmatrix} I_r & 0 \\ 0 & 0 \end{bmatrix}M^{\mathrm{T}}, \; A^{\mathrm{T}} = N^{\mathrm{T}}\begin{bmatrix} A_1^{\mathrm{T}} & 0 \\ 0 & I_{n-r} \end{bmatrix}M^{\mathrm{T}}.$$

In that case, the stability of the pair (E, A) depends entirely on A_1 or equivalently on A_1^{T}. This can be shown from Theorem 2.3.1. Consequently, the stability of the pair (E, A) is equivalent to that of the pair $\left(E^{\mathrm{T}}, A^{\mathrm{T}}\right)$. This proves the equivalence between items (i) and (ii).

The proof that (i) is equivalent to (iii) results directly from the equality between the following sets

$$\Xi = \left\{X \in \mathbb{R}^{n\times n} : \ E^{\mathrm{T}}X = X^{\mathrm{T}}E, \ E^{\mathrm{T}}X \geq 0, \operatorname{rank} E^{\mathrm{T}}X = r \right\},$$
$$\Lambda = \left\{X = PE + E_0Q : P \in \mathbb{R}^{n\times n}, \ P > 0, \ Q \in \mathbb{R}^{(n-r)\times n}\right\}.$$

where $E_0 \in \mathbb{R}^{n\times(n-r)}$ is a matrix of full-column rank such that $E^{\mathrm{T}}E_0 = 0$, and $r = \operatorname{rank} E$. This completes the proof of Theorem 5.4.1. □

Remark 5.4.2 Notice that for the case $E = I_n$, we get $E_0 = 0_n$ and $\mathcal{LMI}$ conditions given in Theorem 5.4.1 are reduced to $\mathcal{LMI}$ conditions given in Theorem 2.3.1.

On the other hand, for $\alpha = 1$, the obtained $\mathcal{LMI}$ admissibility conditions of Theorem 5.4.1 are only the admissibility conditions for singular systems. From this viewpoint, the result in Theorem 5.4.1 is a generalization of the results obtained in previous works for integer-order singular systems (see [8]).

Next result is an immediate corollary of Lemma 2.4.3 and Theorem 5.4.1. It is just to simplify the notations in this book.

Theorem 5.4.2 *Let* $\theta = (1-\alpha)\frac{\pi}{2}$. *The unforced singular linear fractional-order linear system of* (5.7) *is admissible if and only if one of the following equivalent statements holds (i) there exist matrices* $X \in \mathbb{R}^{n\times n}0$ *and* $Y \in \mathbb{R}^{(n-r)\times n}$ *such that the following* $\mathcal{LMI}$ *holds*

$$\operatorname{Her}\left\{e^{j\theta}A^{\mathrm{T}}(XE + E_0Y)\right\} < 0, \tag{5.55}$$

where $E_0 \in \mathbb{R}^{n\times(n-r)}$ *is any matrix of full column rank and verifies* $E^TE_0 = 0$.

(ii) there exist matrices $X \in \mathbb{R}^{n\times n}0$ *and* $Y \in \mathbb{R}^{(n-r)\times n}$ *such that the following* $\mathcal{LMI}$ *holds*

$$\operatorname{Her}\left\{e^{-j\theta}A(XE^T + E_0Y)\right\} < 0, \tag{5.56}$$

where $E_0 \in \mathbb{R}^{n\times(n-r)}$ *is any matrix of full column rank and verifies* $EE_0 = 0$.

(iii) there exist a matrix $P \in \mathbb{R}^{n\times n}$ *such that*

$$\begin{aligned} E^TP = P^T \geq 0, \\ \operatorname{Her}(e^{j\theta}A^TP) < 0. \end{aligned} \tag{5.57}$$

(iv) there exist a matrix $P \in \mathbb{R}^{n\times n}$ *such that*

$$\begin{aligned} EP = P^TE^T \geq 0, \\ \operatorname{Her}(e^{-j\theta}AP) < 0. \end{aligned} \tag{5.58}$$

Example 5.4.1 Consider the linear singular fractional-order system (5.7) with the following data:

$$E = \begin{bmatrix} 1 & 2 & 3 \\ 1 & 2 & 3 \\ 1 & 2 & 3 \end{bmatrix}, \ A = \begin{bmatrix} -1 & 2 & 0 \\ -3 & -4 & 1 \\ 0 & 0 & -2 \end{bmatrix}, \ \alpha = 1.5. \tag{5.59}$$

The system (5.59) is admissible. Indeed,

$$det(s^{\alpha}E - A) = 38s^{\alpha} + 20,$$

which is not identically zero and impulse free with $s^{\alpha} = \lambda$. The finite mode of the pair(E, A) lives in the stability region since it is equal to $-\frac{10}{19}$. This can also be verified by solving the linear matrix inequality (5.43) in Theorem 5.4.1 which is feasible, and concludes that the singular fractional-order system (5.59) is admissible.

A feasible solution solution of (5.43) is as follows

$$X = \begin{bmatrix} 4.0117 & -1.5249 & -1.3670 \\ -1.5249 & 1.5061 & -0.4056 \\ -1.3670 & -0.4056 & 1.4156 \end{bmatrix}, \ Y = \begin{bmatrix} 0.5736 & 0.3607 \\ 1.1342 & 1.3679 \\ 1.3225 & 1.9889 \end{bmatrix}.$$

The state response of system (5.59) is shown in Fig. 5.2 which confirms that it is asymptotically stable and its states converge to zero.

To present a new criterion guaranteeing the admissibility of the system (5.7), the following lemma will be used.

Lemma 5.4.5 *[13] There are four real matrices X_1, X_2, X_3, X_4 of appropriate dimension such that*

$$\text{sym}\left\{\Theta \otimes \begin{bmatrix} X_1 & X_2 \\ X_3 & X_4 \end{bmatrix}\right\} < 0, \tag{5.60}$$

if and only if

$$\text{sym}\left\{\begin{bmatrix} \Theta \otimes X_1 & \Theta \otimes X_2 \\ \Theta \otimes X_3 & \Theta \otimes X_4 \end{bmatrix}\right\} < 0, \tag{5.61}$$

with Θ is defined in (5.39).

Proof For the proof, see Appendix D.

□

Theorem 5.4.3 *[13] System* (5.7) *is admissible if and only if there exist matrices $P_1 \in \mathbb{R}^{r \times r}$, $P_2 \in \mathbb{R}^{(n-r) \times r}$ and non-singular matrix $P_3 \in \mathbb{R}^{(n-r)\times(n-r)}$ satisfying*

$$P_1 > 0, \tag{5.62}$$

$$\text{sym}\left\{\Theta \otimes (MANP)\right\} < 0, \tag{5.63}$$

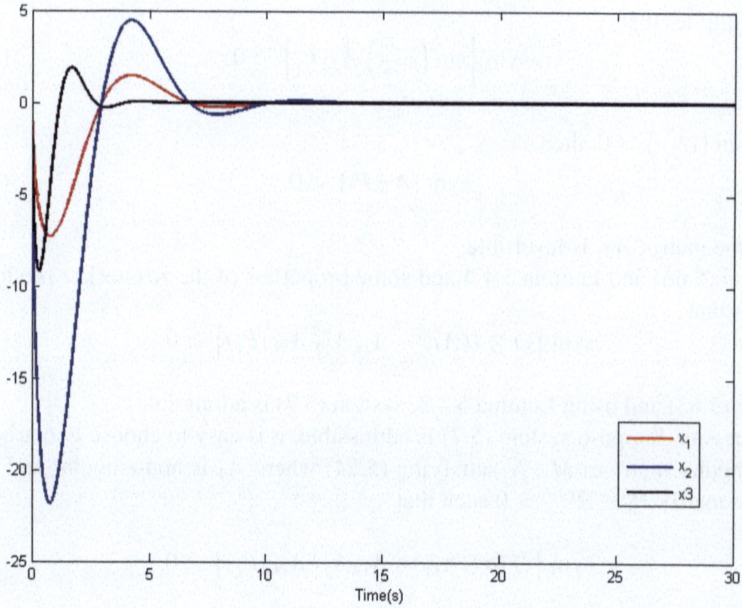

Fig. 5.2 Time response of $x(t)$ with $x(0) = [-1, -3, -2]^{\mathrm{T}}$ and fractional order $\alpha = 1.5$.

where Θ *is defined in* (5.39), $M, N \in \mathbb{R}^{n \times n}$ *are two arbitrary chosen non-singular matrices satisfying* (5.24), *and*

$$P = \begin{bmatrix} P_1 & 0 \\ P_2 & P_3 \end{bmatrix}. \tag{5.64}$$

Proof Sufficiency. For system (5.7), it is easy to choose two arbitrary non-singular matrices M and N satisfying (5.24). Assume that there exist matrices $P_1 \in \mathbb{R}^{r \times r}$, $P_2 \in \mathbb{R}^{(n-r) \times r}$ and non-singular matrix $P_3 \in \mathbb{R}^{(n-r) \times (n-r)}$ such that (5.62) and (5.63) hold. It then follows from (5.63) and (5.64) that

$$\text{sym}\left\{\Theta \otimes \begin{bmatrix} A_{11}P_1 + A_{12}P_2 & A_{12}P_3 \\ A_{21}P_1 + A_{22}P_2 & A_{22}P_3 \end{bmatrix}\right\} < 0, \tag{5.65}$$

It is re-written by Lemma 5.4.5 as

$$\text{sym}\left\{\begin{bmatrix} \Theta \otimes (A_{11}P_1 + A_{12}P_2) & \Theta \otimes (A_{12}P_3) \\ \Theta \otimes (A_{21}P_1 + A_{22}P_2) & \Theta \otimes (A_{22}P_3) \end{bmatrix}\right\} < 0. \tag{5.66}$$

For the (2, 2) sub-block of the matrix of the left hand side of (5.66), it follows that

$$\text{sym}\left\{\Theta \otimes (A_{22}P_3)\right\} < 0, \tag{5.67}$$

which implies that

$$\mathrm{sym}\left\{\sin\left(\alpha\frac{\pi}{2}\right)A_{22}P_3\right\} < 0. \tag{5.68}$$

Since $\sin\left(\alpha\frac{\pi}{2}\right) > 0$, then

$$\mathrm{sym}\left\{A_{22}P_3\right\} < 0. \tag{5.69}$$

Thus, the matrix A_{22} is invertible.

From (5.66) and Lemma 5.4.4 and some properties of the Kronecker product, it follows that

$$\mathrm{sym}\left\{\Theta \otimes ((A_{11} - A_{12}A_{22}^{-1}A_{21})P_1)\right\} < 0. \tag{5.70}$$

Noting (5.62) and using Lemma 5.4.3, system (5.7) is admissible.

Necessity. Suppose system (5.7) is admissible, it is easy to choose two arbitrary non-singular matrices $M,\ N$ satisfying (5.24) where A_4 is non-singular and there exists a matrix $P_1 \in \mathbb{R}^{r\times r} > 0$ such that

$$\mathrm{sym}\left\{\Theta \otimes ((A_{11} - A_{12}A_{22}^{-1}A_{21})P_1)\right\} < 0. \tag{5.71}$$

Let

$$M_1 = \begin{bmatrix} I_r & -A_{12}A_{22}^{-1} \\ 0 & A_{22}^{-1} \end{bmatrix},$$

$$N_1 = \begin{bmatrix} I_r & 0 \\ -A_{22}^{-1}A_{21} & I_{n-r} \end{bmatrix},$$

$$\bar{X} = \begin{bmatrix} P_1 & 0 \\ 0 & -I_{n-r} \end{bmatrix}.$$

Then, it is easy to see that M1 and N1 are non-singular and

$$M_1MENN_1 = M_1\begin{bmatrix} I_r & 0 \\ 0 & 0 \end{bmatrix}N_1 = \begin{bmatrix} I_r & 0 \\ 0 & 0 \end{bmatrix},$$

$$M_1MANN_1 = M_1\begin{bmatrix} A_{11} & A_{12} \\ A_{21} & A_{22} \end{bmatrix}N_1 = \begin{bmatrix} A_{11} - A_{12}A_{22}^{-1}A_{21} & 0 \\ 0 & I_{n-r} \end{bmatrix}.$$

Therefore, it is easy to see

$$\mathrm{sym}\left\{\Theta \otimes (M_1MANN_1\bar{X})\right\} < 0. \tag{5.72}$$

Pre- and post-multiplying (5.72) by $I_2 \otimes M_1^{-1}$ and $I_2 \otimes M_1^{-T}$, respectively, results in

$$\mathrm{sym}\left\{\Theta \otimes (MANN_1\bar{X}M_1^{-T})\right\} < 0. \tag{5.73}$$

It is easy to calculate

$$M_1^{-T} = \begin{bmatrix} I_r & 0 \\ A_{12}^{\mathrm{T}} & A_{22}^{\mathrm{T}} \end{bmatrix}.$$

Then, we obtain

$$N_1 \bar{X} M_1^{-T} = \begin{bmatrix} P_1 & 0 \\ -A_{22}^{-1} A_{21} P_1 - A_{12}^{\mathrm{T}} & -A_{22}^{\mathrm{T}} \end{bmatrix}.$$

Let

$$P = \begin{bmatrix} P_1 & 0 \\ -A_{22}^{-1} A_{21} P_1 - A_{12}^{\mathrm{T}} & -A_{22}^{\mathrm{T}} \end{bmatrix}.$$

the matrices P_1 and P satisfy (5.62) and (5.63), respectively. This completes the proof. □

Remark 5.4.3 The two Theorems 5.4.1 and 5.4.3 require the simple full rank decomposition of the singular matrix E, without involving the decompositions of the system matrix A and the input matrix B. However, Theorem 5.4.1 involves fewer solved variables in the linear matrix inequality ($\mathcal{LMI}$) formulation, making it effectively solvable using the $\mathcal{LMI}$ toolbox.

Corollary 5.4.1 *[13] System* (5.7) *is admissible if and only if there exist matrices* $P_1 \in \mathbb{R}^{r\times r}$, $P_2 \in \mathbb{R}^{(n-r)\times r}$, *and non-singular matrix* $P3 \in \mathbb{R}^{(n-r)\times(n-r)}$ *satisfying*

$$P_1 > 0, \tag{5.74}$$

$$\mathrm{sym}\left\{\Theta \otimes (ANPM^{-T})\right\} < 0, \tag{5.75}$$

where Θ *is defined in* (5.39), $M, N \in \mathbb{R}^{n\times n}$ *are two arbitrary chosen non-singular matrices satisfying* (5.24), *and*

$$P = \begin{bmatrix} P_1 & 0 \\ P_2 & P_3 \end{bmatrix}. \tag{5.76}$$

Proof By pre- and post-multiplying (5.63) by $I_2 \otimes M^{-\mathrm{T}}$ and $I_2 \otimes M^{-1}$ respectively, we get that (5.63) is equivalent to

$$\mathrm{sym}\left\{(I_2 \otimes M^{-1})(\Theta \otimes (MANP))(I_2 \otimes M^{-\mathrm{T}})\right\} < 0.$$

Using the properties of the Kronecker product, the last inequality is the same as

$$\mathrm{sym}\left\{(I_2.\Theta.I_2) \otimes (M^{-1}.MANPM^{-\mathrm{T}})\right\} < 0,$$

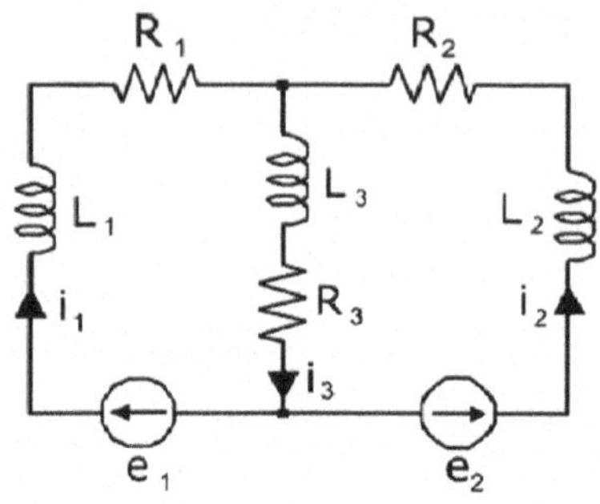

Fig. 5.3 Electronic network

which is equivalent to

$$sym\left\{\Theta \otimes (ANPM^{-T})\right\} < 0.$$

This completes the proof. □

Example 5.4.2 [13] Consider the electrical circuit shown in Fig. 5.3 with given resistances R_1, R_2, R_3, inductances L_1, L_2, L_3 and source voltages e_1 and e_2 [14]. Using the Kirchhoff's laws we write the following equations for the electrical circuit:

$$e_1 = R_1 i_1 + L_1 D^\alpha i_1 + R_3 i_3 + L_3 D^\alpha i_3,$$

$$e_2 = R_2 i_2 + L_2 D^\alpha i_2 + R_3 i_3 + L_3 D^\alpha i_3,$$

$$i_1 + i_2 - i_3 = 0.$$

Then, the following SFOS is obtained:

$$ED^\alpha x(t) = Ax(t) + Bu(t), \tag{5.77}$$

where

$$E = \begin{bmatrix} L_1 & 0 & L_3 \\ 0 & L_2 & L_3 \\ 0 & 0 & 0 \end{bmatrix}, \ A = \begin{bmatrix} -R_1 & 0 & -R_3 \\ 0 & -R_2 & -R_3 \\ 1 & 1 & -1 \end{bmatrix}, \ B = \begin{bmatrix} 1 & 0 \\ 0 & 1 \\ 0 & 0 \end{bmatrix},$$

$$x(t) = \begin{bmatrix} i_1 \\ i_2 \\ i_3 \end{bmatrix}, \ u(t) = \begin{bmatrix} e_1 \\ e_2 \end{bmatrix}, \ \alpha = 1.2.$$

For system (5.77) with $u(t) = 0$, setting $L_1 = L_2 = L_3 = 1, R_1 = 1,\ R_2 = 3,\ R_3 = 1$, it is easy to verify that $\det(s^{1.2}E - A)$ is not identically zero. So, system (5.77) is regular. It is easy to calculate $\deg(\det(sE - A)) = \deg(3s2 + 14s + 15) = 2 =$

$rank(E)$. Thus, system (5.77) is impulse-free. The roots λ of the polynomial $\det(sE - A) = 0$ are $-3, -\frac{5}{3}$ satisfying

$$\left|\arg(\lambda) > \alpha\frac{\pi}{2}\right|.$$

Therefore, the system (5.77) is asymptotically stable. Consequently, system (5.77) is admissible.

Given the initial conditions

$$x(0) = \begin{bmatrix} 2 & 1 & 0.3789 \end{bmatrix}^{\mathrm{T}},$$

the state variable responses of system (5.77) in Example 5.4.2 are shown in Fig. 5.4.

By using $\mathcal{LMI}$ toolbox, inequalities (5.62) and (5.64) in Theorem 5.4.3 are feasible. One feasible solution is given by

$$P_1 = \begin{bmatrix} 12.4797 & 4.1076 \\ 4.1076 & 12.4797 \end{bmatrix}, \; P_2 = \begin{bmatrix} 6.4149 & -1.9571 \end{bmatrix}, \; P3 = [-4.0751].$$

The following theorem gives an other result assuring the admissibility of system (5.7) in terms of non-strict $\mathcal{LMI}$s, which will be used for the control problem.

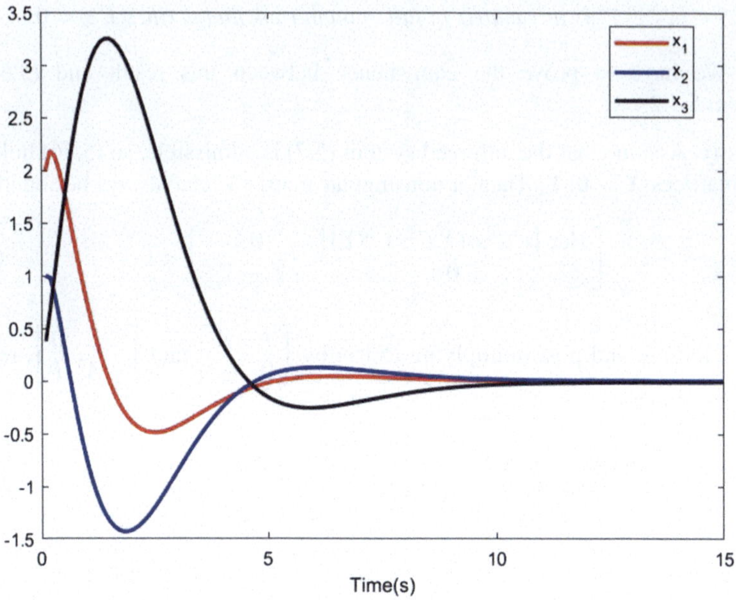

Fig. 5.4 State responses of SFOS (5.77) in Example 5.4.2

Theorem 5.4.4 *[15] System* (5.7) *with fractional-order* α *belonging to* $1 \leq \alpha < 2$ *is admissible, if and only if there exists nonsingular matrix* $P \in \mathbb{R}^{n\times n}$ *such that*

$$EP = P^T E^T \geq 0,$$
$$\text{sym}\{\Gamma \otimes (AP)\} < 0, \tag{5.78}$$

where

$$\Gamma = \begin{bmatrix} \sin\left(\alpha\frac{\pi}{2}\right) & -\cos\left(\alpha\frac{\pi}{2}\right) \\ \cos\left(\alpha\frac{\pi}{2}\right) & \sin\left(\alpha\frac{\pi}{2}\right) \end{bmatrix} \tag{5.79}$$

Proof The proof is similar to that of Theorem 5.4.6; it is omitted. □

The following theorem gives a new formulation of admissibility conditions for the unforced singular fractional-order system (5.7).

Theorem 5.4.5 *[16] Let* $\theta = (1-\alpha)\frac{\pi}{2}$. *The unforced singular fractional-order system* (5.7) *is admissible if and only if the following statement holds. There exist matrices* $X \in \mathbb{R}^{n\times n} > 0$, $Y \in \mathbb{R}^{n-r\times n}$, $V \in \mathbb{R}^{n\times n}$, *and* $U \in \mathbb{C}^{n\times n}$ *such that:*

$$\begin{bmatrix} \text{Her}\{AU\} & (e^{-j\theta}(XE^{\mathrm{T}} + SY))^* + AV - U^* \\ (*) & -V - V^{\mathrm{T}} \end{bmatrix} < 0, \tag{5.80}$$

where $S \in \mathbb{R}^{n\times(n-r)}$ *is any matrix of full column rank and verifies* $ES = 0$.

Proof We have to prove the equivalence between this result and (5.56) in Theorem 5.4.2.

Necessity. Assume that the unforced system (5.7) is admissible, so (5.56) holds for some matrices $X > 0$, Y. Then, a nonsingular matrix V can always be found such that

$$\begin{bmatrix} \text{Her}\left\{e^{-j\theta}A(XE^{\mathrm{T}} + SY)\right\} & 0 \\ 0 & -V - V^{\mathrm{T}} \end{bmatrix} < 0. \tag{5.81}$$

Pre-multiplying and post-multiplying (5.81) by $\begin{bmatrix} I & -A \\ 0 & I \end{bmatrix}$ and $\begin{bmatrix} I & 0 \\ -A^{\mathrm{T}} & I \end{bmatrix}$, respectively, we obtain

$$\begin{bmatrix} \text{Her}\left\{e^{-j\theta}A(XE^{\mathrm{T}} + SY) - AVA^{\mathrm{T}}\right\} & AV + AV^{\mathrm{T}} \\ VA^{\mathrm{T}} + V^{\mathrm{T}}A^{\mathrm{T}} & -V - V^{\mathrm{T}} \end{bmatrix} < 0. \tag{5.82}$$

Then by setting $U = e^{-j\theta}(XE^{\mathrm{T}} + SY) - VA^{\mathrm{T}}$, we get

$$\begin{bmatrix} AU + U^*A^{\mathrm{T}} & AV + AV^{\mathrm{T}} \\ VA^{\mathrm{T}} + V^{\mathrm{T}}A^{\mathrm{T}} & -V - V^{\mathrm{T}} \end{bmatrix} < 0 \tag{5.83}$$

and, from $AV^{\mathrm{T}} = \left(e^{-j\theta}(XE^{\mathrm{T}} + SY)\right)^* - U^*$, (5.80) is obtained.

Sufficiency. To obtain (5.56), pre-multiply and post-multiply (5.80) by $\begin{bmatrix} I & A \end{bmatrix}$ and $\begin{bmatrix} I & A \end{bmatrix}^{\mathrm{T}}$, respectively. This means that (5.80) implies the admissibility of the system (5.7). This ends the proof. □

Remark 5.4.4 Note that the introduction of matrix variables U and V benefits to the control design, since the gain matrix depends on U and not on X. This new representation is more suitable for the $\mathcal{H}_\infty$ control design in comparison to the conditions (5.43), (5.56) in Theorems 5.4.1 and 5.4.2 respectively which cannot be directly used to design controllers, since they are numerically intractable due to involving the inversion of the matrix $XE^T + SY$. Also, for $\alpha = 1$, the condition (5.80) coincides with the condition (9) in Theorem 1 in [17].

5.4.2 *Admissibility of Singular Fractional-Order Systems: Case* $0 < \alpha \leq 1$

Here, we focus with the admissibility problem of the system (5.7) with $0 < \alpha \leq 1$. Different derivations for the system (5.7) to be admissible were derived. The results are a necessary and sufficient condition involving strict and non-strict $\mathcal{LMI}$s. A preliminary result in terms of non-strict $\mathcal{LMI}$s is proposed in [15].

Theorem 5.4.6 *[15] System* (5.7) *with fractional-order* α *belonging to* $0 < \alpha < 1$ *is admissible, if and only if there exist nonsingular matrices*

$$P_{k1} \in \mathbb{R}^{n\times n},\ P_{k2} \in \mathbb{R}^{n\times n},\ k = 1, 2$$

such that

$$\sum_{i=1}^{2}\sum_{j=1}^{2} \mathrm{sym}\left\{\Theta_{ij} \otimes (AP_{ij})\right\} < 0, \tag{5.84}$$

$$\begin{bmatrix} EP_{11} & EP_{12} \\ -EP_{12} & EP_{11} \end{bmatrix} \geq 0,$$

$$\begin{bmatrix} EP_{21} & EP_{22} \\ -EP_{22} & EP_{21} \end{bmatrix} \geq 0, \tag{5.85}$$

$$\begin{aligned} EP_{11} &= P_{11}^T E^T \geq 0, \\ EP_{21} &= P_{21}^T E^T \geq 0, \end{aligned} \tag{5.86}$$

$$\begin{aligned} EP_{12} + P_{12}^T E^T &= 0, \\ EP_{22} + P_{22}^T E^T &= 0, \end{aligned} \tag{5.87}$$

where Θ_{ij}, $i = 1, 2$ *and* $j = 1, 2$ *satisfy* (2.24).

Proof Sufficiency. Since $\mathrm{rank}(E) = r \leq n$, there exist nonsingular matrices G and H such that

$$GEH = \begin{bmatrix} I_r & 0 \\ 0 & 0 \end{bmatrix}. \tag{5.88}$$

Denote

$$GAH = \begin{bmatrix} A_1 & A_2 \\ A_3 & A_4 \end{bmatrix},$$

$$\begin{aligned} H^{-1}P_{k1}G^{-1} &= \begin{bmatrix} P_{k11} & P_{k12} \\ P_{k13} & P_{k14} \end{bmatrix}, \\ H^{-1}P_{k2}G^{-1} &= \begin{bmatrix} P_{k21} & P_{k22} \\ P_{k23} & P_{k24} \end{bmatrix}, \\ &\qquad k = 1,\ 2. \end{aligned} \tag{5.89}$$

From (5.86)–(5.87) and using the expressions in (5.88) and (5.89), it is easy to get

$$P_{k12} = 0,\ P_{k22} = 0,\ k = 1,\ 2,$$

Premultiplying and postmultiplying inequality (5.84) by $I_2 \otimes G$ and $(I_2 \otimes G)^{\mathrm{T}}$, respectively,we obtain

$$\mathrm{sym}\,\{A_4(P_{114} + P_{214} + P_{124} + P_{224})\} < 0, \tag{5.90}$$

which implies that the matrix A_4 is non-singular. According to Lemma 5.3.4, system (5.7) is regular and impulse-free.

The next step is devoted to the asymptotic stability of the system. Since A_4 is non-singular, we set

$$\hat{G} = \begin{bmatrix} I_r & -A_2A_4^{-1} \\ 0 & A_4^{-1} \end{bmatrix} G.$$

It is easy to see that

$$
\begin{aligned}
\hat{G}EH &= \begin{bmatrix} I_r & 0 \\ 0 & 0 \end{bmatrix}, \\
\hat{G}AH &= \begin{bmatrix} \hat{A}_1 & 0 \\ \hat{A}_3 & I_{n-r} \end{bmatrix}, \\
H^{-1}P_{k1}\hat{G}^{\mathrm{T}} &= \begin{bmatrix} \hat{P}_{k11} & \hat{P}_{k12} \\ \hat{P}_{k13} & \hat{P}_{k14} \end{bmatrix}, \\
H^{-1}P_{k2}\hat{G}^{\mathrm{T}} &= \begin{bmatrix} \hat{P}_{k21} & \hat{P}_{k22} \\ \hat{P}_{k23} & \hat{P}_{k24} \end{bmatrix}, \\
k &= 1,\ 2,
\end{aligned} \tag{5.91}
$$

where

$$\hat{A}_1 = A_1 - A_2A_4^{-1}A_3,\ \hat{A}_3 = A_4^{-1}A_3.$$

From (5.84)–(5.87) and using the expression in (5.91), we obtain

$$
\begin{aligned}
\hat{P}_{k11} = \hat{P}_{k11}^{\mathrm{T}} &> 0, \\
\hat{P}_{k21} + \hat{P}_{k21}^{\mathrm{T}} &= 0, \\
k &= 1,\ 2, \\
\sum_{i=1}^{2}\sum_{j=1}^{2} \mathrm{sym}\left\{\Theta_{ij} \otimes (\hat{A}_1\hat{P}_{ij1})\right\} &< 0, \\
\begin{bmatrix} \hat{P}_{111} & \hat{P}_{121} \\ -\hat{P}_{121} & \hat{P}_{111} \end{bmatrix} &> 0, \\
\begin{bmatrix} \hat{P}_{211} & \hat{P}_{221} \\ -\hat{P}_{221} & \hat{P}_{211} \end{bmatrix} &> 0.
\end{aligned} \tag{5.92}
$$

Define

$$\begin{bmatrix} \eta_1(t) \\ \eta_2(t) \end{bmatrix} = H^{-1}\eta(t);$$

system (5.7) is the restricted system equivalent to

$$
\begin{aligned}
D^{\alpha}\eta_1(t) &= \hat{A}_1\eta_1(t), \\
0 &= \hat{A}_3\eta_1(t) + \eta_2(t), \\
0 &< \alpha < 1.
\end{aligned} \tag{5.93}
$$

By Lemma 2.4.2, we can derive from (5.92) that

$$\left|\arg\left(\operatorname{spec}(\hat{A}_1)\right)\right| > \alpha\frac{\pi}{2},$$

which is equivalent to

$$\lim_{t\to+\infty} \|\eta_1(t)\| = 0,$$

therefore

$$\lim_{t\to+\infty} \|\eta_2(t)\| = \lim_{t\to+\infty} \left\|\hat{A}_3\eta_1(t)\right\| = 0.$$

Consequently, according to Definition 5.4.1, the system (5.7) is admissible.

Necessity. Since the pair (E, A) is admissible, there exist nonsingular matrices L and R such that

$$LER = \begin{bmatrix} I_r & 0 \\ 0 & 0 \end{bmatrix}, \ LAR = \begin{bmatrix} \bar{A}_1 & 0 \\ 0 & I_{n-r} \end{bmatrix}. \tag{5.94}$$

By Lemma 2.4.2, there exist two real symmetric positive definite matrices $\bar{P}_{k11}$, $k = 1, 2$, and two skew-symmetric matrices $\bar{P}_{k21}$, $k = 1,\ 2$, such that

$$\sum_{i=1}^{2}\sum_{j=1}^{2}\{\Theta_{ij} \otimes (\bar{A}_1\bar{P}_{ij1})\} < 0, \tag{5.95}$$

$$\begin{bmatrix} \bar{P}_{111} & \bar{P}_{121} \\ -\bar{P}_{121} & \bar{P}_{111} \end{bmatrix} > 0,$$

$$\begin{bmatrix} \bar{P}_{211} & \bar{P}_{221} \\ -\bar{P}_{221} & \bar{P}_{211} \end{bmatrix} > 0. \tag{5.96}$$

Define

$$P_{k1} = R\begin{bmatrix} \bar{P}_{k11} & 0 \\ 0 & I \end{bmatrix}L^{-T},$$

$$P_{k2} = R\begin{bmatrix} \bar{P}_{k21} & 0 \\ 0 & 0 \end{bmatrix}L^{-T}, \tag{5.97}$$

$$k = 1,\ 2. \tag{5.98}$$

The verification of conditions (5.84)–(5.87) can be done effortlessly, which leads us to the conclusion that the proof is now finished. □

Remark 5.4.5 The statements of Theorem 5.4.6 summarize the existing results on the the stability of a larger class of dynamic linear systems. Indeed,

- For $E = I_n$, it can easily be shown that conditions (5.84)–(5.87) in Theorem 5.4.6 are equivalent to the conditions (2.22)–(2.24) in Lemma 2.4.2.

- For $\alpha = 1$, conditions (5.84)–(5.87) in Theorem 5.4.6 imply

$$E(P_{11} + P_{21}) = (P_{11} + P_{21})^{\mathrm{T}} E^{\mathrm{T}} \geq 0, \tag{5.99}$$
$$\mathrm{sym}\,\{A(P_{11} + P_{21})\} < 0. \tag{5.100}$$

  These last inequalities coincide with the admissibility conditions (2.14)–(2.15) in Theorem 2.1 for a standard singular system in [8].

- For $E = I_n$ and $\alpha = 1$, conditions (5.84)–(5.87) in Theorem 5.4.6 imply the following inequalities

$$\mathrm{sym}\,\{A(P_{11} + P_{21})\} < 0, \tag{5.101}$$
$$P_{11} + P_{21} > 0, \tag{5.102}$$

  which are the conditions for the asymptomatic stability of linear system according to Lyapunov stability theory in [18].

From Lemma 2.4.1, the following theorem can be obtained easily.

Theorem 5.4.7 *[10] The singular fractional-order linear continuous system* (5.7) *with order* $0 < \alpha \leq 1$ *is admissible if and only if there exist positive matrices* $X_1 = X_1^* \in \mathbb{C}^{n\times n}$ *and* $X_2 = X_2^* \in \mathbb{C}^{n\times n}$ *such that*

1. *The matrix* A_4 *is invertible.*
2. *The following* $\mathcal{LMI}$ *satisfied*

$$\tilde{A}(rX_1 + \bar{r}X_2) + (\bar{r}X_1 + rX_2)\tilde{A}^T < 0, \tag{5.103}$$

where

$$r = \mathrm{e}^{j(1-\alpha)\frac{\pi}{2}},\ \tilde{A} = A_1 - A_2A_4^{-1}A_3, \tag{5.104}$$

with $A_1,\ A_2,\ A_3,\ A_4$ *are defined in* (5.24).

Proof The result of this theorem is a direct consequence of Lemmas 5.4.2 and 2.4.1. □

Theorem 5.4.8 *[19] The singular fractional-order linear continuous system* (5.7) *with order* $0 < \alpha \leq 1$ *is admissible if and only if there exist matrices* $X = X^* \in \mathbb{C}^{n\times n} > 0$, $Y \in \mathbb{R}^{n\times(n-r)}$ *such that*

$$\mathrm{sym}\left\{A\left(\left(zX + \bar{z}\bar{X}\right)E^T + E_0Y^T\right)\right\} < 0 \tag{5.105}$$

where E_0 *is any matrix of full rank column that satisfies* $EE_0 = 0$ *and* $z = e^{j(1-\alpha)\frac{\pi}{2}}$.

Proof Sufficiency - Assume that the inequality (5.105) holds for some matrices $X = X^* \in \mathbb{C}^{n\times n} > 0$ and $Y \in \mathbb{R}^{n\times(n-r)}$. Since rank $E = r < n$, invertible matrices M, N can always be found such that

$$MEN = \begin{bmatrix} I_r & 0 \\ 0 & 0 \end{bmatrix}. \tag{5.106}$$

The matrices M and N are not unique and can be obtained either by rank decomposition or by singular value decomposition.

Let

$$MAN = \begin{bmatrix} A_{11} & A_{12} \\ A_{21} & A_{22} \end{bmatrix}, E_0 = N \begin{bmatrix} 0 \\ I_{n-r} \end{bmatrix}. \tag{5.107}$$

Then from (5.106), (5.107) and the invertibility of the matrix N, we have

$$EE_0 = M^{-1} \begin{bmatrix} I_r & 0 \\ 0 & 0 \end{bmatrix} N^{-1} N \begin{bmatrix} 0 \\ I_{n-r} \end{bmatrix} = 0_{n\times(n-r)}.$$

Let us now define matrices X and Y as

$$X = N \begin{bmatrix} X_{11} & X_{12} \\ X_{21} & X_{22} \end{bmatrix} N^{\mathrm{T}}, Y = M^{-1} \begin{bmatrix} Y_1 \\ Y_2 \end{bmatrix}. \tag{5.108}$$

Tacking account (5.106), (5.107) and (5.108) into (5.105), we get $\Sigma + \Sigma^{\mathrm{T}} < 0$ with

$$\begin{aligned} \Sigma &= A\left(\left(zX + \bar{z}\overline{X}\right)E^{\mathrm{T}} + E_0 Y^{\mathrm{T}}\right) \\ &= M^{-1} \times \begin{bmatrix} A_{11}\left(zX_{11} + \bar{z}\overline{X_{11}}\right) + A_{12}\left(zX_{21} + \bar{z}\overline{X_{21}} + Y_1^{\mathrm{T}}\right) & A_{12}Y_2^{\mathrm{T}} \\ A_{21}\left(zX_{11} + \bar{z}\overline{X_{11}}\right) + A_{22}\left(zX_{21} + \bar{z}\overline{X_{21}} + Y_1^{\mathrm{T}}\right) & A_{22}Y_2^{\mathrm{T}} \end{bmatrix} M^{-T} \end{aligned}$$

which implies

$$A_{22}Y_2^{\mathrm{T}} + Y_2 A_{22}^{\mathrm{T}} < 0.$$

Therefore the matrix A_{22} is invertible which means that the system (5.7) is regular and impulse free. Since the system (5.7) is regular and impulse free, invertible matrices L and R can always be found such that

$$LER = \begin{bmatrix} I_r & 0 \\ 0 & 0 \end{bmatrix}, LAR = \begin{bmatrix} A_1 & 0 \\ 0 & I_{n-r} \end{bmatrix}. \tag{5.109}$$

Let now,

$$X = R \begin{bmatrix} X_{11} & X_{12} \\ X_{21} & X_{22} \end{bmatrix} R^{\mathrm{T}}, Y = L^{-1} \begin{bmatrix} Y_1 \\ Y_2 \end{bmatrix}, E_0 = R \begin{bmatrix} 0 \\ I_{n-r} \end{bmatrix}. \tag{5.110}$$

From (5.109) and (5.110) and the invertibility of the matrix L, the condition (5.105) is equivalent to

$$\begin{bmatrix} \Omega_{11} & \Omega_{12} \\ \Omega_{12}^{\mathrm{T}} & \Omega_{22} \end{bmatrix} < 0,$$

where

$$\Omega_{11} = A_1 \left(zX_{11} + \overline{z}\overline{X}_{11}\right) + \left(zX_{11} + \overline{z}\overline{X}_{11}\right)^{\mathrm{T}} A_1^{\mathrm{T}},$$
$$\Omega_{12} = \left(zX_{21} + \overline{z}\overline{X}_{21}\right)^{\mathrm{T}} + Y_1,$$
$$\Omega_{22} = Y_2 + Y_2^{\mathrm{T}}.$$

It follows that $\Omega_{11} < 0$. From Theorem 2.4.1, we deduce that the system $D^{\alpha}x_1(t) = A_1x(t)$ is stable, and consequently the system (5.7) is stable. We conclude that the system (5.7) is admissible since it is regular, impulse free and stable.

Necessity - Assume that the system (5.7) is admissible. Then invertible matrices can always be found such that (5.109) holds and the system $D^{\alpha}x_1(t) = A_1x(t)$ is stable. Then according to Theorem 2.4.1, there exist a matrix $X_1 = X_1^* \in \mathbb{C}^{r\times r} > 0$ such that

$$A_1 \left(zX_1 + \overline{z}\overline{X}_1\right) + \left(zX_1 + \overline{z}\overline{X}_1\right)^{\mathrm{T}} A_1^{\mathrm{T}} < 0, \tag{5.111}$$

with $z = e^{j(1-\alpha)\frac{\pi}{2}}$.

Let matrices,

$$X = R \begin{bmatrix} X_1 & 0 \\ 0 & I_{n-r} \end{bmatrix} R^{\mathrm{T}}, \; Y = L^{-1} \begin{bmatrix} 0 \\ -I_{n-r} \end{bmatrix}, \; E_0 = R \begin{bmatrix} 0 \\ I_{n-r} \end{bmatrix}. \tag{5.112}$$

It is easy to show that $X = X^*$ and the matrix E_0 is of full rank column and satisfies $EE_0 = 0$.

From (5.109) and (5.112) and some computation, we have

$$\mathrm{sym}\left\{A\left(\left(zX + \overline{z}\overline{X}\right)E^{\mathrm{T}} + E_0Y^{\mathrm{T}}\right)\right\} = L^{-1} \begin{bmatrix} \mathrm{sym}\left\{A_1\left(zX_1 + \overline{z}\overline{X}_1\right)\right\} & 0 \\ 0 & -2I_{n-r} \end{bmatrix} L^{-T}$$

Tacking into account (5.111), we get

$$\mathrm{sym}\left\{A\left(\left(zX + \overline{z}\overline{X}\right)E^{\mathrm{T}} + E_0Y^{\mathrm{T}}\right)\right\} < 0,$$

which completes the proof of the Theorem 5.4.8. □

Note that, since the admissibility of the pair (E, A) is equivalent to the admissibility of the pair $\left(E^{\mathrm{T}}, A^{\mathrm{T}}\right)$, we have

Corollary 5.4.2 *[19] The singular fractional-order linear continuous system* (5.7) *with order* $0 < \alpha \leq 1$ *is admissible if and only if there exist matrices* $X = X^* \in \mathbb{C}^{n\times n} > 0$, $Y \in \mathbb{R}^{n\times(n-r)}$ *such that*

$$\mathrm{sym}\left\{A^T\left(\left(zX+\overline{zX}\right)E+E_0Y^T\right)\right\}<0 \tag{5.113}$$

where E_0 *is any matrix of full rank column that satisfies* $E^TE_0=0$ *and* $z=e^{j(1-\alpha)\frac{\pi}{2}}$.

Remark 5.4.6 Let remark that for $E=I_n$, $E_0=0$ and the $\mathcal{LMI}$ condition (5.105) in Theorem 5.4.8 is the same as the $\mathcal{LMI}$ condition of stability for FOS given in Theorem 2.4.1.

If we take $\alpha=1$, (5.105) becomes

$$\mathrm{sym}\left\{A\left(\left(X+\overline{X}\right)E^{\mathrm{T}}+E_0Y^{\mathrm{T}}\right)\right\}<0$$

with $X+\bar{X}>0$ as $X>0$ which means that we recover the admissibility condition for singular system (see Theorem 2.2 in [8]).

Now, if we take $E=I_n$ and $\alpha=1$ we obtain the Lyapunov stability for the system $\dot{x}(t)=Ax(t)$. According to this discussion, we deduce that the derived result in Theorem 5.4.8 is more general.

Example 5.4.3 Consider the linear singular fractional-order system (5.7) with $\alpha=0.3$ and

$$E=\begin{bmatrix}1&2&3\\1&2&3\\1&0&0\end{bmatrix},\ A=\begin{bmatrix}1&3&2\\1&2&1\\-1&0.5&0.3\end{bmatrix}. \tag{5.114}$$

A feasible solution of the linear matrix inequality (5.105) of Theorem 5.4.8 is as follows:

$$X=\begin{bmatrix}73.9517+0.0000i & 6.0022-2.9893i & 9.0033-4.4840i\\ 6.0022+2.9893i & 84.8527+0.0000i & 3.0206+0.0000i\\ 9.0033+4.4840i & 3.0206-0.0000i & 87.3698+0.0000i\end{bmatrix},\ Y=\begin{bmatrix}825.7043\\663.8281\\-82.5359\end{bmatrix}$$

We conclude that the singular fractional-order system (5.114) is admissible. The state responses of the selected system is shown in Fig. 5.5, which confirms that the system is asymptotically stable and its states converge to zero.

Example 5.4.4 Now, we will consider the same system taken in Example 1 in [11]

$$E=\begin{bmatrix}1&1\\8&8\end{bmatrix},\ A=\begin{bmatrix}-1&0\\1&-9\end{bmatrix},$$

with $\alpha=0.5$.

Result of Corollary 5.4.2 is applied. $\mathcal{LMI}$ (5.113) leads to

$$X=\begin{bmatrix}0.9882+0.0000i & -0.1228-0.0000i\\ -0.1228+0.0000i & 0.0209+0.0000i\end{bmatrix},\ Y=\begin{bmatrix}0.5053\\0.0518\end{bmatrix}.$$

An equivalent version of Theorem 5.4.8 is given in the following Theorem.

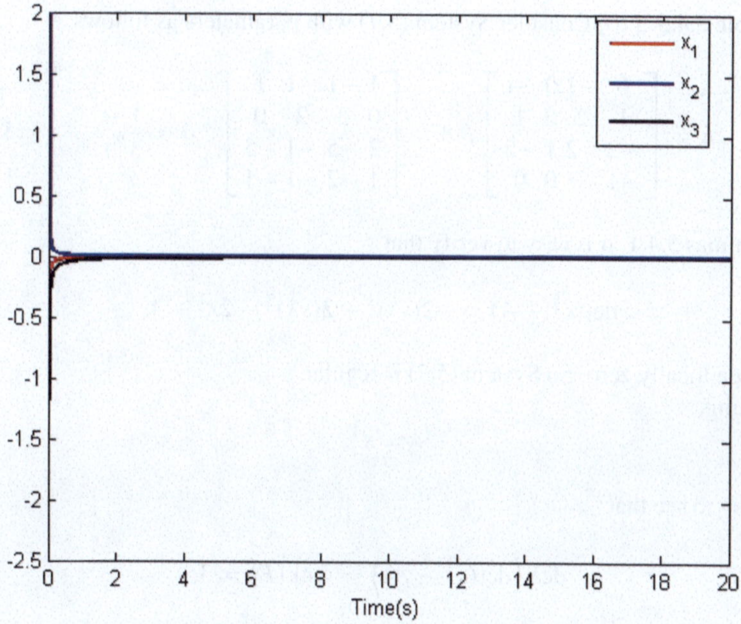

Fig. 5.5 State responses of the selected system (5.114)

Theorem 5.4.9 *[20] The singular fractional-order linear continuous system* (5.7) *with order* $0 < \alpha \leq 1$ *is admissible if and only if there exist matrices* $X_1, X_2 \in \mathbb{R}^{n\times n}$, $Q \in \mathbb{R}^{(n-r)\times n}$, *such that*

$$\begin{bmatrix} X_1 & X_2 \\ -X_2 & X_1 \end{bmatrix} > 0 \tag{5.115}$$

$$\text{sym}\left\{A((\sin(\alpha\frac{\pi}{2})X_1 - \cos(\alpha\frac{\pi}{2})X_2)E^T + SQ)\right\} < 0 \tag{5.116}$$

where E_0 *is any matrix of full rank column that satisfies* $ES = 0$.

Proof By setting in Theorem 5.4.8, $X = Z_1 + jZ_2$ and after a few computations, it is easy to see that the $\mathcal{LMI}$ (5.105) is equivalent to that (5.116) with $X_1 = 2Z_1$ and $X_2 = 2Z_2$. Moreover, according to Lemma 2.4.3, $X > 0$ is equivalent to $\begin{bmatrix} X_1 & X_2 \\ -X_2 & X_1 \end{bmatrix} > 0$. This completes the proof. □

Example 5.4.5 [20] Consider System (5.7) with parameters as follows:

$$E = \begin{bmatrix} 0 & -1 & 0 & -1 \\ 1 & 2 & 0 & 1 \\ -2 & -2 & 1 & -2 \\ -1 & -1 & 0 & 0 \end{bmatrix}, \ A = \begin{bmatrix} 1 & -1 & -1 & 1 \\ 0 & 3 & 2 & 0 \\ 2 & -5 & -1 & -5 \\ 1 & -2 & -1 & -1 \end{bmatrix}, \ \alpha = \frac{1}{3}. \tag{5.117}$$

By Lemmas 5.4.1, it is easy to verify that

$$\det(s^{\frac{1}{3}} - A) = -2(s^{\frac{1}{3}})^3 + 2(s^{\frac{1}{3}})^2 - 2s^{\frac{1}{3}} - 4,$$

is not identically zero. So System (5.7) is regular.
By setting,

$$\lambda = s^{\frac{1}{3}},$$

it is easy to see that

$$\deg\left(\det(s^{\frac{1}{3}} - A)\right) = \mathrm{rank}(E) = 3,$$

i.e.System (5.7) is impulse-free. The three roots λ of the polynomial $\det(s^{\frac{1}{3}} - A) = 0$ are $1 \pm \mathrm{i}$ and -1. All their roots satisfy $\arg(\lambda) > \frac{\pi}{6}$, which shows that System (5.7) is stable. So, we can obtain that System (5.7) is admissible.

Solving the $\mathcal{LMIs}$ (5.115) and (5.116) in Theorem 5.4.9, the feasible solutions are obtained as follows:

$$X = \begin{bmatrix} 0.7589 & 0.1248 & 0.0045 & 0.0773 \\ 0.1248 & 0.7936 & 0.0196 & -0.0909 \\ 0.0045 & 0.0196 & 0.6474 & 0.1614 \\ 0.0773 & -0.0909 & 0.1614 & 0.2295 \end{bmatrix},$$

$$Y = \begin{bmatrix} 0 & -0.1940 & -0.1326 & 0.0712 \\ 0.1940 & 0 & -0.1934 & 0.1929 \\ 0.1326 & 0.1934 & 0 & 0.0609 \\ -0.0712 & -0.1929 & -0.0609 & 0 \end{bmatrix},$$

$$Q = \begin{bmatrix} -0.1105 & 2.5768 & -2.4126 & -1.4127 \end{bmatrix}.$$

The criteria presented in Theorem 5.4.9 establishes the basis of null space of the singular matrix E. The following theorem is limited to the simple full-rank decomposition of the singular matrix E, without incorporating the decompositions of the system matrix A and the input matrix B.

Theorem 5.4.10 *[20] The singular fractional-order linear continuous system* (5.7) *with order* $0 < \alpha \leq 1$ *is admissible if and only if there exist matrices* $X_1, X_2 \in \mathbb{R}^{r\times r}$, $X_3 \in \mathbb{R}^{(n-r)\times r}$, $X_4 \in \mathbb{R}^{(n-r)\times(n-r)}$, *such that*

$$\begin{bmatrix} X_1 & X_2 \\ -X_2 & X_1 \end{bmatrix} > 0 \tag{5.118}$$

$$\text{sym}\left\{\sin(\alpha\frac{\pi}{2})MANX - \cos(\alpha\frac{\pi}{2})MANY\right\} < 0 \tag{5.119}$$

where

$$X = \begin{bmatrix} X_1 & 0 \\ X_3 & X_4 \end{bmatrix}, \; Y = \begin{bmatrix} X_2 & 0 \\ 0 & 0 \end{bmatrix}, \tag{5.120}$$

and $M, \; N \in \mathbb{R}^{n\times n}$ *are two arbitrary chosen nonsingular matrices satisfying* (5.24).

Proof Sufficiency. For system (5.7), it is easy to choose two arbitrary nonsingular matrices M and N satisfying (5.24). Assume that there exist matrices X_1, X_2, X_3, and X_4 such that (5.118), (5.119), (5.120) hold. Then,considering (5.119), and using the expression in (5.120), we have

$$\begin{bmatrix} \Phi_1 & \Phi_2 \\ \Phi_2^{\mathrm{T}} & \Phi_3 \end{bmatrix} < 0, \tag{5.121}$$

where

$$\begin{aligned}
\Phi_1 &= \text{sym}\left\{A_1(\sin(\alpha\frac{\pi}{2})X_1 - \cos(\alpha\frac{\pi}{2})X_2) + \sin(\alpha\frac{\pi}{2})A_2X_3\right\}, \\
\Phi_2 &= \sin(\alpha\frac{\pi}{2})A_2X_4 + (\sin(\alpha\frac{\pi}{2})X_1 + \cos(\alpha\frac{\pi}{2})X_2)A_3^{\mathrm{T}} + \sin(\alpha\frac{\pi}{2})X_3^{\mathrm{T}}A_4^{\mathrm{T}}, \\
\Phi_3 &= \sin(\alpha\frac{\pi}{2})A_4X_4 + \sin(\alpha\frac{\pi}{2})X_4^{\mathrm{T}}A_4^{\mathrm{T}}.
\end{aligned}$$

Since, $0 < \alpha \leq 1, \; \sin(\alpha\frac{\pi}{2}) > 0$, then the 2-2 block in (5.121) implies

$$A_4X_4 + X_4^{\mathrm{T}}A_4^{\mathrm{T}} < 0.$$

Thus, A_4X_4 is invertible which entails that the matrix A_4 is nonsingular. According to Lemma 5.3.4, the system (5.7) is regular and impulse-free. One defines

$$\Phi = \begin{bmatrix} A_1(\sin(\alpha\frac{\pi}{2})X_1 - \cos(\alpha\frac{\pi}{2})X_2) + \sin(\alpha\frac{\pi}{2})X_3^{\mathrm{T}}A_2^{\mathrm{T}} & \sin(\alpha\frac{\pi}{2})A_2X_4 \\ A_3(\sin(\alpha\frac{\pi}{2})X_1 - \cos(\alpha\frac{\pi}{2})X_2) + \sin(\alpha\frac{\pi}{2})A_4X_3 & \sin(\alpha\frac{\pi}{2})A_4X_4 \end{bmatrix},$$

and considering (5.121), it is easy to see that

$$\Phi + \Phi^{\mathrm{T}} = \begin{bmatrix} \Phi_1 & \Phi_2 \\ \Phi_2^{\mathrm{T}} & \Phi_3 \end{bmatrix} < 0.$$

Therefore, by Lemma 5.4.4, it ensues that

$$\mathrm{sym}\left\{(A_1 - A_2A_4^{-1}A_3)(\sin(\alpha\frac{\pi}{2})X_1 - \cos(\alpha\frac{\pi}{2})X_2) + \sin(\alpha\frac{\pi}{2})X_1\right\} < 0 \quad (5.122)$$

Noting (5.118) and using Lemma 5.4.2, we get the stability of the pair (E, A). This together with the regularity and the non-impulsiveness of the pair (E, A) gives that the system (5.7) is admissible.

Necessity. Suppose that the system (5.7) is admissible. According to Lemma 5.4.2, from the full rank decomposition of descriptor matrix E, for the two arbitrary chosen nonsingular matrices M and N satisfying (5.24), where A_4 is nonsingular, and

$$\left|\arg(\mathrm{spec}(A_1 - A_2A_4^{-1}A_3))\right| > \alpha\frac{\pi}{2}.$$

Noting the above inequality and employing Lemma 2.4.4, it follows that there exist matrices X_1, X_2 such that (5.118) and (5.122) hold.

Let

$$M_1 = \begin{bmatrix} I_r & -A_2A_4^{-1} \\ 0 & A_4^{-1} \end{bmatrix},$$
$$N_1 = \begin{bmatrix} I_r & 0 \\ -A_4^{-1}A_3 & I_{n-r} \end{bmatrix},$$
$$\bar{X} = \begin{bmatrix} \sin\left(\alpha\frac{\pi}{2}\right)X_1 - \cos\left(\alpha\frac{\pi}{2}\right)X_2 & 0 \\ 0 & I_{n-r} \end{bmatrix}.$$

It is clear that M_1 and N_1 are nonsingular and

$$M_1MENN_1 = M_1\begin{bmatrix} I_r & 0 \\ 0 & 0 \end{bmatrix}N_1 = \begin{bmatrix} I_r & 0 \\ 0 & 0 \end{bmatrix},$$
$$M_1MANN_1 = M_1\begin{bmatrix} A_1 & A_2 \\ A_3 & A_4 \end{bmatrix}N_1 = \begin{bmatrix} A_1 - A_2A_4^{-1}A_3 & 0 \\ 0 & I_{n-r} \end{bmatrix}.$$

Hence, it is easy to see

$$\mathrm{sym}\left\{M_1MANN_1\bar{X}\right\} < 0. \quad (5.123)$$

Pre-multiplying and post-multiplying (5.123) by M_1^{-1} and $M_1^{-\mathrm{T}}$, respectively, we obtain

$$\mathrm{sym}\left\{MANN_1\bar{X}M_1^{-\mathrm{T}}\right\} < 0.$$

Consider

$$N_1\bar{X}M_1^{-\mathrm{T}} = \begin{bmatrix} I_r & 0 \\ -A_4^{-1}A_3 & I_{n-r} \end{bmatrix}\begin{bmatrix} \sin\left(\alpha\frac{\pi}{2}\right)X_1 - \cos\left(\alpha\frac{\pi}{2}\right)X_2 & 0 \\ 0 & I_{n-r} \end{bmatrix}\begin{bmatrix} I_r & 0 \\ A_4^{-\mathrm{T}}A_2^{\mathrm{T}} & I_{n-r} \end{bmatrix},$$

$$= \begin{bmatrix} \sin\left(\alpha\frac{\pi}{2}\right)X_1 - \cos\left(\alpha\frac{\pi}{2}\right)X_2 & 0 \\ -A_4^{-1}A_3\left(\sin\left(\alpha\frac{\pi}{2}\right)X_1 - \cos\left(\alpha\frac{\pi}{2}\right)\right) - A_4^{-\mathrm{T}}A_2^{\mathrm{T}} & -I_{n-r} \end{bmatrix},$$

$$= \begin{bmatrix} \sin\left(\alpha\frac{\pi}{2}\right)X_1 & 0 \\ -A_4^{-1}A_3\left(\sin\left(\alpha\frac{\pi}{2}\right)X_1 - \cos\left(\alpha\frac{\pi}{2}\right)\right) - A_4^{-\mathrm{T}}A_2^{\mathrm{T}} & -I_{n-r} \end{bmatrix}$$

$$+ \begin{bmatrix} -\cos\left(\alpha\frac{\pi}{2}\right)X_2 & 0 \\ 0 & 0 \end{bmatrix}.$$

Let

$$X = \begin{bmatrix} X_1 & 0 \\ X_3 & X_4 \end{bmatrix},$$

$$Y = \begin{bmatrix} X_2 & 0 \\ 0 & 0 \end{bmatrix},$$

$$X_3 = \sin\left(\alpha\frac{\pi}{2}\right)^{-1}\left(-A_4^{-1}A_3\left(\sin\left(\alpha\frac{\pi}{2}\right)X_1 - \cos\left(\alpha\frac{\pi}{2}\right)\right) - A_4^{-\mathrm{T}}A_2^{\mathrm{T}}\right),$$

$$X_4 = -\sin\left(\alpha\frac{\pi}{2}\right)^{-1} I_{n-r}.$$

Therefore, the matrices X_1, X_2, $X$and Y satisfy (5.118) and (5.119), respectively.This ends the proof. □

In the following, we present some results regarding the admissibility of a singular fractional-order system, expressed in the form of a Linear Matrix Inequality $\mathcal{LMI}$ which contains equality constraint, and the number of variables to solve reduced by half compared to the result in Theorem 5.4.6.

Theorem 5.4.11 *[20] The unforced SFOS in* (5.7) *with order* $0 < \alpha \leq 1$ *is admissible if and only if there exist matrices* $X, Y \in \mathbb{R}^{n\times n}$, *such that*

$$\begin{bmatrix} EX & EY \\ -EY & EX \end{bmatrix} = \begin{bmatrix} EX & EY \\ -EY & EX \end{bmatrix}^T \geq 0 \tag{5.124}$$

$$\mathrm{sym}\left\{A(\sin(\alpha\frac{\pi}{2})X - \cos(\alpha\frac{\pi}{2})Y)\right\} < 0 \tag{5.125}$$

Proof Sufficiency. For the system (5.7), it is always possible to choose two nonsingular matrices M and N satisfying (5.24). Write

$$N^{-1}XM^{\mathrm{T}} = \begin{bmatrix} X_1 & X_2 \\ X_3 & X_4 \end{bmatrix},\ N^{-1}YM^{\mathrm{T}} = \begin{bmatrix} Y_1 & Y_2 \\ Y_3 & Y_4 \end{bmatrix} \tag{5.126}$$

where the partition is compatible with that of A in (5.24). By (5.124), it can be shown that $X_2 = Y_2 = 0$, $X_1 = X_1^{\mathrm{T}} \geq 0$ and $Y_1^{\mathrm{T}} = -Y_1$.

Considering (5.125), and using the expressions in (5.126), we have

$$\begin{bmatrix} U_1 & U_2 \\ U_2^{\mathrm{T}} & U_3 \end{bmatrix} < 0 \tag{5.127}$$

where

$$\begin{aligned}
U_1 &= \mathrm{sym}\left\{A_1(\sin(\alpha\frac{\pi}{2})X_1 - \cos(\alpha\frac{\pi}{2})Y_1) + A_2(\sin(\alpha\frac{\pi}{2})X_3 - \cos(\alpha\frac{\pi}{2})Y_3)\right\} \\
U_2 &= A_2(\sin(\alpha\frac{\pi}{2})X_4 - \cos(\alpha\frac{\pi}{2})Y_4) + (\sin(\alpha\frac{\pi}{2})X_1^{\mathrm{T}} - \cos(\alpha\frac{\pi}{2})Y_1^{\mathrm{T}})A_3^{\mathrm{T}} \\
&\quad + (\sin(\alpha\frac{\pi}{2})X_3^{\mathrm{T}} - \cos(\alpha\frac{\pi}{2})Y_3^{\mathrm{T}})A_4^{\mathrm{T}} \\
U_3 &= \mathrm{sym}\left\{A_4(\sin(\alpha\frac{\pi}{2})X_4 - \cos(\alpha\frac{\pi}{2})Y_4)\right\}
\end{aligned}$$

Then, the 2-2 block in (5.127) gives

$$\mathrm{sym}\left\{A_4(\sin(\alpha\frac{\pi}{2})X_4 - \cos(\alpha\frac{\pi}{2})Y_4)\right\} < 0 \tag{5.128}$$

So, the matrix $A_4(\sin(\alpha\frac{\pi}{2})X_4 - \cos(\alpha\frac{\pi}{2})Y_4)$ is invertible, which implies A_4 is invertible too. Thus, by Lemma 5.3.4, the system (5.7) is impulse-free and regular. To show the stability of the pair (E, A), one define

$$\mathcal{N} = \begin{bmatrix} \mathcal{P} & \mathcal{X} \\ \mathcal{Y} & \mathcal{Z} \end{bmatrix} \tag{5.129}$$

with

$$\begin{aligned}
\mathcal{P} &= A_1(\sin(\alpha\frac{\pi}{2})X_1 - \cos(\alpha\frac{\pi}{2})Y_1) + (\sin(\alpha\frac{\pi}{2})X_3 - \cos(\alpha\frac{\pi}{2})Y_3)^{\mathrm{T}}A_2^{\mathrm{T}} \\
\mathcal{X} &= A_2(\sin(\alpha\frac{\pi}{2})X_4 - \cos(\alpha\frac{\pi}{2})Y_4) \\
\mathcal{Y} &= A_3(\sin(\alpha\frac{\pi}{2})X_1 - \cos(\alpha\frac{\pi}{2})Y_1) + A_4(\sin(\alpha\frac{\pi}{2})X_3 - \cos(\alpha\frac{\pi}{2})Y_3) \\
\mathcal{Z} &= A_4(\sin(\alpha\frac{\pi}{2})X_4 - \cos(\alpha\frac{\pi}{2})Y_4)
\end{aligned}$$

Considering (5.127), it is easy to see that

$$\mathcal{N} + \mathcal{N}^{\mathrm{T}} = \begin{bmatrix} U_1 & U_2 \\ U_2^{\mathrm{T}} & U_3 \end{bmatrix} < 0 \tag{5.130}$$

As a consequence of Lemma 5.4.4, it follows that

$$\text{sym}\left\{(A_1 - A_2 A_4^{-1} A_3)(\sin(\alpha\frac{\pi}{2})X_1 - \cos(\alpha\frac{\pi}{2})Y_1)\right\} < 0 \tag{5.131}$$

Noting (5.131) and employing Lemmas 2.4.4, 2.2.2 and 5.4.2, we have that the system (5.7) is asymptotically stable. This together with the regularity and non-impulsiveness of the system (5.7) gives that system (5.7) is admissible.

Necessity. Suppose that the system (5.7) is admissible. According to Lemma 5.4.1, from the full rank decomposition of descriptor matrix E, for the two arbitrary chosen nonsingular matrices P and Q satisfying (5.27) and $|\arg(\text{spec}(A_1))| > \alpha\frac{\pi}{2}$. Taking into account the above inequality and applying Lemma 2.4.4, it can be seen that there exist matrices X_1, Y_1 such that such that (2.39) holds and

$$\text{sym}\left\{A_1(\sin(\alpha\frac{\pi}{2})X_1 - \cos(\alpha\frac{\pi}{2})X_2)\right\} < 0 \tag{5.132}$$

By setting

$$X = Q\begin{bmatrix} X_1 & 0 \\ 0 & -I_{n-r} \end{bmatrix} P^{-\mathrm{T}},\ Y = Q\begin{bmatrix} Y_1 & 0 \\ 0 & 0 \end{bmatrix} P^{-\mathrm{T}}, \tag{5.133}$$

it is easy to check that conditions (5.124) and (5.125) are satisfied. This completes the proof. □

Theorem 5.4.12 *[21] The unforced SFOS in* (5.7) *with order* $0 < \alpha \le 1$ *is admissible if and only if the following equivalent statements hold:*
(i) There exist $X_1 \in \mathbb{C}^{n\times n}$, $X_2 \in \mathbb{C}^{n\times n}$ *such that*

$$EX_1 \ge 0,\ EX_2 \ge 0, \tag{5.134}$$

$$\text{sym}\{zAX_1 + \bar{z}AX_2\} < 0. \tag{5.135}$$

(ii) There exists $X \in \mathbb{C}^{n\times n}$ *such that*

$$EX \ge 0, \tag{5.136}$$

$$\text{sym}\left\{zAX + \bar{z}A\bar{X}\right\} < 0, \tag{5.137}$$

where $z = e^{j(1-\alpha)\frac{\pi}{2}}$.

Proof First, we prove the equivalence between admissibility and (ii).
(Sufficiency) Suppose the inequalities (5.136) and (5.137) are satisfied for some matrix X, since E is singular, there exist invertible matrices M and N satisfying (5.24). Let

$$N^{-1}XM^{\mathrm{T}} = \begin{bmatrix} X_{11} & X_{12} \\ X_{21} & X_{22} \end{bmatrix}, \tag{5.138}$$

the inequality (5.136) is then equivalent to

$$\begin{aligned} EX &= M^{-1}\begin{bmatrix} I & 0 \\ 0 & 0 \end{bmatrix} N^{-1}N \begin{bmatrix} X_{11} & X_{12} \\ X_{21} & X_{22} \end{bmatrix} M^{-T} \\ &= M^{-1}\begin{bmatrix} X_{11} & X_{12} \\ 0 & 0 \end{bmatrix} M^{-T} \geq 0 \end{aligned} \tag{5.139}$$

It is easy to show that $X_{11} \geq 0,\ X_{12} = 0$.
Considering (5.138), and using the expressions in (5.24), we have

$$\begin{aligned} \mathrm{sym}\left\{zAX + \bar{z}A\bar{X}\right\} = \mathrm{sym}\Bigg\{ & zM^{-1}\begin{bmatrix} A_{11}X_{11} + A_{12}X_{21} & A_{12}X_{22} \\ A_{21}X_{11} + A_{22}X_{21} & A_{22}X_{22} \end{bmatrix} M^{-\mathrm{T}} \\ & + \bar{z}M^{-1}\begin{bmatrix} A_{11}\bar{X}_{11} + A_{12}\bar{X}_{21} & A_{12}\bar{X}_{22} \\ A_{21}\bar{X}_{11} + A_{22}\bar{X}_{21} & A_{22}\bar{X}_{22} \end{bmatrix} M^{-\mathrm{T}} \Bigg\} < 0 \end{aligned} \tag{5.140}$$

Then, it induces A_{22} invertible. By Lemma 5.3.4, the system (5.7) is then regular and impulse free. Therefore, invertible matrices L and R can always be defined such that

$$LER = \begin{bmatrix} I_r & 0 \\ 0 & 0 \end{bmatrix},\ LAR = \begin{bmatrix} A_1 & 0 \\ 0 & I_{n-r} \end{bmatrix}. \tag{5.141}$$

Let

$$X = R\begin{bmatrix} X_{11} & 0 \\ X_{21} & X_{22} \end{bmatrix} L^{-\mathrm{T}} \tag{5.142}$$

From (5.141) and (5.142), we get that the condition (5.137) is equivalent to

$$\begin{bmatrix} \Psi_{11} & \Psi_{21}^{\mathrm{T}} \\ \Psi_{21} & \Psi_{22} \end{bmatrix} < 0, \tag{5.143}$$

with

$$\begin{aligned} \Psi_{11} &= \mathrm{sym}\left\{zA_1X_{11} + \bar{z}A_1\bar{X}_{11}\right\}, \\ \Psi_{21} &= \mathrm{sym}\left\{zX_{21} + \bar{z}\bar{X}_{21}\right\}, \\ \Psi_{22} &= \mathrm{sym}\left\{zX_{22} + \bar{z}\bar{X}_{22}\right\}, \end{aligned} \tag{5.144}$$

It follows that $\Psi_{11} < 0$, which implies that the system $D^{\alpha}x_1(t) = A_1x_1(t)$ is asymptotically stable. According to Lemma 5.4.1, the system (5.7) is also is asymptotically stable. Therefore, We conclude that system (5.7) is admissible since it is regular, impulse free and asymptotically stable.

(Necessity) Under the assumption that the system (5.7) is admissible, there exist invertible matrices such that (5.24) holds and $D^{\alpha}x_1(t) = A_1x_1(t)$ is asymptotically stable. So, by Theorem 2.4.1, there exists a matrix $X_1 = X_1^* \in \mathbb{C}^{r\times r} > 0$ such that

$$\mathrm{sym}\left\{A_1(zX_1 + \bar{z}\bar{X}_1)\right\} < 0. \tag{5.145}$$

Let

$$X = R\begin{bmatrix} X_1 & 0 \\ 0 & -I \end{bmatrix} L^{-\mathrm{T}} \tag{5.146}$$

It is easy to show that $EX \geq 0$, then we get

$$\mathrm{sym}\left\{L^{-1}\begin{bmatrix} A_1(zX_1 + \bar{z}\bar{X}_1) & 0 \\ 0 & -2I \end{bmatrix} L^{-\mathrm{T}}\right\} < 0 \tag{5.147}$$

and this ends the proof of the equivalence between (ii) and the admissibility of the system (5.7). Afterward, the equivalence of (i) and (ii) is given.
$(ii) \Rightarrow (i)$ is obvious: Take $X_1 = X$ and $X_2 = \bar{X}$.
$(i) \Rightarrow (ii)$ can be proved as follows. $\exists X_1, X_2$ such that $EX_1 \geq 0$, $EX_2 \geq 0$ and

$$zAX_1 + \bar{z}X_1^*A^{\mathrm{T}} + \bar{z}AX_2 + zX_2^*A^{\mathrm{T}} < 0 \tag{5.148}$$

This last inequality is equivalent to

$$\overline{zAX_1 + \bar{z}X_1^*A^{\mathrm{T}} + \bar{z}AX_2 + zX_2^*A^{\mathrm{T}}} < 0 \tag{5.149}$$

This implies that $\exists X_1, X_2$ such that $EX_1 \geq 0$, $EX_2 \geq 0$ and

$$zAX_1 + \bar{z}X_1^*A^{\mathrm{T}} + \bar{z}AX_2 + zX_2^*A^{\mathrm{T}} + \overline{zAX_1 + \bar{z}X_1^*A^{\mathrm{T}} + \bar{z}AX_2 + zX_2^*A^{\mathrm{T}}} < 0 \tag{5.150}$$

After few computations, this last inequality is equivalent to relation (5.135) with

$$X = X_1 + \bar{X}_2.$$

□

5.5 Robust Admissibility Analysis

In this section, the robust admissibility of uncertain linear SFO systems is investigated for the fractional-order α belonging to $1 \le \alpha < 2$ and $0 < \alpha \le 1$.

Considering the following uncertain linear SFO continuous-time system described by

$$\begin{aligned} ED^{\alpha}x(t) &= (A + \Delta A)x(t) + (B + \Delta B)u(t),\ 0 < \alpha < 2, \\ y(t) &= Cx(t). \end{aligned} \tag{5.151}$$

where α is the time fractional derivative order, $x(t) \in \mathbb{R}^n$ is the pseudo-state, $u(t) \in \mathbb{R}^m$ is the control input and $y(t) \in \mathbb{R}^p$ is the output. $A \in \mathbb{R}^{n\times n}$, $B \in \mathbb{R}^{n\times m}$ and $C \in \mathbb{R}^{p\times n}$ with rank $C = p$ and matrix $E \in \mathbb{R}^{n\times n}$ is singular with rank $E = r < n$. ΔA and ΔB are time-invariant matrices representing norm-bounded parameter uncertainties, and assumed to be of the following form

$$[\Delta A \ \ \Delta B] = MF(\sigma)\,[N_1 \ \ N_2] \tag{5.152}$$

where $M \in \mathbb{R}^{n\times h}$, $N_1 \in \mathbb{R}^{q\times n}$ and $N_2 \in \mathbb{R}^{q\times m}$ are known real constant matrices. The uncertain matrix $F(\sigma) \in \mathbb{R}^{h\times q}$ satisfies

$$F(\sigma)^{\mathrm{T}}F(\sigma) \le I_q \tag{5.153}$$

with $\sigma \in \Xi$, where Ξ is a compact set in $\mathbb{R}$. Furthermore, it is assumed that given any matrix $F : F^{\mathrm{T}}F \le I_q$, there exists $\sigma \in \Xi$ such that $F = F(\sigma)$. ΔA and ΔB are said to be allowable if both (5.152) and (5.153) hold.

The unforced SFO system of the uncertain singular fractional-order system (5.151) can be written as

$$ED^{\alpha}x(t) = (A + \Delta A)x(t) \tag{5.154}$$

To investigate the robust stability problem for the uncertain singular system in (5.154), we have to consider not only stability robustness, but also regularity and non-impulsiveness of (5.154). In view of this, we first introduce the following definition.

Definition 5.5.1 [22] The uncertain singular system in (5.154) is said to be robustly admissible if it is regular, impulse-free and stable for all allowable uncertainties.

Case one: $1 \le \ < 2$

The purpose is to develop conditions for robust admissibility of continuous singular fractional-order systems with order $1 \le \alpha < 2$. Similar to the study of stability robustness of singular integer order systems [8], we introduce the following definition of robust admissibility for the uncertain continuous singular system in (5.154).

Definition 5.5.2 [22] The uncertain continuous singular system in (5.154) with order $1 \leq \alpha < 2$ is said to be robustly admissible if there exist matrices $P > 0$ and Q such that

$$\mathrm{sym}\left\{\Theta \otimes (A + \Delta A)(PE^{\mathrm{T}} + SQ)\right\} < 0 \tag{5.155}$$

for all allowable $F(\sigma)$, where $S \in \mathbb{R}^{n\times(n-r)}$ is any matrix with full column rank and satisfies $ES = 0$.

In view of Definition 5.5.2, necessary and sufficient $\mathcal{LMI}$ conditions for robust admissibility are presented.

Theorem 5.5.1 *[22] The uncertain system* (5.154) *with* $1 \leq \alpha < 2$ *is admissible for all allowable uncertainties* ΔA *if and only if the following equivalent statements hold.*
(i) There exist matrices $P \in \mathbb{R}^{n\times n} > 0$, $Q \in \mathbb{R}^{(n-r)\times n}$ *and a scalar* $\epsilon > 0$ *satisfying the following* $\mathcal{LMI}$

$$\begin{bmatrix} \mathrm{sym}\left\{\Theta \otimes (A(PE^{T} + SQ)\right\} + \epsilon I_2 \otimes MM^{T} & * \\ \Theta \otimes (N_1(PE^{T} + SQ)) & -\epsilon I_{2q} \end{bmatrix} < 0 \tag{5.156}$$

where $S \in \mathbb{R}^{n\times(n-r)}$ *is any matrix of full column rank and satisfies* $ES = 0$.
(ii) There exist matrices $P \in \mathbb{R}^{n\times n} > 0$, $Q \in \mathbb{R}^{(n-r)\times n}$ *and a scalar* $\epsilon > 0$ *satisfying the following* $\mathcal{LMI}$

$$\begin{bmatrix} \mathrm{sym}\left\{\Theta \otimes (A^{T}(PE + SQ)\right\} + \epsilon I_2 \otimes N_1^{T}N_1 & * \\ \Theta \otimes (M^{T}(PE + SQ)) & -\epsilon I_{2h} \end{bmatrix} < 0 \tag{5.157}$$

where $S \in \mathbb{R}^{n\times(n-r)}$ *is any matrix of full column rank and satisfies* $E^{T}S = 0$.

Proof Sufficiency: Suppose that there exist matrices $P > 0$, Q and a scalar $\epsilon > 0$ such that the $\mathcal{LMI}$ (5.110) holds.

$$\begin{aligned}\mathrm{sym}\left\{\Theta \otimes ((A + \Delta A)(PE^{\mathrm{T}} + SQ))\right\} &= \mathrm{sym}\left\{\Theta \otimes (A(PE^{\mathrm{T}} + SQ))\right\} \\ &\quad + \mathrm{sym}\left\{\Theta \otimes (\Delta A(PE^{\mathrm{T}} + SQ))\right\}\end{aligned} \tag{5.158}$$

Note that (5.153) implies

$$F(\sigma)F(\sigma)^{\mathrm{T}} \leq I \tag{5.159}$$

It follows from (5.159) that

$$(I_2 \otimes F(\sigma))(I_2 \otimes F(\sigma))^{\mathrm{T}} = I_2 \otimes (F(\sigma)F(\sigma)^{\mathrm{T}}) \le I \tag{5.160}$$

It follows from (5.160) and Lemma 3.4.1 that for any real scalar $\varepsilon > 0$

$$\begin{aligned}
&\mathrm{sym}\left\{\Theta \otimes (\Delta A(PE^{\mathrm{T}} + SQ))\right\} = \\
&\mathrm{sym}\left\{(I_2 \otimes (MF(\sigma)))(\Theta \otimes (N_1(PE^{\mathrm{T}} + SQ)))\right\} \\
&\le \varepsilon(I_2 \otimes (MF(\sigma)))(I_2 \otimes (MF(\sigma)))^{\mathrm{T}} \\
&+ \frac{1}{\varepsilon}(\Theta \otimes (N_1(PE^{\mathrm{T}} + SQ)))^{\mathrm{T}}(\Theta \otimes (N_1(PE^{\mathrm{T}} + SQ))) \\
&\le \ \varepsilon(I_2 \otimes M)(I_2 \otimes F(\sigma))(I_2 \otimes F(\sigma))^{\mathrm{T}}(I_2 \otimes M)^{\mathrm{T}} \\
&+ \frac{1}{\varepsilon}(\Theta \otimes (N_1(PE^{\mathrm{T}} + SQ)))^{\mathrm{T}}(\Theta \otimes (N_1(PE^{\mathrm{T}} + SQ))) \\
&\le \ \varepsilon(I_2 \otimes M)(I_2 \otimes M)^{\mathrm{T}} \\
&+ \frac{1}{\varepsilon}(\Theta \otimes (N_1(PE^{\mathrm{T}} + SQ)))^{\mathrm{T}}(\Theta \otimes (N_1(PE^{\mathrm{T}} + SQ) \\
&\le \varepsilon(I_2 \otimes MM^{\mathrm{T}}) + \frac{1}{\varepsilon}(\Theta \otimes (N_1(PE^{\mathrm{T}} + SQ)))^{\mathrm{T}} \\
&\times (\Theta \otimes (N_1(PE^{\mathrm{T}} + SQ)
\end{aligned} \tag{5.161}$$

Substituting (5.161) into (5.158), we get

$$\begin{aligned}
\mathrm{sym}\left\{\Theta \otimes ((A + \Delta A)(PE^{\mathrm{T}} + SQ))\right\} \le \mathrm{sym}\left\{\Theta \otimes (A(PE^{\mathrm{T}} + SQ))\right\} + \varepsilon(I_2 \otimes MM^{\mathrm{T}}) \\
+ \frac{1}{\varepsilon}(\Theta \otimes (N_1(PE^{\mathrm{T}} + SQ)))^{\mathrm{T}}(\Theta \otimes (N_1(PE^{\mathrm{T}} + SQ)
\end{aligned} \tag{5.162}$$

Taking into account (5.162) and using the Schur complement of (5.156), we obtain From condition (5.153), Lemma 3.4.1 and by using the Schur complement [23] of (5.156), we obtain

$$\mathrm{sym}\left\{\Theta \otimes ((A + \Delta A)(PE^{\mathrm{T}} + SQ))\right\} < 0 \tag{5.163}$$

It follows from the above inequality (5.163) and Theorem 5.4.1 that the uncertain SFOS (5.154) is admissible.

Necessity: Suppose that the uncertain SFO system (5.154) is admissible. It follows from Theorem 5.4.1, that there exist matrices $P > 0$ and Q such that

$$\mathrm{sym}\left\{\Theta \otimes (A + \Delta A)(PE^{\mathrm{T}} + SQ)\right\} < 0 \tag{5.164}$$

Inequality (5.164) is the same as

$$\mathrm{sym}\left\{\Theta \otimes (A(PE^{\mathrm{T}} + SQ))\right\} + \mathrm{sym}\left\{\Theta \otimes (\Delta A(PE^{\mathrm{T}} + SQ))\right\} < 0 \tag{5.165}$$

Define $\tilde{F} = I_2 \otimes F(\sigma)$. It follows from (5.153) that

$$\tilde{F}^{\mathrm{T}}(\sigma)\tilde{F}(\sigma) = (I_2 \otimes F^{\mathrm{T}}(\sigma)F(\sigma)) \leq I \tag{5.166}$$

Inequality (5.165) implies that for all $\xi \in \mathbb{R}^{2n}$, $\xi \neq 0$

$$\xi^{\mathrm{T}}(\mathrm{sym}\left\{\Theta \otimes (A(PE^{\mathrm{T}} + SQ))\right\} + \mathrm{sym}\left\{\Theta \otimes (\Delta A(PE^{\mathrm{T}} + SQ))\right\})\xi < 0 \tag{5.167}$$

i.e.,

$$\xi^{\mathrm{T}}(\mathrm{sym}\left\{\Theta \otimes (A(PE^{\mathrm{T}} + SQ))\right\})\xi + \xi^{\mathrm{T}}(\mathrm{sym}\left\{\Theta \otimes (\Delta A(PE^{\mathrm{T}} + SQ))\right\})\xi < 0 \tag{5.168}$$

Let

$$Z = \mathrm{sym}\left\{\Theta \otimes (A(PE^{\mathrm{T}} + SQ))\right\},\ G = PE^{\mathrm{T}} + SQ \tag{5.169}$$

Then, (5.168) can be rewritten as

$$\begin{aligned} \xi^{\mathrm{T}} Z\xi &< -2\xi^{\mathrm{T}}(\Theta \otimes (\Delta AG))\xi \\ &< -2\xi^{\mathrm{T}}(I_2 \otimes M)(I_2 \otimes F(\sigma))(\Theta \otimes N_1 G)\xi \end{aligned} \tag{5.170}$$

It follows from (5.166) and (5.169) that

$$\xi^{\mathrm{T}} Z\xi < -2\xi^{\mathrm{T}}(I_2 \otimes M)\tilde{F}(\Theta \otimes N_1 G)\xi \tag{5.171}$$

Hence

$$\xi^{\mathrm{T}} Z\xi < -2\max\left\{\xi^{\mathrm{T}}(I_2 \otimes M)\tilde{F}(\Theta \otimes (N_1 G))\xi / \tilde{F}^{\mathrm{T}}\tilde{F} \leq I\right\} < 0 \tag{5.172}$$

which implies

$$\left(\xi^{\mathrm{T}} Z\xi\right)^2 > 4 \times \left(\max\left\{\xi^{\mathrm{T}}(I_2 \otimes M)\tilde{F}(\Theta \otimes (N_1 G))\xi / \tilde{F}^{\mathrm{T}}\tilde{F} \leq I\right\}\right)^2 \tag{5.173}$$

Define

$$\Sigma = \left\{\xi^{\mathrm{T}}(I_2 \otimes M)\tilde{F}(\Theta \otimes (N_1 G))\xi / \tilde{F}^{\mathrm{T}}\tilde{F} \leq I\right\}$$

Since the set Σ is symmetric with respect to the real axis then

$$\begin{aligned}&\left(\max\left\{\xi^{\mathrm{T}}(I_2\otimes M)\tilde{F}(\Theta\otimes(N_1G))\xi/\tilde{F}^{\mathrm{T}}\tilde{F}\leq I\right\}\right)^2\\&=\max\left\{(\xi^{\mathrm{T}}(I_2\otimes M)\tilde{F}(\Theta\otimes(N_1G)\xi))^2/\tilde{F}^{\mathrm{T}}\tilde{F}\leq I\right\}\end{aligned}\tag{5.174}$$

By substituting (5.174) into (5.173), then applying Lemma 3.4.3, we can obtain

$$\left(\xi^{\mathrm{T}}Z\xi\right)^2>4\left(\xi^{\mathrm{T}}(I_2\otimes M)(I_2\otimes M^{\mathrm{T}})\xi\right)\times\left(\xi^{\mathrm{T}}(\Theta\otimes N_1G)^{\mathrm{T}}(\Theta\otimes N_1G)\xi\right)\tag{5.175}$$

Thus, by Lemma 3.4.4, there exists a scalar $\varepsilon>0$ such that

$$\varepsilon^2I_2\otimes(MM^{\mathrm{T}})+\varepsilon Z+(\Theta\otimes N_1G)^{\mathrm{T}}(\Theta\otimes N_1G)<0\tag{5.176}$$

That is,

$$Z+\varepsilon I_2\otimes(MM^{\mathrm{T}})+\frac{1}{\varepsilon}(\Theta\otimes N_1G)^{\mathrm{T}}(\Theta\otimes N_1G)<0\tag{5.177}$$

By the Schur complement Lemma 4.2.4, (5.177) together with (5.169) means that the $\mathcal{LMI}$ (5.156) holds.

The equivalence between (i) and (ii) is due to the duality criterion. This ends the proof. □

Remark 5.5.1 For the nominal system of (5.154), i.e. $\Delta A=0$, the $\mathcal{LMI}$ conditions (5.156) becomes

$$\begin{bmatrix}\operatorname{sym}\left\{\Theta\otimes(A(PE^{\mathrm{T}}+SQ)\right\} & *\\ 0 & -\epsilon I_{2q}\end{bmatrix}<0\tag{5.178}$$

The $\mathcal{LMI}$ conditions (5.178) is equivalent to $\operatorname{sym}\left\{\Theta\otimes(A(PE^{\mathrm{T}}+SQ)\right\}<0$ which is the necessary and sufficient admissibility condition derived in [24] for nominal SFOS.

Example 5.5.1 [22] Consider the singular fractional-order uncertain linear systems (5.109) with the fractional order $\alpha=1.50$ and

$$E=\begin{bmatrix}1&2&3\\1&2&3\\1&2&3\end{bmatrix},\ A=\begin{bmatrix}-1&2&0\\-3&-4&1\\0&0&-2\end{bmatrix}$$

$$M=\begin{bmatrix}0.1&0\\0.1&0.3\\0.2&0\end{bmatrix},\ N=\begin{bmatrix}0.2&0.1&0\end{bmatrix},$$

with fractional-order derivative $\alpha=1.5$.

By using the Matlab LMI Control Toolbox, it is found that the $\mathcal{LMI}$s given in (5.156) in Theorem 5.5.1 are feasible. The solution is given by

$$P = \begin{bmatrix} 0.9872 & -0.1460 & -0.2190 \\ -0.1460 & 0.7682 & -0.4379 \\ -0.2190 & -0.4379 & 0.4033 \end{bmatrix},$$

$$Q = \begin{bmatrix} -0.2622 & 0.0158 & -0.0880 \\ 0.0740 & -0.0571 & 0.0570 \end{bmatrix},$$

and

$$\xi = 1.0538.$$

Therefore, we have that the uncertain singular system in this example is robustly admissible.

Case two: $0 < \alpha < 1$

Similar to the study of stability robustness of singular integer order systems, the following definitions of quadratic admissibility [25] and generalized quadratic stability [26] of singular fractional-order systems are introduced.

Definition 5.5.3 [20] The uncertain singular fractional-order system in (5.154) is said to be quadratically admissible if there exist matrices $X, Y \in \mathbb{R}^{n\times n}$ such that

$$\begin{bmatrix} EX & EY \\ -EY & EX \end{bmatrix} = \begin{bmatrix} X^{\mathrm{T}}E^{\mathrm{T}} & -Y^{\mathrm{T}}E^{\mathrm{T}} \\ Y^{\mathrm{T}}E^{\mathrm{T}} & X^{\mathrm{T}}E^{\mathrm{T}} \end{bmatrix} \geq 0, \tag{5.179}$$

$$\mathrm{sym}\left\{(A + \Delta A)(\sin(\alpha\frac{\pi}{2})X - \cos(\alpha\frac{\pi}{2})Y)\right\} < 0, \tag{5.180}$$

for all allowable uncertainty ΔA.

Definition 5.5.4 [20] The uncertain singular fractional-order system in (5.154) is said to be generalized quadratically stable if there exist matrices $X, Y \in \mathbb{R}^{n\times n}$ and $Q \in \mathbb{R}^{(n-r)\times n}$ such that

$$\begin{bmatrix} X & Y \\ -Y & X \end{bmatrix} > 0 \tag{5.181}$$

$$\mathrm{sym}\left\{(A + \Delta A)\left((\sin(\alpha\frac{\pi}{2})X - \cos(\alpha\frac{\pi}{2})Y)E^{\mathrm{T}} + SQ\right)\right\} < 0, \tag{5.182}$$

where $S \in \mathbb{R}^{n\times(n-r)}$ is any matrix of full column rank and satisfies $ES = 0$.

According to Definition 5.5.3 and applying Lemmas 3.4.2 and 4.2.4, the following result is easily obtained.

Theorem 5.5.2 *[20] The uncertain singular fractional-order system in* (5.154) *is quadratically admissible if and only if there exist matrices* $X, Y \in \mathbb{R}^{n\times n}$ *and a scalar* $\epsilon > 0$ *such that*

$$\begin{bmatrix} EX & EY \\ -EY & EX \end{bmatrix} = \begin{bmatrix} X^T E^T & -Y^T E^T \\ Y^T E^T & X^T E^T \end{bmatrix} \geq 0 \tag{5.183}$$

$$\begin{bmatrix} \mathrm{sym}\,\{A\mathcal{P}\} + \epsilon M M^T & \mathcal{P}^T N_1^T \\ N_1 \mathcal{P} & -\epsilon I_n \end{bmatrix} < 0 \tag{5.184}$$

where $\mathcal{P} = \sin(\alpha\frac{\pi}{2})X - \cos(\alpha\frac{\pi}{2})Y$.

According to Definition 5.5.4 and applying Lemmas 3.4.2 and 4.2.4, the following result is easily obtained.

Theorem 5.5.3 *[20] The uncertain singular fractional-order system in* (5.154) *is generalized quadratically admissible if and only if there exist matrices* $X, Y \in \mathbb{R}^{n\times n}$, $Q \in \mathbb{R}^{(n-r)\times n}$ *and a scalar* $\epsilon > 0$ *such that*

$$\begin{bmatrix} X & Y \\ -Y & X \end{bmatrix} > 0 \tag{5.185}$$

$$\begin{bmatrix} \mathcal{W} & (\mathcal{P}E^T + SQ)^T N_1^T \\ N_1(\mathcal{P}E^T + SQ) & -\epsilon I_n \end{bmatrix} < 0 \tag{5.186}$$

where

$$\mathcal{P} = \sin(\alpha\tfrac{\pi}{2})X - \cos(\alpha\tfrac{\pi}{2})Y$$
$$\mathcal{W} = \mathrm{sym}\left\{A(\mathcal{P}E^T + SQ)\right\} + \epsilon M M^T$$

and $S \in \mathbb{R}^{n\times(n-r)}$ *is any matrix of full column rank and satisfies* $ES = 0$.

Theorem 5.5.4 *[20] System* (5.154) *is quadratically admissible if and only if it is generalized quadratically stable.*

Proof Sufficiency. Suppose that System (5.154) is generalized quadratically stable, then it follows from Theorem 5.5.3 that there exist matrices $X, Y \in \mathbb{R}^{n\times n}$ and $Q \in \mathbb{R}^{(n-r)\times n}$ such that (5.185) and (5.186) hold.

Setting $\hat{X} = XE^T + \frac{1}{\sin(\alpha\frac{\pi}{2})}SQ$, $\hat{Y} = YE^T$, where $S \in \mathbb{R}^{n\times(n-r)}$ is any matrix of full column rank and satisfies $ES = 0$.

It is easy to check $\hat{X}$ and $\hat{Y}$ satisfy (5.183) and (5.184). Then by the use of the result of Theorem 5.5.2, it follows that System (5.154) is quadratically admissible. Necessity. Suppose that System (5.154) is quadratically admissible. Then from Theorem 5.5.2, there exist matrices $\hat{X}, \hat{Y} \in \mathbb{R}^{n\times n}$ and a positive scalar ϵ such that (5.183) and (5.184) hold, and there exist two unitary matrices U and V such that

$$\hat{E} = UEV = \begin{bmatrix} I_r & 0 \\ 0 & 0 \end{bmatrix}.$$

Setting

$$\hat{X} = V^{-1}XU^{\mathrm{T}} = \begin{bmatrix} \hat{X}_1 & \hat{X}_2 \\ \hat{X}_3 & \hat{X}_4 \end{bmatrix}, \; \hat{Y} = V^{-1}YU^{\mathrm{T}} = \begin{bmatrix} \hat{Y}_1 & \hat{Y}_2 \\ \hat{Y}_3 & \hat{Y}_4 \end{bmatrix},$$

and let

$$\hat{A} = UAV, \; \hat{M} = UM, \; \hat{N}_1 = UN_1V.$$

Pre-multiplying and post-multiplying (5.183) and (5.184) by $\begin{bmatrix} U & 0 \\ 0 & U \end{bmatrix}$ and its transpose, respectively, we have

$$\begin{bmatrix} \hat{E}\hat{X} & \hat{E}\hat{Y} \\ -\hat{E}\hat{Y} & \hat{E}\hat{X} \end{bmatrix} = \begin{bmatrix} \hat{X}^{\mathrm{T}}\hat{E}^{\mathrm{T}} & -\hat{Y}^{\mathrm{T}}\hat{E}^{\mathrm{T}} \\ \hat{Y}^{\mathrm{T}}\hat{E}^{\mathrm{T}} & \hat{X}^{\mathrm{T}}\hat{E}^{\mathrm{T}} \end{bmatrix} \geq 0, \tag{5.187}$$

$$\begin{bmatrix} \mathrm{sym}\left\{\hat{A}\hat{\mathcal{P}}\right\} + \epsilon\hat{M}\hat{M}^{\mathrm{T}} & \hat{\mathcal{P}}^{\mathrm{T}}\hat{N}_1^{\mathrm{T}} \\ \hat{N}_1\mathcal{P} & -\epsilon I_n \end{bmatrix} < 0, \tag{5.188}$$

where

$$\hat{\mathcal{P}} = \sin(\alpha\frac{\pi}{2})\hat{X} - \cos(\alpha\frac{\pi}{2})\hat{Y}. \tag{5.189}$$

By (5.187), it can be shown, in one hand, that

$$\hat{X}_2 = \hat{Y}_2 = 0, \tag{5.190}$$

and in the other hand that

$$\begin{bmatrix} \hat{X}_1 & 0 & \hat{Y}_1 & 0 \\ 0 & 0 & 0 & 0 \\ -\hat{Y}_1 & 0 & \hat{X}_1 & 0 \\ 0 & 0 & 0 & 0 \end{bmatrix} \geq 0. \tag{5.191}$$

By pre- and post- multiplying the inequality (5.191) by

$$\begin{bmatrix} I_r & 0 & 0 & 0 \\ 0 & 0 & I_r & 0 \\ 0 & 0 & 0 & 0 \\ 0 & 0 & 0 & 0 \end{bmatrix}$$

and its transpose, we get

$$\begin{bmatrix} \hat{X}_1 & \hat{Y}_1 & 0 & 0 \\ -\hat{Y}_1 & \hat{X}_1 & 0 & 0 \\ 0 & 0 & 0 & 0 \\ 0 & 0 & 0 & 0 \end{bmatrix} \geq 0 \tag{5.192}$$

which implies

$$\begin{bmatrix} \hat{X}_1 & \hat{Y}_1 \\ -\hat{Y}_1 & \hat{X}_1 \end{bmatrix} > 0 \tag{5.193}$$

Setting

$$\mathcal{X} = \begin{bmatrix} \hat{X}_1 & 0 \\ 0 & I_{n-r} \end{bmatrix}, \mathcal{Y} = \begin{bmatrix} \hat{Y}_1 & 0 \\ 0 & 0 \end{bmatrix}, \hat{S} = \begin{bmatrix} 0 \\ I_{n-r} \end{bmatrix}, \tag{5.194}$$

$$\hat{Q} = \left[\sin(\alpha\tfrac{\pi}{2})\hat{X}_3 - \cos(\alpha\tfrac{\pi}{2})\hat{Y}_3 \; \sin(\alpha\tfrac{\pi}{2})\hat{X}_4 - \cos(\alpha\tfrac{\pi}{2})\hat{Y}_4 - \sin(\alpha\tfrac{\pi}{2})I_{n-r} \right]. \tag{5.195}$$

Therefore, the inequality (5.188) becomes

$$\begin{bmatrix} \text{sym}\left\{\hat{A}(\sin(\alpha\frac{\pi}{2})\mathcal{X} - \cos(\alpha\frac{\pi}{2})\mathcal{Y} + \hat{S}\hat{Q})\right\} + \epsilon\hat{M}\hat{M}^{\mathrm{T}} & (\sin(\alpha\frac{\pi}{2})\hat{X}_1 - \cos(\alpha\frac{\pi}{2})\hat{Y}_1)^{\mathrm{T}}\hat{N}_1^{\mathrm{T}} \\ \hat{N}_1(\sin(\alpha\frac{\pi}{2})\hat{X}_1 - \cos(\alpha\frac{\pi}{2})\hat{Y}_1) & -\epsilon I \end{bmatrix} < 0 \tag{5.196}$$

Considering the definition of the matrix S in Definition 5.5.4, we can write $S = V\hat{S}H$, where H is any $n - r$ order nonsingular matrix. Denoting

$$X = V\mathcal{X}V^{\mathrm{T}},\ Y = V\mathcal{Y}V^{\mathrm{T}},\ Q = H^{-1}\hat{Q}U^{-\mathrm{T}}, \tag{5.197}$$

and substituting

$$\begin{aligned} \hat{N}_1 &= N_1U, \qquad \hat{Q} = HQU^{\mathrm{T}}, \qquad \hat{S} = V^{-1}SH^{-1}, \\ \mathcal{X} &= V^{-1}XV^{-\mathrm{T}},\ \mathcal{Y} = V^{-1}YV^{-\mathrm{T}},\ \hat{P} = UP, \end{aligned} \tag{5.198}$$

into (5.196) gives

$$\begin{bmatrix} U & 0 \\ 0 & U \end{bmatrix} \begin{bmatrix} \mathcal{W} & (\mathcal{P}E^{\mathrm{T}} + SQ)^{\mathrm{T}}N_1^{\mathrm{T}} \\ N_1(\mathcal{P}E^{\mathrm{T}} + SQ) & -\epsilon I_n \end{bmatrix} \begin{bmatrix} U^{\mathrm{T}} & 0 \\ 0 & U^{\mathrm{T}} \end{bmatrix} < 0, \tag{5.199}$$

where

$$\begin{aligned} \mathcal{P} &= \sin(\alpha\tfrac{\pi}{2})X - \cos(\alpha\tfrac{\pi}{2})Y, \\ \mathcal{W} &= \text{sym}\left\{A(\mathcal{P}E^{\mathrm{T}} + SQ)\right\} + \epsilon MM^{\mathrm{T}}. \end{aligned}$$

Therefore, matrices X, Y, Q, and scalar $\epsilon > 0$ satisfy (5.185) and (5.186). This completes the proof. □

5.6 Summary

This chapter addresses the issue of admissibility for singular fractional-order systems in both cases $(0, 1)$ and $(1, 2)$. Extensions of fundamental concepts from integer-order singular systems to fractional-order singular systems are provided, including definitions of regularity, impulse-freeness, and admissibility. The study also focuses on the robust admissibility of fractional-order singular systems, taking into account uncertainties in system parameters. Robust admissibility ensures that the system remains stable and consistent under all permissible perturbations. Both necessary and sufficient conditions for admissibility and robust admissibility are established. A framework based on Linear Matrix Inequalities (LMIs) is proposed, offering a computationally efficient method for verifying these properties. These LMI-based conditions serve as a powerful tool for the analysis and design of fractional-order singular systems. Notably, the stability results developed in this chapter will play a crucial role in addressing the stabilization problem, as discussed in the following chapter.

References

1. Kaczorek, T.: Singular fractional linear systems and electrical circuits. Int. J. Math. Comput. Sci. **21**, 379–384 (2011)
2. N'Doye, I., Darouach, M., Zasadzinski, M., Radhy, N.: Robust stabilisation of uncertain descriptor fractional-order systems. Automatica **49**, 1907–1913 (2013). June
3. N'Doye, I., Zasadzinski, M., Darouach, M., Radhy, N.: Stabilization of singular fractional-order systems: an lmi approach. In: 18th Mediterranean Conference on Control and Automation. Marrakech, Morocco (2010)
4. Farges, C., Moze, M., Sabatier, J.: Pseudo-state feedback stabilization of commensurate fractional-order systems. Automatica **46**, 1730–1734 (2010)
5. Huang, X., Wang, Z., Li, Z.Y., Lu, J.: Design of fuzzy state feedback controller for robust stabilization of uncertain fractional-order chaotic system. J. Franklin Inst. **351**, 5480–5493 (2014)
6. Li, C., Wang, J.: Robust stability and stabilization of fractional order interval systems with coupling relationships: the $0<\alpha<1$ case. J. Franklin Inst. **349**, 2406–2419 (2012)
7. Sabatier, J., Moze, M., Farges, C.: $\mathcal{LMI}$ conditions for fractional order systems. Comput. Math. Appl. **59**, 1594–1609 (2010)
8. Xu, S.Y., Lam, J.: Robust Control and Filtering of Singular Systems. Springer, Berlin (2006)
9. Dai, L.: Singular Control Systems. Springer, Berlin (1989)
10. Song, X., Liu, L., Wang, Z.: Stabilization of singular fractional-order systems: a linear matrix inequality approach. In: The IEEE International Conference on Automation and Logistics, Zhengzhou, China, August 2012
11. Yu, Y., Jiao, Z., Sun, C.Y.: Sufficient and necessary condition of admissibility for fractional-order singular system. Acta Automatica Sinica **39**, 2160–2164 (2013)
12. Marir, S., Chadli, M., Bouagada, D.: New admissibility conditions for singular linear continuous-time fractional-order systems. J. Franklin Inst. **354**, 752–766 (2017)
13. Zhang, X., Zhao, Z.N., Wang, Q.: Static and dynamic output feedback stabilisation of descriptor fractional order systems. IET Control Theory Appl. **14** (2019). https://doi.org/10.1049/iet-cta.2019.0694

14. Kaczorek, T.: Descriptor fractional linear systems with regular pencils. Asian J. Control **15**(4), 1051–1064 (2013)
15. Liu, Y., Cui, L., Duan, D.: Dynamic output feedback stabilization of singular fractional-order systems. Math. Problems Eng. **2016**, 1–8 (2016). https://doi.org/10.1155/2016/9694780
16. Marir, S., Chadli, M., Basin, M.V.: Bounded real lemma for singular linear continuous-time fractional-order systems. Automatica **135** (2022). https://doi.org/10.1016/j.automatica.2021.109962
17. Chadli, M., Shi, P., Feng, Z., Lam, J.: New bounded real lemma formulation and $\mathcal{H}_\infty$ control for continuous-time descriptor systems. Asian J. Control **20**(1), 1–7, 2018. https://doi.org/10.1002lasjc.1606
18. Anderson, B.D.O., Moore, J.B.: Linear Optimal Control. Englewood Cliffs, Prentice-Hall, NJ (1990)
19. Marir, S., Chadli, M., Bouagada, D.: A novel approach of admissibility for singular linear continuous-time fractional-order systems. Int. J. Control Autom. Syst. (2017). https://doi.org/10.1007/s12555-016-0003-0
20. Xuefeng, Z., Chen, Y.Q.: Admissibility and robust stabilization of continuous linear singular fractional order systems with the fractional order α: the $0<\alpha<1$ case. ISA Trans. **82**, 42–50 (2018)
21. Zhang, X.F., Yan, Y.: Admissibility of fractional order descriptor systems based on complex variables: an lmi approach. Fractal and Fractional **4** (2020). https://doi.org/10.3390/fractalfract4010008
22. Marir, S., Chadli, M.: Robust admissibility and stabilization of uncertain singular fractional-order linear time-invariant systems. IEEE/CAA J. Automatica Sinica **6**(3), 685–692 (2019). https://doi.org/10.1109/JAS.2019.1911480. May
23. Masubuchi, I., Kamitane, Y., Ohara, N., Suda, A.: H_∞ control for descriptor systems: a matrix inequalities approach. Automatica **33**, 669–673 (1997)
24. Luenberger, D.G.: Time-invariant descriptor systems. Automatica **14**, 473–480 (1978)
25. Lin, C., Chen, B., Zhuang, H., Zhang, L.: A direct method of static output feedback design for t-s fuzzy systems. In: Proceedings of the 11th IEEE International Conference on Control and Automation, pp. 445–448, Taichung, June 2014
26. Xuefeng, Z.: Stability and stabilization of singular systems: strict lmi sufficient conditions. In: Proceedings of the 10th World Congress on Intelligent Control and Automation, July 6–8, Beijing, China (2012)

Chapter 6
Stabilization of Singular Fractional-Order Systems

6.1 Introduction

In this chapter, we shall deal with the stabilization problem for singular fractional-order systems. The purpose is the design of controllers such that the closed-loop system is regular, stable and impulse-free. Both state feedback and output feedback controllers are considered in the case $1 \leq \alpha < 2$ and observer-based controller for the case $0 < \alpha \leq 1$. Based on the admissibility conditions presented in the previous chapter, necessary and sufficient conditions for the existence of stabilizing controllers are obtained and the stabilizing controllers design can be formulated.

6.2 State Feedback Controller

Consider singular system (5.6). It is assumed that $r = \text{rank}\, E < n$ and the system is regular.

Let the state feedback control have the form of

$$u(t) = -LD^{\alpha}x(t) + Kx(t), \tag{6.1}$$

where $L \in \mathbb{R}^{m \times n}$ and $K \in \mathbb{R}^{m \times n}$ are constant matrices.

The closed-loop system formed by (5.6) and (6.1) is

$$\begin{aligned} (E + BL)D^{\alpha}x(t) &= (A + BK)x(t) \\ y(t) &= Cx(t) \end{aligned} \tag{6.2}$$

S. Marir and M. Chadli, *Control of Singular Fractional Order Systems: LMI Approach*, Studies in Systems, Decision and Control 292,
https://doi.org/10.1007/978-3-031-87382-9_6

6.2.1 Normalization of Singular Fractional-Order System

The definition of a singular integer-order system normalizable given in [1] is suitable for a singular fractional-order system.

Definition 6.2.1 [1] The system (5.6) is called normalizable if a feedback control (6.1) may be chosen such that the closed-loop system (6.2) is normal, i.e.,

$$\det(E + BL) \neq 0 \tag{6.3}$$

When the condition (6.3) is satisfied, the closed-loop system (6.2) would become

$$\begin{aligned} D^{\alpha}x(t) &= (E + BL)^{-1}(A + BK)x(t) \\ y(t) &= Cx(t) \end{aligned} \tag{6.4}$$

The following lemma generalizes well known results of integer-order singular systems to fractional-order singular systems.

Lemma 6.2.1 *[2] System* (5.6) *is normalizable if and only if*

$$\operatorname{rank}\begin{bmatrix} E & B \end{bmatrix} = n \tag{6.5}$$

Proof Necessity. Assume that the system (5.6) is normalizable. By Definition 6.2.1, there exists a matrix $L \in \mathbb{R}^{m\times n}$ such that $\det(E + BL) \neq 0$, i.e.,

$$\operatorname{rank}(E + BL) = \operatorname{rank}\left(\begin{bmatrix} E & B \end{bmatrix}\begin{bmatrix} I \\ L \end{bmatrix}\right) = n. \tag{6.6}$$

We know that

$$\operatorname{rank}\begin{bmatrix} E & B \end{bmatrix} \leq \min(n, n + m) = n \tag{6.7}$$

and

$$\operatorname{rank}\begin{bmatrix} I \\ L \end{bmatrix} \leq \min(n + m, n) = n \tag{6.8}$$

In the other hand we have

$$\operatorname{rank}(E + BL) \leq \min\left(\operatorname{rank}\begin{bmatrix} E & B \end{bmatrix}, \operatorname{rank}\begin{bmatrix} I \\ L \end{bmatrix}\right) \tag{6.9}$$

Thus,

$$\operatorname{rank}\begin{bmatrix} E & B \end{bmatrix} = n.$$

Sufficiency. If $\operatorname{rank}\begin{bmatrix} E & B \end{bmatrix} = n$, from the matrix rank decomposition via singular values, there exist two nonsingular matrices Q and P satisfying

$$QEP = \begin{bmatrix} I & 0 \\ 0 & 0 \end{bmatrix} \tag{6.10}$$

Denoting $QB = \begin{bmatrix} B_1 \\ B_2 \end{bmatrix}$, we have

$$\begin{aligned} \operatorname{rank}\begin{bmatrix} E & B \end{bmatrix} &= \operatorname{rank}\begin{bmatrix} QEP & QB \end{bmatrix} \\ &= \operatorname{rank}\begin{bmatrix} I_r & 0 & B_1 \\ 0 & 0 & B_2 \end{bmatrix} \end{aligned}$$

and I_r is invertible, so

$$\operatorname{rank}\begin{bmatrix} E & B \end{bmatrix} = r + \operatorname{rank} B_2 \tag{6.11}$$

We deduce that $\operatorname{rank} B_2 = n - r$, indicating $B_2 B_2^{\mathrm{T}} > 0$ is symmetric positive definite.
Let $L = \begin{bmatrix} 0 & B_2^{\mathrm{T}} \end{bmatrix} P^{-1} \in \mathbb{R}^{m\times n}$.
Then,

$$\begin{aligned} \det(E + BL) &= \det\left(Q^{-1}\begin{bmatrix} I_r & 0 \\ 0 & 0 \end{bmatrix} P^{-1} + Q^{-1}\begin{bmatrix} B_1 \\ B_2 \end{bmatrix}\begin{bmatrix} 0 & B_2^{\mathrm{T}} \end{bmatrix} P^{-1} \right) \\ &= \det\left(Q^{-1}\begin{bmatrix} I_r & B_1 B_2^{\mathrm{T}} \\ 0 & B_2 B_2^{\mathrm{T}} \end{bmatrix} P^{-1} \right) \\ &= \det\left(Q^{-1}\right)\det\left(P^{-1}\right)\det\left(B_2 B_2^{\mathrm{T}}\right) \\ &\neq 0 \end{aligned}$$

In the sequel, we shall need the following Lemma.

Lemma 6.2.2 *[2] Let $A \in \mathbb{C}^{n\times n}$ be a complex matrix. Then A is nonsingular if and only if there exists a nonsingular matrix $X \in \mathbb{C}^{n\times n}$ such that*

$$AX + X^* A^* < 0 \tag{6.12}$$

Proof Sufficiency. Assume that there exists a nonsingular matrix $X \in \mathbb{C}^{n\times n}$ such that (6.12) is satisfied and A be singular. Then for any $x \in \mathcal{N}(A^*)$[1] we have

$$x^*\left(AX + X^* A^*\right)x = 0$$

[1] Consider a linear map represented as a $m \times n$ matrix A with coefficients in a field $\mathbb{K}$ (typically $\mathbb{R}$ or $\mathbb{C}$), that is operating on column vectors x with n components over $\mathbb{K}$.

The kernel of this linear map is the set of solutions to the equation $Ax = 0$, where 0 is understood as the zero vector. The dimension of the kernel of A is called the nullity of A. In set-builder notation, $\mathcal{N}(A) = \mathrm{Null}(A) = \ker(A) = \{\mathbf{x} \in \mathbb{K}^n \mid A\mathbf{x} = \mathbf{0}\}$.

which is in contradiction with (6.12). This concludes that the matrix A must be invertible.

Necessity. Assume that the matrix A is invertible, so the matrix A^{-1} is well defined. By denoting $X = -A^{-1}$, it is easy to show that

$$AX + X^* A^* = -2I < 0$$

The following theorem gives necessary and sufficient conditions in the $\mathcal{LMI}$ formulation for the existence of gain matrix L making the system (5.6) normalizable.

Theorem 6.2.1 *[2] The system* (5.6) *is normalizable if and only if there exist a nonsingular matrix P and a matrix Y such that the following $\mathcal{LMI}$ holds*

$$\text{sym}\,\{EP + BY\} < 0 \tag{6.13}$$

In this case, the gain matrix L is given by

$$L = YP^{-1} \tag{6.14}$$

Proof System (5.6) is normalizable if and only if there exists a gain matrix L such that the real matrix $E + BL$ is nonsingular. By applying the result of Lemma 6.2.2 to the matrix $E + BL$ we obtain the following condition, there exists a nonsingular matrix P satisfying

$$(E + BL)\,P + P^{\mathrm{T}}\,(E + BL)^{\mathrm{T}} < 0 \tag{6.15}$$

By taking $Y = LP$, the inequality (6.15) is exactly (6.13). □

The following section will be devoted to the design of the gain matrix K in (6.1) such that the obtained normalized system is asymptotically stable.

6.2.2 Stabilization of the Normalized System

Assume that the system (5.6) is normalizable. We may choose appropriate feedback control (6.1) such that its closed-loop system is normal, enabling the application of results in normal system theory to singular ones. As long as condition (6.3) is satisfied, the closed-loop system (6.2) would become

$$D^{\alpha}x(t) = (A_1 + B_1 K)\,x(t) \tag{6.16}$$

where $E_1 = (E + BL)^{-1}$, $A_1 = E_1 A$, $B_1 = E_1 B$ stand for the new resulting nominal matrices.

The design of the gain matrix K to ensure singular fractional-order system (5.6) stable is treated for both cases $1 \leq \alpha < 2$ and $0 < \alpha < 1$.

1. **Case one: $0 < \alpha < 1$**
   Applying the result of Theorem 6.6.1 to a system with zero uncertainties, the computation of the controller is summarized in the following statement.

Theorem 6.2.2 *Assume that the system* (5.6) *is normalizable, then there exists a gain matrix K such that the singular fractional-order system* (5.6) *with fractional-order derivative $0 < \alpha < 1$ controlled by the control* (6.1) *is asymptotically stable if there exist matrices $P_0 \in \mathbb{R}^{n\times n} > 0$ and $X \in \mathbb{R}^{n\times n}$ such that*

$$\sum_{i=1}^{2} \text{sym}\,\{\Theta_{i1} \otimes (A_1 P_0 + B_1 X)\} < 0 \tag{6.17}$$

$\Theta_{i1}(i = 1, 2)$ *satisfy* (2.24) *and matrices P and Y are determined by $\mathcal{LMI}$* (6.14). *Moreover, the asymptotically stabilizing state-feedback gain matrix is given by*

$$K = X P_0^{-1} \tag{6.18}$$

Proof Under the assumption that the system (5.6) is normalizable from Theorem 6.2.1, there exist a gain matrix L such that the system (5.6) can be rewritten in the form (6.16).

Suppose that the $\mathcal{LMI}$ (6.17) holds for some matrices P_0 and X.

By setting

$$P_{11} = P_{21} = P_0$$

and

$$P_{12} = P_{22} = 0$$

and defining the matrix

$$X = K P_0$$

it easily follows from (6.17)

$$\sum_{i=1}^{2}\sum_{j=1}^{2} \text{sym}\left\{\Theta_{ij} \otimes \left((A_1 + B_1 K)\, P_{ij}\right)\right\} < 0 \tag{6.19}$$

From Lemma 2.4.2, inequality (6.19) means that the system (6.16) is asymptotically stable.

In this case, the gain matrix K is deduced by $K = X P_0^{-1}$ which ends the proof. □

Now, we provide necessary and sufficient condition for stabilization by direct application of the Theorem 3.2.3 to the obtained normalized system (6.16).

Theorem 6.2.3 *Assume that the system* (5.6) *is normalizable, then there exists a gain matrix K such that the singular fractional-order system* (5.6) *with fractional-order derivative* $0 < \alpha < 1$ *controlled by the control* (6.1) *is asymptotically stable if and only if there exist matrices* $P_0 \in \mathbb{C}^{n \times n} > 0$ *and* $X \in \mathbb{R}^{n \times n}$ *such that*

$$(r P_0 + \bar{r} \bar{P}_0)^T A_1^T + A_1 (r P_0 + \bar{r} \bar{P}_0) + X^T B_1^T + B_1 X < 0 \tag{6.20}$$

where $r = e^{j(1-\alpha)\frac{\pi}{2}}$ *and matrices P and Y are determined by* $\mathcal{LMI}$ (6.14).

Moreover, the asymptotically stabilizing state-feedback gain matrix K is given by

$$K = X (r P_0 + \bar{r} \bar{P}_0)^{-1} \tag{6.21}$$

Example 6.2.1 Consider the stabilization by normalization of the following singular fractional-order system described in (5.6) for the fractional-order α belonging to $0 < \alpha < 1$ and with parameters as follows

$$E = \begin{bmatrix} -1 & -1 & -1 & -2 \\ 4 & 3 & 2 & 5 \\ -6 & 0 & 2 & 2 \\ 1 & 1 & 1 & 2 \end{bmatrix}, \quad A = \begin{bmatrix} -2 & -4 & -2 & -6 \\ 9 & 15 & 5 & 18 \\ -18 & -4 & 2 & -4 \\ 2 & 5 & 2 & 6 \end{bmatrix},$$

$$B = \begin{bmatrix} 10 & 1 \\ -1 & 0 \\ 0 & 1 \\ 2 & -1 \end{bmatrix}, \quad \alpha = 0.5$$

According to the characteristic polynomial,

$$\det(\lambda E - A) = -4\lambda^3 + 28\lambda^2 - 64\lambda + 40,$$

the system is regular, impulse free but it is not asymptotically stable since its finite modes $\{1, 3+i, 3-i\}$ don't satisfy the condition $|\arg(\lambda)| > \alpha\frac{\pi}{2}$. Indeed, the time response of the system with control $u(t) = 0$ is shown in Fig. 6.1, which shows that its states are not convergent. Therefore, the system is not admissible. The purpose of this example is the design of a state feedback control law such that the resulting normalized fractional-order system is asymptotically stable.

The solution of the $\mathcal{LMI}$ (6.14) with respect to P and Y gives

$$P = \begin{bmatrix} 499.5611 & -54.8165 & 314.6902 & -123.6835 \\ -54.8165 & 371.1135 & 112.5457 & -306.8706 \\ 314.6902 & 112.5457 & 663.9565 & -515.6155 \\ -123.6835 & -306.8706 & -515.6155 & 595.9295 \end{bmatrix},$$

$$Y = 10^3 \begin{bmatrix} 0.1153 & -0.1949 & 0.2159 & -0.1554 \\ -0.9032 & 0.0261 & 0.14023 & 0.1396 \end{bmatrix},$$

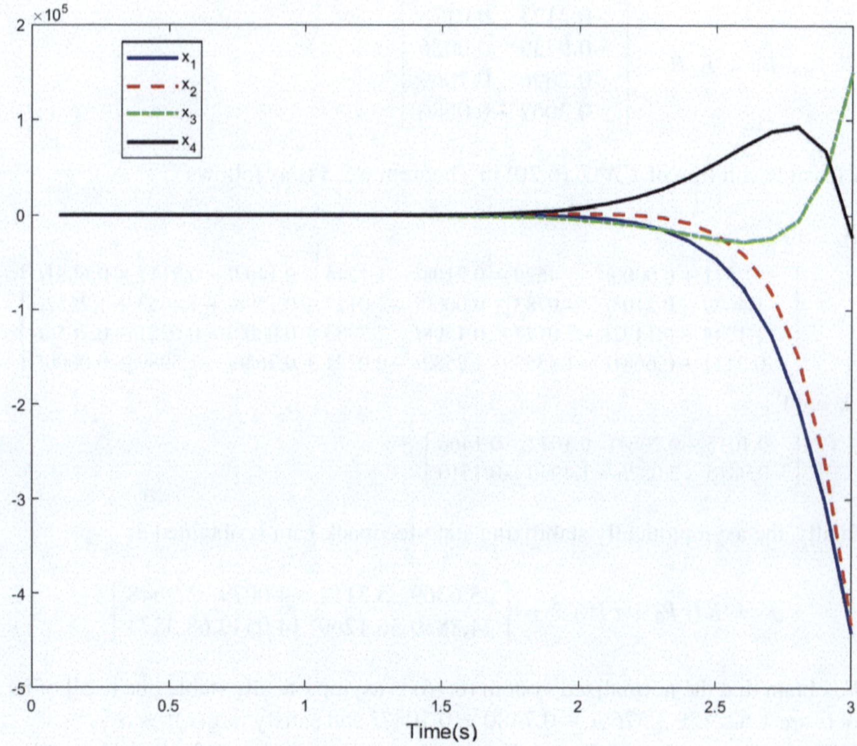

Fig. 6.1 Time response of the system in Example 6.2.1 with $u(t) = 0$

and the gain matrix L given by

$$L = YP^{-1} = \begin{bmatrix} 0.2295 & -2.1202 & -1.3337 & -2.4590 \\ -9.7768 & 13.5771 & 25.8475 & 27.5606 \end{bmatrix}.$$

This means that the system can be normalized by the full state controller. Using the full state derivative controller, we get the new matrices

$$E_1 = (E + BL)^{-1} = \begin{bmatrix} 0.0528 & 0.1197 & -0.1329 & -0.0979 \\ -0.0586 & -0.2757 & -0.2709 & -0.3320 \\ 0.0839 & -0.1460 & -0.3348 & -0.3578 \\ -0.0240 & 0.3280 & 0.4028 & 0.4373 \end{bmatrix},$$

$$A_1 = E_1 A = \begin{bmatrix} 3.1676 & 1.6263 & 0.0314 & 1.7819 \\ 1.8477 & -4.4774 & -2.4670 & -5.5193 \\ 3.8294 & -2.9756 & -2.2832 & -3.9393 \\ -3.3745 & 5.5923 & 3.3684 & 7.0617 \end{bmatrix},$$

$$B_1 = E_1 B = \begin{bmatrix} 0.2123 & 0.0178 \\ -0.9739 & 0.0026 \\ 0.2696 & 0.1069 \\ 0.3067 & -0.0586 \end{bmatrix}.$$

A feasible solution of $\mathcal{LMI}$ (6.20) in Theorem 6.2.3 is as follows

$$\begin{aligned} P_0 &= 10^3 \\ &\times \begin{bmatrix} 0.7511+0.0000i & -1.4679+0.2108i & 1.1244-0.1492i & 0.2111-0.6681i \\ -1.4679-0.2108i & 4.0783+0.0000i & -2.0127+0.1598i & -1.4553+1.2582i \\ 1.1244+0.1492i & -2.0127-0.1598i & 2.7353+0.0000i & -0.0221-0.7656i \\ 0.2111+0.6681i & -1.4553-1.2582i & -0.0221+0.7656i & 1.5989+0.0000i \end{bmatrix} \\ X &= 10^5 \\ &\times \begin{bmatrix} 0.1015 & -0.2064 & 0.1072 & 0.1466 \\ -0.9815 & 2.0876 & -1.4940 & -0.1510 \end{bmatrix} \end{aligned}$$

Finally, the asymptotically stabilizing state-feedback gain is obtained as

$$K = X(rP_0 + \bar{r}\bar{P}_0)^{-1} = \begin{bmatrix} 25.6369 & 5.3112 & -4.0079 & 3.2648 \\ 24.8830 & 56.1269 & 14.0549 & 68.3574 \end{bmatrix}.$$

We obtain that the normalized system (6.16) is asymptotically stable due to all of its roots are $1.6221 \pm 3.3769i$, $-0.7480 \pm 0.1047i$ and satisfy $|\arg(\lambda)| > \alpha\frac{\pi}{2}$.

The state responses of the stabilized system is given in Fig. 6.2 which shows that is asymptotically stable and its states converge to zero.

2. **Case one:** $1 \leq \alpha < 2$
   Applying the result of Theorem 2 in [2] to a system with zero uncertainties, the computation of the controller is summarized in the following statement.

Theorem 6.2.4 *Assume that the system* (5.6) *is normalizable, then there exists a gain matrix K such that the singular fractional-order system* (5.6) *with fractional-order derivative* $1 \leq \alpha < 2$ *controlled by the control* (6.1) *is asymptotically stable if and only if there exist matrices* $P_0 \in \mathbb{R}^{n\times n} > 0$ *and* $X \in \mathbb{R}^{n\times n}$ *such that*

$$\mathrm{sym}\{\Theta \otimes (A_1 P_0 + B_1 X)\} < 0 \tag{6.22}$$

with

$$\Theta = \begin{bmatrix} \sin(\alpha\frac{\pi}{2}) & \cos(\alpha\frac{\pi}{2}) \\ -\cos(\alpha\frac{\pi}{2}) & \sin(\alpha\frac{\pi}{2}) \end{bmatrix}$$

and matrices P and Y are determined by $\mathcal{LMI}$ (6.14) where A_1 and B_1 are defined in (6.16).

Moreover, the asymptotically stabilizing state feedback gain matrix is given by

$$K = XP_0^{-1} \tag{6.23}$$

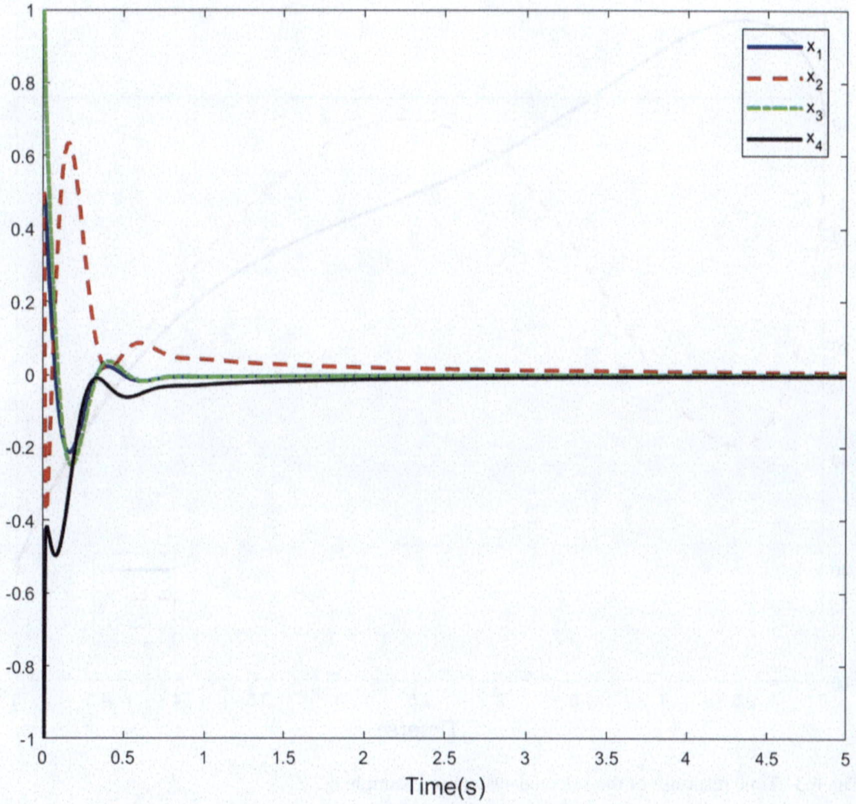

Fig. 6.2 Time response of the normalized system in Example 6.2.1

Proof The result of this theorem is deduced by direct application of Theorem 2.3.1 and Remark 2.3.1 to the obtained normalized system (6.16). □

To show the effectiveness of the result in Theorem 6.2.4, we provide the following numerical example.

Example 6.2.2 Consider a continuous singular system in (5.6) with parameters as follows:

$$E = \begin{bmatrix} 1 & 2 & 1 \\ 3 & 2 & 1 \\ 0 & 0 & 0 \end{bmatrix}, \; A = \begin{bmatrix} 1 & 1.5 & 1 \\ -2.5 & 1.5 & 1 \\ 1 & 2.4 & 1 \end{bmatrix}, \; B = \begin{bmatrix} 1 & 1 \\ 2.5 & 1 \\ 2 & 1 \end{bmatrix}, \; \alpha = 1.5. \tag{6.24}$$

It is easy to see that the singular fractional-order system regular, impulse-free but unstable, with finite modes $\frac{9}{4}$ and $\frac{-7}{4}$ as shown in Fig. 6.3.

The purpose of this system is the design of a state-feedback control law such that the resulting normalized fractional-order system is asymptotically stable.

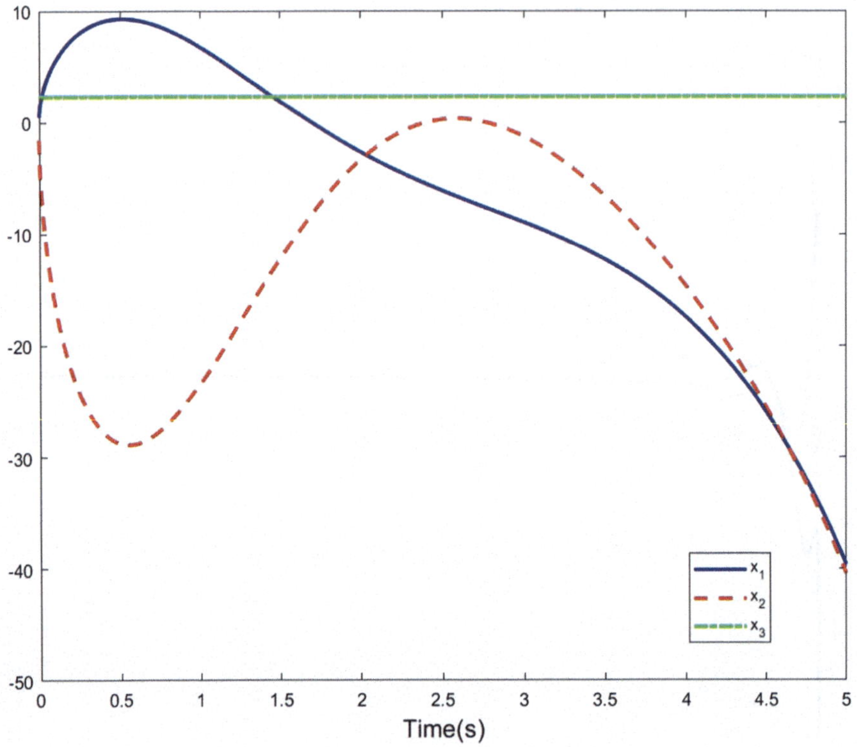

Fig. 6.3 Time response of the selected system in Example 6.2.2

The solution of the $\mathcal{LMI}$ (6.13) with respect to P and Y gives

$$P = \begin{bmatrix} 70.9234 & -18.8981 & 17.0083 \\ -18.8981 & 61.4744 & 11.3389 \\ 17.0083 & 11.3389 & 101.1603 \end{bmatrix},$$

$$Y = \begin{bmatrix} -34.9618 & 29.0138 & -20.8438 \\ -56.7773 & -190.3140 & 3.8637 \end{bmatrix},$$

and the gain matrix L is given by

$$L = YP^{-1} = \begin{bmatrix} -0.3388 & 0.4037 & -0.1943 \\ -2.0262 & -3.8686 & 0.8125 \end{bmatrix}.$$

This means that the system can be normalized by the full state controller. Using the full state derivative controller, we obtain the new matrices

$$E_1 = \begin{bmatrix} -0.9014 & 1.0564 & 0.1348 \\ 0.8877 & -0.9257 & -0.4916 \\ 0.6612 & 0.0531 & -0.3313 \end{bmatrix},$$

$$A_1 = \begin{bmatrix} -3.4077 & 0.5560 & 0.2898 \\ -2.7104 & -1.2368 & -0.5296 \\ 0.1972 & 0.2763 & 0.3830 \end{bmatrix},$$

$$B_1 = \begin{bmatrix} 2.0092 & 0.2898 \\ -2.4097 & -0.5296 \\ 0.1313 & 0.3830 \end{bmatrix}.$$

A feasible solution of the $\mathcal{LMI}$ (6.22) is as follows

$$P_0 = \begin{bmatrix} 94.2535 & 8.9466 & 17.3120 \\ 8.9466 & 131.2259 & 6.6775 \\ 17.3120 & 6.6775 & 120.4298 \end{bmatrix},$$

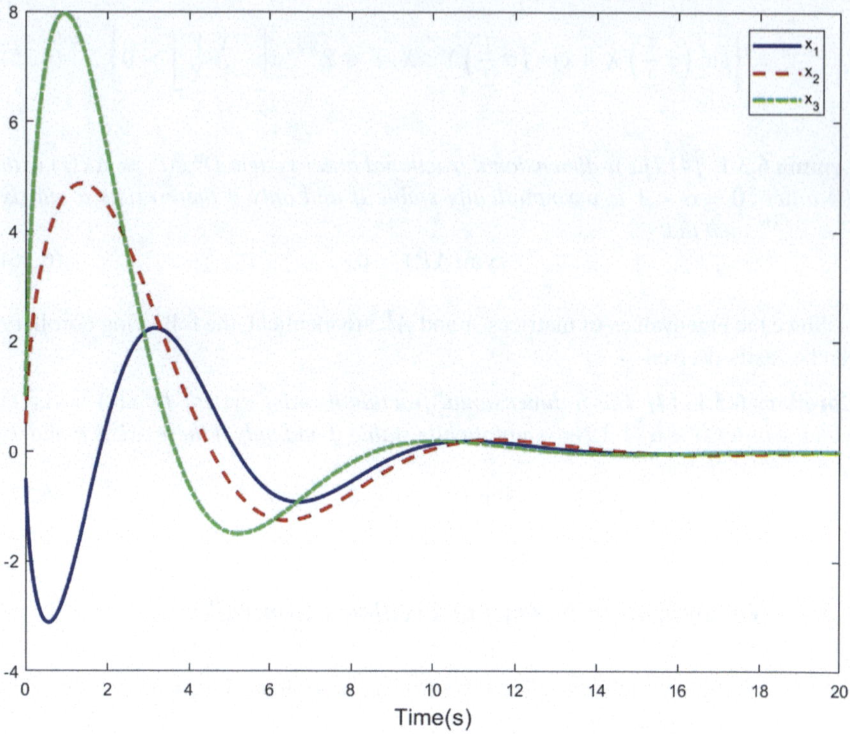

Fig. 6.4 Time response of the normalized system in Example 6.2.2

$$X = \begin{bmatrix} 127.0115 & -4.6988 & 67.8249 \\ -123.1565 & -96.5116 & -370.0503 \end{bmatrix},$$

Finally, the asymptotically stabilizing state-feedback gain is obtained as

$$K = XP_0^{-1} = \begin{bmatrix} 1.2903 & -0.1434 & 0.3857 \\ -0.7157 & -0.5371 & -2.9401 \end{bmatrix}.$$

The state responses of the selected system is given in Fig. 6.4 which shows that is asymptotically stable and its states converge to zero.

6.3 Output Feedback Control Synthesis of SFOS

As the state variables are rarely available in practical applications, an output feedback controller is often applied. It is practically and theoretically appealing to normalize and stabilize singular fractional order systems by output feedback technique [3].

In order to further explore the main results, a valuable lemma will be presented for subsequent use. In the sequel, $\mathbb{P}_\alpha^{n\times n}$ is defined as

$$\mathbb{P}_\alpha^{n\times n} = \left\{ \sin\left(\alpha\frac{\pi}{2}\right) X + \cos\left(\alpha\frac{\pi}{2}\right) Y : X, Y \in \mathbb{R}^{n\times n}, \begin{bmatrix} X & Y \\ -Y & X \end{bmatrix} > 0 \right\}. \tag{6.25}$$

Lemma 6.3.1 *[4] The n-dimensional fractional order system $D^\alpha x(t) = Ax(t)$ with the order $0 < \alpha < 1$ is asymptotically stable if and only if there exists a matrix $P \in \mathbb{P}_\alpha^{n\times n}$ such that*

$$\mathrm{sym}(AP) < 0. \tag{6.26}$$

Since the eigenvalues of matrices A and A^T are identical, the following corollary can be easily derived.

Corollary 6.3.1 *[4] The n-dimensional fractional order system $D^\alpha x(t) = Ax(t)$ with the order $0 < \alpha < 1$ is asymptotically stable if and only if there exists a matrix $Q \in \mathbb{P}_\alpha^{n\times n}$ such that*

$$\mathrm{sym}(A^T Q) < 0. \tag{6.27}$$

6.3.1 Normalization by Output Feedback Controller

Let us consider the output feedback controller for system (5.6), with $\alpha \in (0, 1) \cup (1, 2)$, in the following form

$$u(t) = -LD^\alpha y(t) + v(t). \tag{6.28}$$

The resulting closed-loop system can be expressed as

$$\begin{cases} (E + BLC)D^{\alpha}x(t) = Ax(t) + Bv(t), \\ y(t) = Cx(t), \end{cases} \tag{6.29}$$

where $L \in \mathbb{R}^{m \times p}$ is the gain matrix to normalize the singular system (5.6) and $v(t) \in \mathbb{R}^m$ is the virtual control input with output feedback, which is constructed to stabilize the normalized system (6.29). The objective of this study is to design suitable L and v to guarantee the admissibility of resulting closed-loop control systems. Before giving such a condition, we first introduce the following result, which will be used in the sequel.

Lemma 6.3.2 *[5] Let $\mathcal{X}$ and $\mathcal{Y}$ be any given real matrices of appropriate dimensions. Then, for any scalar $\epsilon > 0$,*

$$\mathcal{X}^T\mathcal{Y} + \mathcal{Y}^T\mathcal{X} \leq \epsilon^{-1}\mathcal{X}^T\mathcal{X} + \epsilon\mathcal{Y}^T\mathcal{Y}.$$

Similar to the conventional integer order case [1], the system (6.29) is normalized if and only if $\det(E + BLC) \neq 0$. Hence, the following theorem is given to get the needed L.

Theorem 6.3.1 *[3] The system (6.29) is normalized if there exist a scalar $\epsilon \in (0, 1)$ and a matrix $L \in \mathbb{R}^{m \times p}$ such that*

$$\begin{bmatrix} E^TE + \text{sym}\left\{E^TBLC + BLC\right\} & I_n \\ I_n & \epsilon I_n \end{bmatrix} > 0. \tag{6.30}$$

Proof Suppose (6.30) holds for some matrix L. According to Schur complement operation, an equivalent condition can be derived

$$E^TE + \text{sym}\left\{E^TBLC + BLC\right\} - \frac{1}{\epsilon}I_n > 0 \tag{6.31}$$

By Lemma 6.3.2, the following inequality holds for any scalar $\epsilon > 0$

$$\text{sym}\left\{BLC\right\} \leq \epsilon C^TL^TB^TBLC + \frac{1}{\epsilon}I_n. \tag{6.32}$$

It is easy to see that the inequality (6.32) implies

$$E^TE + \text{sym}\left\{E^TBLC + BLC\right\} - \frac{1}{\epsilon}I_n \leq E^TE + \text{sym}\left\{E^TBLC\right\} + \epsilon(BLC)^TBLC. \tag{6.33}$$

In the other hand, it is noted that for any scalar $0 < \epsilon < 1$, we have

$$\epsilon(BLC)^TBLC \leq (BLC)^TBLC. \tag{6.34}$$

Using (6.34), we can verify the following inequality

$$E^T E + \text{sym}\left\{E^T BLC\right\} + \epsilon(BLC)^T BLC \leq E^T E + \text{sym}\left\{E^T BLC\right\} + (BLC)^T BLC. \tag{6.35}$$

From the inequalities (6.33) and (6.35), we obtain

$$E^T E + \text{sym}\left\{E^T BLC + BLC\right\} - \frac{1}{\epsilon} I_n \leq E^T E + \text{sym}\left\{E^T BLC\right\} + (BLC)^T BLC. \tag{6.36}$$

According to (6.31), we deduce the inequality

$$E^T E + \text{sym}\left\{E^T BLC\right\} + (BLC)^T BLC > 0. \tag{6.37}$$

Then, from (6.37), it follows

$$(E + BLC)^T (E + BLC) > 0. \tag{6.38}$$

which is equivalent to

$$\det(E + BLC) \neq 0.$$

This completes the proof. □

Remark 6.3.1 Due to the application of (6.13) in Theorem 6.2.1 to calculate L as that of [2] leads to o bilinear matrix inequality ($\mathcal{BMI}$), sufficient condition but not necessary is proposed in Theorem 6.3.1, which can be solved easily by $\mathcal{LMI}$ toolbox, while the one in [2] cannot.

Defining

$$E_1 = E + BLC, \; A_1 = E_1^{-1} A, \; B_1 = E_1^{-1} B, \tag{6.39}$$

the normalized system in turn becomes

$$\begin{cases} D^\alpha x(t) = A_1 x(t) + B_1 v(t) \\ y(t) = C x(t) \end{cases} \tag{6.40}$$

6.3.2 *Static Output Feedback Control Technique*

A relevant lemma will now be presented, which is crucial for addressing the decoupling issue of matrix variables, before proceeding further.

Lemma 6.3.3 *[6] Let Ψ, A and B be given matrices with appropriate dimensions, thus*

$$\begin{cases} \Psi < 0 \\ \Psi + \text{sym}(AB^{T)} < 0 \end{cases} \tag{6.41}$$

hold if and only if there exists a matrix G of appropriate dimensions that satisfies

$$\begin{bmatrix} \Psi & A + BG^T \\ * & -\operatorname{sym}(G) \end{bmatrix} < 0. \tag{6.42}$$

If the controller (6.28) is designed with the static output control law

$$v(t) = Ky(t), \tag{6.43}$$

The solvable stability condition of the resulting closed-loop control system can be obtained from the following theorem.

Theorem 6.3.2 *[3] Designing the controller* (6.43) *for the normalized system* (6.40) *with* $0 < \alpha < 1$, *the resulting closed-loop control system is asymptotically stable, if there exist matrices* $Z \in \mathbb{P}_\alpha^{n\times n}$, $G \in \mathbb{R}^{m\times m}$, *and* $H \in \mathbb{R}^{m\times p}$, *such that*

$$\begin{bmatrix} \operatorname{sym}\left(Z^T(A_1 + B_1K_0)\right) & Z^TB_1 + C^TH^T - K_0^TG^T \\ * & -\operatorname{sym}(G) \end{bmatrix} < 0 \tag{6.44}$$

and the controller gain is given by

$$K = G^{-1}H, \tag{6.45}$$

where K_0 *is an intermediate matrix, which is derived from* $K_0 = QP^{-1}$.

The matrices $P \in \mathbb{P}_\alpha^{n\times n}$ *and* $Q \in \mathbb{R}^{m\times n}$ *satisfy the following* $\mathcal{LMI}$

$$\operatorname{sym}(A_1P + B_1Q) < 0. \tag{6.46}$$

where A_1 and B_1 are defined in (6.39)

Proof Using Theorem 3.2.3, the inequality (6.46) is equivalent to the existence of a virtual state feedback controller

$$v(t) = K_0x(t)$$

for the system in (6.40), which leads to

$$D^\alpha x(t) = (A_1 + B_1K_0)x(t). \tag{6.47}$$

The system (6.47) is asymptotically stable and K_0 is determined by $K_0P = Q$. By taking

$$GK = H,$$

the inequality (6.44) becomes

$$\begin{bmatrix} \operatorname{sym}\left(Z^T(A_1 + B_1K_0)\right) & Z^TB_1 + C^TK^TG^T - K_0^TG^T \\ * & -\operatorname{sym}(G) \end{bmatrix} < 0 \tag{6.48}$$

Based on Lemma 6.3.3, the inequality (6.48) is equivalent to

$$\Psi + \mathrm{sym}\left(ab^T\right) < 0, \tag{6.49}$$

with $\Psi = \mathrm{sym}\left(Z^T(A_1 + B_1 K_0)\right)$, $a = Z^T B_1$, $b = C^T K^T - K_0^T$. After some computations, we get

$$\mathrm{sym}\left(Z^T\left(A_1 + B_1 K C\right)\right) < 0, \tag{6.50}$$

which means that the system (6.40) is stabilized by the output feedback controller $v(t) = Ky(t)$. This completes the proof of Theorem 6.3.2. □

As mentioned in [3], the same result can be obtained easily for the case $1 < \alpha < 2$ in the following theorem.

Theorem 6.3.3 *Designing the controller* (6.43) *for the normalized system* (6.40) *with* $1 < \alpha < 2$, *the resulting closed-loop control system is asymptotically stable, if there exist matrices* $Z \in \mathbb{R}^{n\times n} > 0$, $G \in \mathbb{R}^{m\times m}$, *and* $H \in \mathbb{R}^{m\times p}$, *such that*

$$\begin{bmatrix} \mathrm{sym}\left(Z^T(A_1 + B_1 K_0)\right) & Z^T B_1 + C^T H^T - K_0^T G^T \\ * & -\mathrm{sym}(G) \end{bmatrix} < 0 \tag{6.51}$$

and the controller gain is given by

$$K = G^{-1} H, \tag{6.52}$$

where K_0 *is an intermediate matrix, which is derived from* $K_0 = QP^{-1}$.

The matrices $P \in \mathbb{R}^{n\times n} > 0$ *and* $Q \in \mathbb{R}^{m\times n}$ *satisfy the following* $\mathcal{LMI}$

$$\mathrm{sym}\left(A_1 P + B_1 Q\right) < 0. \tag{6.53}$$

Proof The proof is practically the same as the case $0 < \alpha < 1$.

Example 6.3.1 [3] Consider the controlled singular fractional order system (5.6) with parameters as follows

$$E = \begin{bmatrix} 1 & 1 & 1 \\ 1 & 1 & 1 \\ 0 & 0 & 1 \end{bmatrix}, \quad A = \begin{bmatrix} 0.1 & -0.3 & 2 \\ 0.5 & -3 & -1 \\ 0.2 & 0.4 & -0.1 \end{bmatrix},$$
$$B = \begin{bmatrix} -1 & 1 \\ 1 & -1 \\ 1 & 1 \end{bmatrix}, \quad C = \begin{bmatrix} 1 & 0 & -1 \\ 0 & 1 & 0 \end{bmatrix}, \quad \alpha = 0.8. \tag{6.54}$$

One can easily observe that the pair (E, A) is unstable since it has two stable finite modes -2.7160, -0.9493 and one unstable finite mode 0.6652. By using the MATLAB LMI toolbox, we find that the linear matrix inequality (6.30) in Theorem 6.3.1 is feasible and the solution gives

$$L = \begin{bmatrix} -0.3865 & 0.0580 \\ -0.0021 & -0.4753 \end{bmatrix}, \ \epsilon = 0.6258. \tag{6.55}$$

This means that the system can be normalized by the output derivative feedback controller. In this case, we obtain the system (6.40) with parameters as follows

$$A_1 = \begin{bmatrix} -0.0730 & -0.1995 & 1.9203 \\ 0.1918 & -0.2558 & -1.4499 \\ 0.1812 & 0.1553 & 0.0296 \end{bmatrix}, \ B_1 = \begin{bmatrix} -1.1826 & 1.0024 \\ 0.6100 & -1.5402 \\ 0.5725 & 0.5378 \end{bmatrix} \tag{6.56}$$

From Theorem 6.3.2, a feasible static output feedback controller (6.51) with the control gain can be obtained

$$K = \begin{bmatrix} -0.8811 & 1.9188 \\ -2.3071 & 1.8966 \end{bmatrix}. \tag{6.57}$$

Setting the initial conditions $x(0) = \begin{bmatrix} -0.4 & -0.8 & 0.4 \end{bmatrix}^T$, the time responses of the resulting closed-loop control system is illustrated in Fig. 6.5, which shows that is asymptotically stable and its states converge to zero.

However, normalizing the system is not always possible (rank $([E\ B]) \neq n$). In such cases, it is necessary to find an alternative method of stabilization.

In the following, a static output feedback controller is proposed, without normalization of the original system, to ensure admissibility of the closed-loop system.

A lemma is essential in this context, serving a crucial role in addressing the decoupling issue concerning matrix variables. Recall that for any matrix $\Pi \in \mathbb{R}^{m\times n}$ with full row rank ($\text{rank}(\Pi) = m$), there exists an singular value decomposition (SVD) of Π as follows:

$$\Pi = U \begin{bmatrix} S & 0 \end{bmatrix} V^{\mathrm{T}},$$

where

$$S = \text{diag}\{\sigma_1, \sigma_2, ..., \sigma_m\}, \ \sigma_1 \geq \sigma_2 \geq ... \geq \sigma_m > 0,$$

and $U \in \mathbb{R}^{m\times m}$, $V \in \mathbb{R}^{n\times n}$ are unitary matrices. In particular, the following lemma holds.

Lemma 6.3.4 *[7] Let $\Pi \in \mathbb{R}^{m\times n}$ with $\text{rank}(\Pi) = m \leq n$ and $X \in \mathbb{R}^{n\times n}$. There exists a matrix $\overline{X}$ satisfying $\Pi X = \overline{X}\Pi$ if and only if*

$$X = V \begin{bmatrix} X_{11} & 0 \\ X_{21} & X_{22} \end{bmatrix} V^T,$$

where $X_{11} \in \mathbb{R}^{m\times m}$, $X_{21} \in \mathbb{R}^{(n-m)\times m}$, $X_{22} \in \mathbb{R}^{(n-m)\times(n-m)}$.

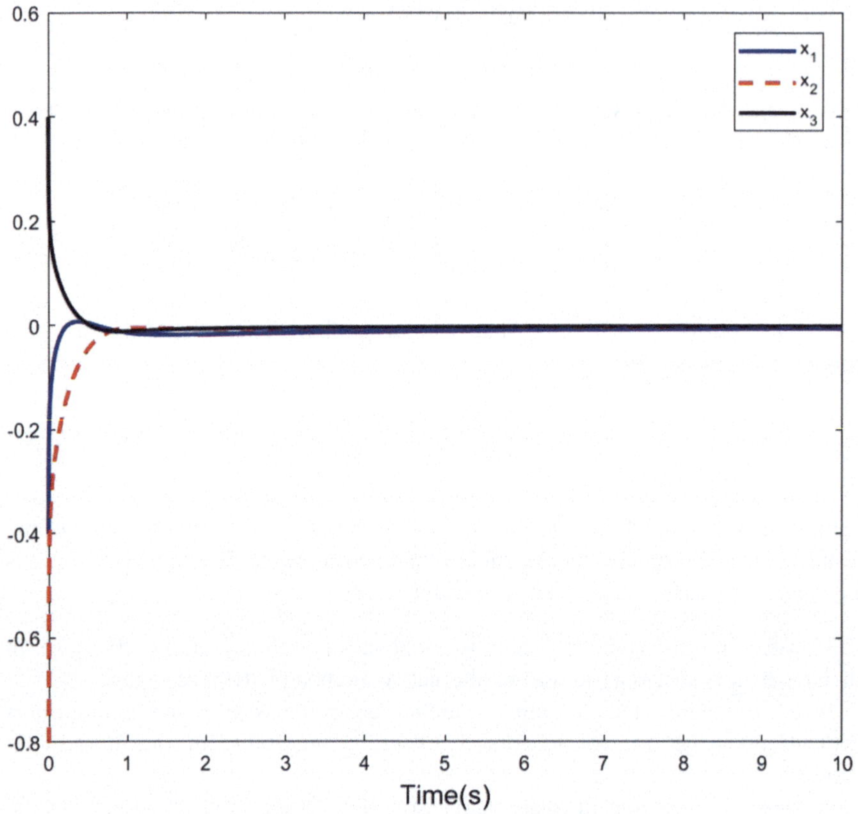

Fig. 6.5 Time response of the selected system in Example 6.3.1 with $v(t) = Ky(t)$

Proof Sufficiency. Suppose that

$$X = V \begin{bmatrix} X_{11} & 0 \\ X_{21} & X_{22} \end{bmatrix} V^{\mathrm{T}},$$

then, we have

$$\begin{aligned} \Pi X &= U \begin{bmatrix} S & 0 \end{bmatrix} V^{\mathrm{T}} V \begin{bmatrix} X_{11} & 0 \\ X_{21} & X_{22} \end{bmatrix} V^{\mathrm{T}}, \\ &= \begin{bmatrix} USX_{11} & 0 \end{bmatrix} V^{\mathrm{T}}, \\ &= \begin{bmatrix} USX_{11}S^{-1}U^{\mathrm{T}}US & 0 \end{bmatrix} V^{\mathrm{T}}, \\ &= USX_{11}S^{-1}U^{\mathrm{T}}U \begin{bmatrix} S & 0 \end{bmatrix} V^{\mathrm{T}}. \end{aligned}$$

By setting

$$\overline{X} = USX_{11}S^{-1}U^{\mathrm{T}},$$

we get that

$$\Pi X = \overline{X}\Pi.$$

Necessity. Let's assume the existence of a matrix $\overline{X}$ satisfying $\Pi X = \overline{X}\Pi$.

For matrix X, one can always find matrices $X_{11} \in \mathbb{R}^{m\times m}$, $X_{12} \in \mathbb{R}^{m\times(n-m)}$, $X_{21} \in \mathbb{R}^{(n-m)\times m}$, $X_{22} \in \mathbb{R}^{(n-m)\times(n-m)}$, such that

$$X = V\begin{bmatrix} X_{11} & X_{12} \\ X_{21} & X_{22} \end{bmatrix} V^{\mathrm{T}}.$$

We just need to show that $X_{21} = 0$.

From $\Pi X = \overline{X}\Pi$, we get

$$U\left[\,S\;0\,\right]V^{\mathrm{T}}V\begin{bmatrix} X_{11} & 0 \\ X_{21} & X_{22} \end{bmatrix} V^{\mathrm{T}} = \overline{X}U\left[\,S\;0\,\right]V^{\mathrm{T}}V. \tag{6.58}$$

The equality (6.58) can be simplified as follows

$$\left[\,USX_{11}\;USX_{12}\,\right] = \left[\,\overline{X}US\;0\,\right]. \tag{6.59}$$

From the equality (6.59), we obtain that

$$USX_{11} = \overline{X}US,$$

which implies that

$$\overline{X} = USX_{11}S^{-1}U^{\mathrm{T}},$$

and

$$USX_{12} = 0,$$

which implies that

$$X_{12} = 0,$$

because matrices U and S are non-singular. Thus, X can be expressed as

$$X = V\begin{bmatrix} X_{11} & 0 \\ X_{21} & X_{22} \end{bmatrix} V^{\mathrm{T}}.$$

This ends the proof. □

Consider the static output feedback controller:

$$u(t) = Ky(t), \tag{6.60}$$

where K is the control gain matrix to be designed. The objective is to design a gain matrix K such that the resulting closed-loop system

$$ED^{\alpha}x(t) = (A + BKC)x(t). \tag{6.61}$$

is admissible. To this end, suppose that M and N are found which satisfy (5.24).The singular value decomposition (SVD) of the matrix CN with full row rank yields

$$CN = U\begin{bmatrix} S & 0 \end{bmatrix}V^{\mathrm{T}}. \tag{6.62}$$

Now we are in a position to state the following result.

Theorem 6.3.4 *[7] System* (6.61), *with* $1 \leq \alpha < 2$, *is admissible if there exist matrices* $P_1 \in \mathbb{R}^{r\times r}$, $P_2 \in \mathbb{R}^{(n-r)\times r}$ *and non-singular matrix* $P_3 \in \mathbb{R}^{(n-r)\times(n-r)}$, $Z \in \mathbb{R}^{m\times p}$ *such that*

$$P_1 > 0, \tag{6.63}$$

$$\mathrm{sym}\left\{\Theta \otimes (ANPM^{-\mathrm{T}} + BZCNM^{-T})\right\} < 0, \tag{6.64}$$

where

$$P = \begin{bmatrix} P_1 & 0 \\ P_2 & P_3 \end{bmatrix}, \tag{6.65}$$

$$V^{\mathrm{T}}PV = \begin{bmatrix} R_1 & 0 \\ R_2 & R_3 \end{bmatrix}, \tag{6.66}$$

with $R_1 \in \mathbb{R}^{p\times p}$, $P_2 \in \mathbb{R}^{(n-p)\times p}$ *and* $R_3 \in \mathbb{R}^{(n-p)\times(n-p)}$. $M, N \in \mathbb{R}^{n\times n}$ *are two arbitrary chosen non-singular matrices satisfying* (5.24), $U \in \mathbb{R}^{p\times p}$, $V \in \mathbb{R}^{n\times n}$ *are defined in* (6.62) *and* $\Theta = \begin{bmatrix} \sin(\alpha\frac{\pi}{2}) & -\cos(\alpha\frac{\pi}{2}) \\ \cos(\alpha\frac{\pi}{2}) & \sin(\alpha\frac{\pi}{2}) \end{bmatrix}$.

The controller gain K is given by

$$K = ZUSR_1^{-1}S^{-1}U^{-1} \tag{6.67}$$

Proof Assume that there exist $P_1 \in \mathbb{R}^{r\times r}$, $P_2 \in \mathbb{R}^{(n-r)\times r}$ and non-singular matrix $P_3 \in \mathbb{R}^{(n-r)\times(n-r)}$, $Z \in \mathbb{R}^{m\times q}$ satisfying (6.63)–(6.67).

Substituting

$$Z = KUSR_1S^{-1}U^{-1}$$

into (6.64) yields

$$\mathrm{sym}\left\{\Theta \otimes (ANPM^{-\mathrm{T}} + BKUSR_1S^{-1}U^{-1}CNM^{-T})\right\} < 0. \tag{6.68}$$

According to Lemma 6.3.4, and the formula (6.66), it is easy to show that the matrix

$$\overline{P} = USR_1S^{-1}U^{-1}$$

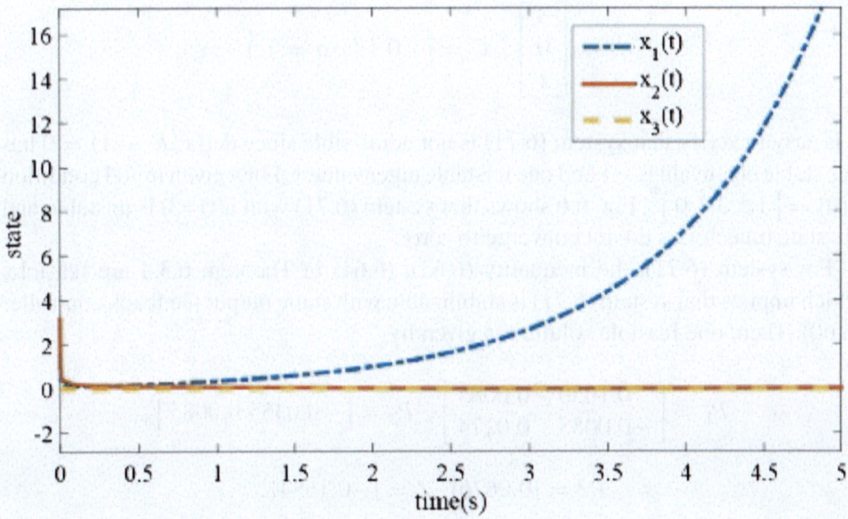

Fig. 6.6 State responses of the SFOS (6.71) with $u(t) = 0$

verify

$$CNP = \overline{P}CN.$$

In this case, (6.68) becomes

$$\text{sym}\left\{\Theta \otimes (ANPM^{-\mathrm{T}} + BKCNPM^{-T})\right\} < 0, \tag{6.69}$$

which is equivalent to

$$\text{sym}\left\{\Theta \otimes (A + BKC)NPM^{-\mathrm{T}}\right\} < 0, \tag{6.70}$$

From (6.63), (6.65) and (6.70), Corollary 5.4.1, system (6.61) is admissible. This completes the proof. □

Example 6.3.2 [7] Consider the descriptor fractional-order system:

$$\begin{aligned} ED^{\alpha} &= Ax(t) + Bu(t), \\ y(t) &= Cx(t), \end{aligned} \tag{6.71}$$

with parameters

$$E = \begin{bmatrix} 3 & 1 & 0 \\ 0 & 5 & 0 \\ 0 & 0 & 0 \end{bmatrix}, \quad A = \begin{bmatrix} 3 & 2 & 0 \\ 0 & -5 & 0 \\ 0 & 0 & -4 \end{bmatrix},$$

$$B = \begin{bmatrix} 3 \\ 0 \\ 1 \end{bmatrix}, \ C = \begin{bmatrix} 1 \ 0 \ 1 \end{bmatrix}, \ \alpha = 1.5.$$

It is easy to verify that system (6.71) is not admissible since $\det(s^\alpha E - A) = 0$ has one stable eigenvalues -1 and one unstable eigenvalue 1. For a given initial condition $x(0) = \begin{bmatrix} 1.5 \ 3.1 \ 0 \end{bmatrix}^{\mathrm{T}}$, Fig. 6.6 shows that system (6.71) with u(t) = 0 is unstable and the state trajectories do not converge to zero.

For system (6.71), the inequality (6.63), (6.64) in Theorem 6.3.4 are feasible, which implies that system (6.71) is stabilizable with static output feedback controller (6.60). Then, one feasible solution is given by

$$P_1 = \begin{bmatrix} 0.1030 & -0.0085 \\ -0.0085 & 0.0274 \end{bmatrix}, \ P_2 = \begin{bmatrix} -0.0355 \ 0.0085 \end{bmatrix},$$

$$P3 = [0.0676], Z = [-0.1654].$$

From (6.67), the gain matrix K is calculated as

$$K = [-1.6047].$$

Given the initial conditions

$$x(0) = \begin{bmatrix} 0.32 \ 0 \ -0.1 \end{bmatrix}^{\mathrm{T}},$$

the state variable responses of closed-loop system (6.61) and the corresponding controller in Example 6.3.2 are shown in Figs. 6.7 and 6.8, respectively, which imply that the stabilisation of system (6.61) is achieved by the static output feedback technique.

It is important to emphasize that the matrix variable P in Theorem 6.3.4 must satisfy the condition that $V^T P V$ takes form

$$\begin{bmatrix} R_1 & 0 \\ R_2 & R_3 \end{bmatrix}.$$

Nevertheless, the existence of this form is not always guaranteed for system (6.61) with matrix C having a full row rank. Hence, the criterion provided in Theorem 6.3.4 is merely sufficient. In the other hand, equality constraint (6.66) in Theorem 6.3.4, may cause a big trouble in using numerical examples to test the validity of the condition. In order to remove equality constraint, the following lemma is given.

Lemma 6.3.5 *[7] Suppose* $\Pi \in \mathbb{R}^{m \times n}$ *with* $m < n$ *and full row rank* ($\mathrm{rank}(\Pi) = m$)*, and its* SVD

$$\Pi = U \begin{bmatrix} S \ 0 \end{bmatrix} V^T. \tag{6.72}$$

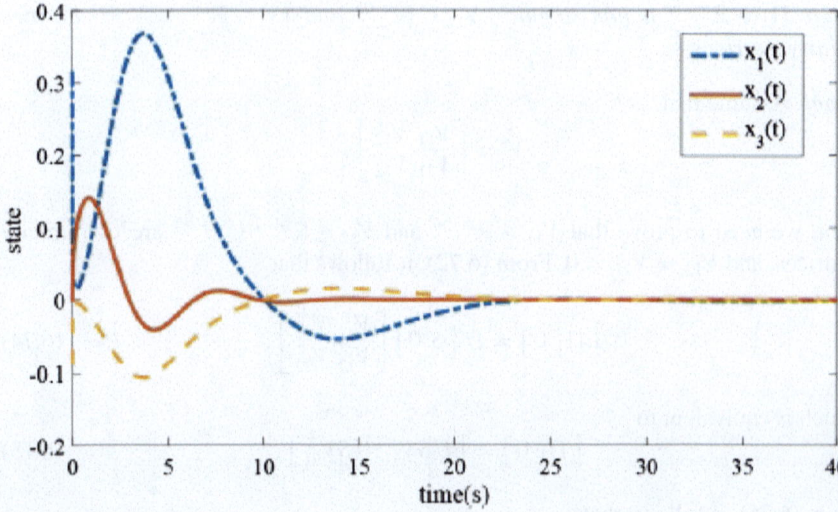

Fig. 6.7 State responses of the closed-loop system (6.61) in Example 6.3.2

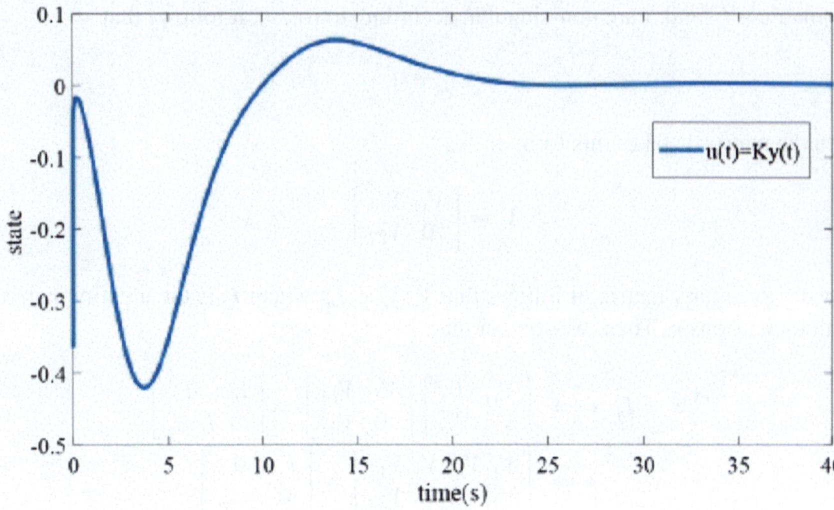

Fig. 6.8 Designed control input $u(t)$ in Example 6.3.2

If Π *has the following form*

$$\Pi = \begin{bmatrix} \Pi_1 & 0 \end{bmatrix},$$

then, V *is given by*

$$V = \begin{bmatrix} V_{11} & 0 \\ 0 & V_{22} \end{bmatrix}, \tag{6.73}$$

where $\Pi_1 \in \mathbb{R}^{m\times m}$ *is non-singular,* $V_{11} \in \mathbb{R}^{m\times m}$ *and* $V_{22} \in \mathbb{R}^{(n-m)\times(n-m)}$ *are two unitary matrices.*

Proof Assume that

$$V = \begin{bmatrix} V_{11} & V_{12} \\ V_{21} & V_{22} \end{bmatrix},$$

then, we need to prove that $V_{11} \in \mathbb{R}^{m\times m}$ and $V_{22} \in \mathbb{R}^{(n-m)\times(n-m)}$ are two unitary matrices, and $V_{12} = V_{21} = 0$. From (6.72), it follows that

$$\begin{bmatrix} \Pi_1 & 0 \end{bmatrix} = U \begin{bmatrix} S & 0 \end{bmatrix} \begin{bmatrix} V_{11}^{\mathrm{T}} & V_{21}^{\mathrm{T}} \\ V_{12}^{\mathrm{T}} & V_{22}^{\mathrm{T}} \end{bmatrix}, \tag{6.74}$$

which is equivalent to

$$\begin{bmatrix} \Pi_1 & 0 \end{bmatrix} = \begin{bmatrix} USV_{11}^{\mathrm{T}} & USV_{21}^{\mathrm{T}} \end{bmatrix}. \tag{6.75}$$

From (6.75), it follows that

$$USV_{21}^{\mathrm{T}} = 0. \tag{6.76}$$

As matrices U and S are non-singular, according to (6.76), it follows that

$$V_{21} = 0,$$

then the matrix V takes this form

$$V = \begin{bmatrix} V_{11} & V_{12} \\ 0 & V_{22} \end{bmatrix}.$$

Since V is unitary matrix, it follows that $V^{\mathrm{T}}V = I_n$, where I_n is the identity matrix with dimensions n. Then, we can get that

$$\begin{aligned} V^{\mathrm{T}}V = I_n \iff & \begin{bmatrix} V_{11}^{\mathrm{T}} & 0 \\ V_{12}^{\mathrm{T}} & V_{22}^{\mathrm{T}} \end{bmatrix} \begin{bmatrix} V_{11} & V_{12} \\ 0 & V_{22} \end{bmatrix} = \begin{bmatrix} I_m & 0 \\ 0 & I_{n-m} \end{bmatrix}, \\ \iff & \begin{bmatrix} V_{11}^{\mathrm{T}}V_{11} & V_{11}^{\mathrm{T}}V_{12} \\ V_{21}^{\mathrm{T}}V_{11} & V_{22}^{\mathrm{T}}V_{22} \end{bmatrix} = \begin{bmatrix} I_m & 0 \\ 0 & I_{n-m} \end{bmatrix}. \end{aligned}$$

The last equality implies that

$$V_{11}^{\mathrm{T}}V_{11} = I_m, \; V_{22}^{\mathrm{T}}V_{22} = I_{n-m},$$

which means that the matrices V_{11} and V_{22} are non-singular and unitary.

We have also

$$V_{11}^{\mathrm{T}}V_{12} = 0.$$

since V_{11} is no-singular, then we gets

$$V_{12} = 0.$$

This completes the proof. □

In the sequel, without loss of generality, suppose $E = \text{diag}(I_r, 0)$ in system (6.61), we make the following assumption.

Assumption 6.3.1
(i) $r = q$, where $r = \text{rank}(E)$, $q = \text{rank}(C)$.
(ii) $C = \begin{bmatrix} C_1 & 0 \end{bmatrix}$, where $C_1 \in \mathbb{R}^{p\times p}$.

Then, combining Lemma 6.3.5, the SVD of the matrix C with full row rank is expressed as

$$C = U\begin{bmatrix} S & 0 \end{bmatrix}\begin{bmatrix} V_1 & 0 \\ 0 & V_2 \end{bmatrix}^{\mathrm{T}}, \tag{6.77}$$

where $V_1 \in \mathbb{R}^{r\times r}$ and $V_2 \in \mathbb{R}^{(n-r)\times(n-r)}$ are two unitary matrices.

Let remark that the Assumption 6.3.1 implies that $r = q = p$.

Under Assumption 6.3.1, a static output feedback stabilisation criterion without equality constraints is presented.

Theorem 6.3.5 *[7] System* (6.61), *with* $1 \le \alpha < 2$, *is admissible if and only if there exist matrices* $P_1 \in \mathbb{R}^{r\times r}$, $P_2 \in \mathbb{R}^{(n-r)\times r}$ *and and non-singular matrix* $P_3 \in \mathbb{R}^{(n-r)\times(n-r)}$, $Z \in \mathbb{R}^{m\times p}$ *such that*

$$P_1 > 0, \tag{6.78}$$

$$\text{sym}\{\Theta \otimes (AP + BZC)\} < 0, \tag{6.79}$$

where

$$P = \begin{bmatrix} P_1 & 0 \\ P_2 & P_3 \end{bmatrix}. \tag{6.80}$$

And the controller gain K is given by

$$K = ZUSV_1^{\mathrm{T}}P_1^{-1}V_1S^{-1}U^{-1}. \tag{6.81}$$

Proof sufficiency: Assume that there exist matrices $P_1 \in \mathbb{R}^{r\times r}$, $P_2 \in \mathbb{R}^{(n-r)\times r}$ and non-singular matrix $P_3 \in \mathbb{R}^{(n-r)\times(n-r)}$, $Z \in \mathbb{R}^{m\times q}$ such that (6.78)–(6.80).

Substituting,

$$Z = KUSV_1^{\mathrm{T}}P_1V_1S^{-1}U^{-1}$$

into (6.79) yields

$$\text{sym}\left\{\Theta \otimes (AP + BKUSV_1^{\mathrm{T}}P_1V_1S^{-1}U^{-1}C\right\} < 0. \tag{6.82}$$

Notice

$$\begin{bmatrix} V_1 & 0 \\ 0 & V_2 \end{bmatrix}^{\mathrm{T}} \begin{bmatrix} P_1 & 0 \\ P_2 & P_3 \end{bmatrix} \begin{bmatrix} V_1 & 0 \\ 0 & V_2 \end{bmatrix} = \begin{bmatrix} V_1^{\mathrm{T}} P_1 V_1 & 0 \\ V_2^{\mathrm{T}} P_2 V_1 & V_2^{\mathrm{T}} P_3 V_2 \end{bmatrix}, \tag{6.83}$$

which implies

$$\begin{bmatrix} P_1 & 0 \\ P_2 & P_3 \end{bmatrix} = \begin{bmatrix} V_1 & 0 \\ 0 & V_2 \end{bmatrix} \begin{bmatrix} V_1^{\mathrm{T}} P_1 V_1 & 0 \\ V_2^{\mathrm{T}} P_2 V_1 & V_2^{\mathrm{T}} P_3 V_2 \end{bmatrix} \begin{bmatrix} V_1 & 0 \\ 0 & V_2 \end{bmatrix}^{\mathrm{T}}. \tag{6.84}$$

From (6.84), and Lemma 6.3.4, it follows that there exists a matrix $\overline{P}$ satisfying

$$CP = \overline{P}C.$$

Using (6.77), we get

$$CP = \overline{P}C \iff \left[USV_1^{\mathrm{T}} P_1 \; 0\right] = \left[\overline{P}USV_1^{\mathrm{T}} \; 0\right],$$

then, we have

$$USV_1^{\mathrm{T}} P_1 = \overline{P}USV_1^{\mathrm{T}},$$

which implies that

$$\overline{P} = USV_1^{\mathrm{T}} P_1 V_1 S^{-1} U^{-1}.$$

Consequently, (6.82) can be rewritten as

$$\mathrm{sym}\left\{\Theta \otimes (A + BKC)NPM^{-\mathrm{T}}\right\} < 0, \tag{6.85}$$

with $N = M = I_n$ because we assumed $E = \mathrm{diag}(I_r, 0)$. Taking into account (6.78) and (6.85), and using Corollary 5.4.1, we deduce that the system (6.61) is admissible. Necessity.Assume that system (6.61) is admissible, there exist matrices $P_1 \in \mathbb{R}^{r\times r}$, $P_2 \in \mathbb{R}^{(n-r)\times r}$ and non-singular matrix $P_3 \in \mathbb{R}^{(n-r)\times(n-r)}$ such that

$$P_1 > 0, \tag{6.86}$$

$$\mathrm{sym}\left\{\Theta \otimes (AP + BKCP) < 0\right\} < 0, \tag{6.87}$$

where P is defined in (6.80). Combining (6.84) and Lemma 6.3.4, there exist $\overline{P}$ such that $CP = \overline{P}C$. Then, (6.87) can be rewritten as

$$\mathrm{sym}\left\{\Theta \otimes (AP + BKUSV_1^{\mathrm{T}} P_1 V_1 S^{-1} U^{-1} C) < 0\right\} < 0. \tag{6.88}$$

By taking

$$Z = KUSV_1^{\mathrm{T}} P_1 V_1 S^{-1} U^{-1},$$

it follows that (6.79) holds. This completes the proof. □

Example 6.3.3 [7] Consider the descriptor fractional-order system:

$$\begin{aligned} ED^{\alpha} &= Ax(t) + Bu(t), \\ y(t) &= Cx(t), \end{aligned} \tag{6.89}$$

with parameters

$$E = \begin{bmatrix} 1 & 0 & 0 \\ 0 & 1 & 0 \\ 0 & 0 & 0 \end{bmatrix}, \; A = \begin{bmatrix} 1 & 2 & 0 \\ 3 & 4 & 0 \\ 0 & 0 & -2 \end{bmatrix},$$

$$B = \begin{bmatrix} 1 \\ 1 \\ 1 \end{bmatrix}, \; C = \begin{bmatrix} 1 & 2 & 0 \\ 0 & 1 & 0 \end{bmatrix}, \; \alpha = 1.2.$$

For given initial condition $x(0) = \begin{bmatrix} 25 & 0.25 & 0 \end{bmatrix}^{\mathrm{T}}$, Fig. 6.9 shows that system (6.89) with u(t) = 0 is unstable and the state trajectories do not converge to zero.

For system (6.89), by using LMI toolbox, we can get that inequalities (6.78) and (6.79) in Theorem 6.3.5 are feasible, which implies that system (6.61) in Example 6.3.3 is stabilised by static output feedback controller (6.60). Then, the feasible solutions are given by

$$P_1 = \begin{bmatrix} 0.9414 & -0.4846 \\ -0.4846 & 0.3785 \end{bmatrix}, \; P_2 = \begin{bmatrix} -0.2940 & -0.3045 \end{bmatrix},$$

$$P_3 = \begin{bmatrix} 0.3009 \end{bmatrix}, Z = \begin{bmatrix} -0.5879 & 0.5668 \end{bmatrix}.$$

From (6.81), the gain matrix K is calculated as

$$K = \begin{bmatrix} -4.2615 & 1.4581 \end{bmatrix}.$$

For given initial conditions

$$x(0) = \begin{bmatrix} 0.25 & 0 & 0.53 \end{bmatrix}^{\mathrm{T}},$$

state responses of closed-loop system (6.61) and the corresponding control input are shown in Figs. 6.10 and 6.11, respectively, which imply that the stabilization of system (6.61) is achieved by the static output feedback technique.

Subsequently, we focus on the system (6.61), where $0 < \alpha < 1$. Similarly to the previous discussions, we assume that M and N are found which satisfy (5.24). An SVD of the matrix CN with full row rank is obtained as in (6.62).

For system (6.61), the following two theorems are developed for the case $0 < \alpha < 1$.

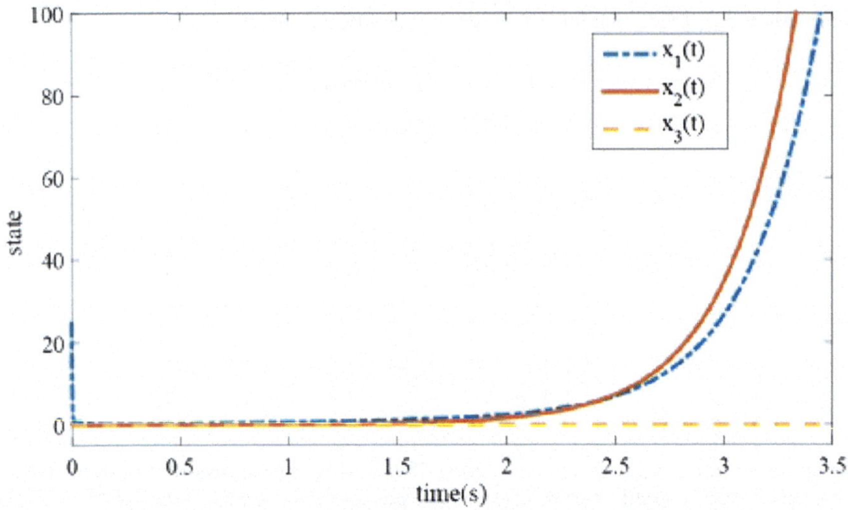

Fig. 6.9 State responses of the SFOS (6.89) with $u(t) = 0$

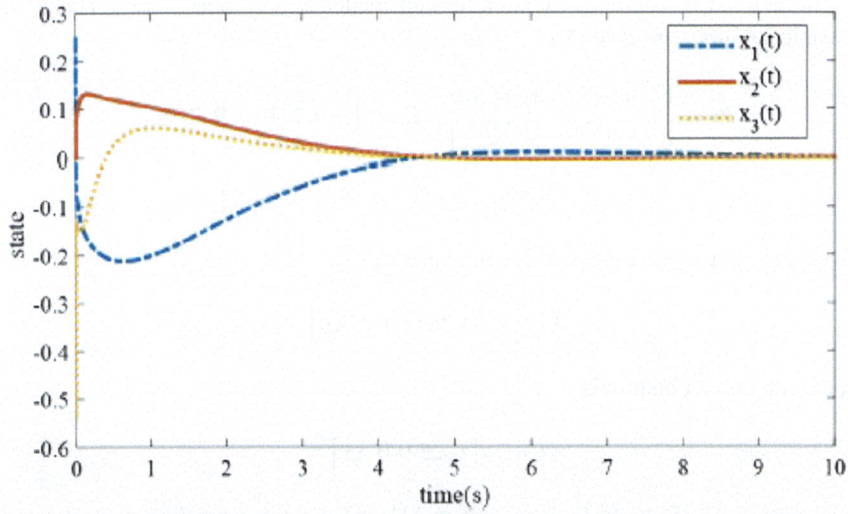

Fig. 6.10 State responses of the closed-loop system (6.61) in Example 6.3.3

Theorem 6.3.6 *[7] System* (6.61), *with* $0 < \alpha < 1$, *is admissible if there exist matrices* $X_1, X_2 \in \mathbb{R}^{r\times r}$, $X_3 \in \mathbb{R}^{(n-r)\times r}$, $Z \in \mathbb{R}^{m\times p}$ *and non-singular matrix* $X_4 \in \mathbb{R}^{(n-r)\times(n-r)}$ *such that*

$$\begin{bmatrix} X_1 & X_2 \\ -X_2 & X_1 \end{bmatrix} < 0, \tag{6.90}$$

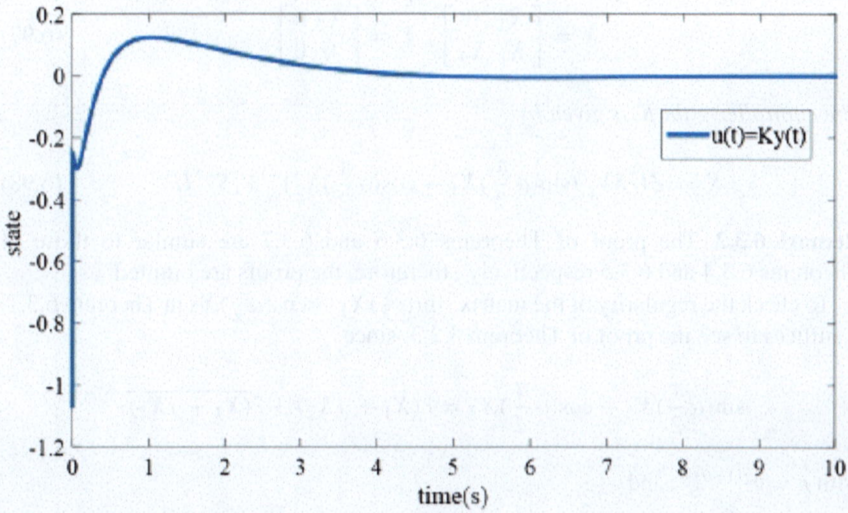

Fig. 6.11 Designed control input $u(t)$ in Example 6.3.3

$$\text{sym}\left\{AN(\sin(\alpha\frac{\pi}{2})X-\cos(\alpha\frac{\pi}{2})Y)M^{-T}+BZCNM^{-T}\right\}<0, \tag{6.91}$$

where

$$X=\begin{bmatrix} X_1 & 0 \\ X_3 & X_4 \end{bmatrix},\ Y=\begin{bmatrix} X_2 & 0 \\ 0 & 0 \end{bmatrix}, \tag{6.92}$$

$$V^T(\sin(\alpha\frac{\pi}{2})X-\cos(\alpha\frac{\pi}{2})Y)V=\begin{bmatrix} \tilde{R}_1 & 0 \\ \tilde{R}_2 & \tilde{R}_3 \end{bmatrix}, \tag{6.93}$$

with $\tilde{R}_1 \in \mathbb{R}^{p\times p}$, $\tilde{R}_2 \in \mathbb{R}^{(n-p)\times p}$, *and* $\tilde{R}_3 \in \mathbb{R}^{(n-p)\times(n-p)}$. *M and N are two matrices satisfying* (5.24). *The controller gain K is given by*

$$K = ZUS\tilde{R}_1^{-1}S^{-1}U^{-1}. \tag{6.94}$$

Theorem 6.3.7 *[7] Under Assumption 6.3.1, system* (6.61), *with* $0<\alpha<1$, *is admissible if and only if there exist matrices* $X_1, X_2 \in \mathbb{R}^{r\times r}$, $X_3 \in \mathbb{R}^{(n-r)\times r}$, $Z \in \mathbb{R}^{m\times p}$ *and non-singular matrix* $X_4 \in \mathbb{R}^{(n-r)\times(n-r)}$ *such that*

$$\begin{bmatrix} X_1 & X_2 \\ -X_2 & X_1 \end{bmatrix}<0, \tag{6.95}$$

$$\text{sym}\left\{AN(\sin(\alpha\frac{\pi}{2})X-\cos(\alpha\frac{\pi}{2})Y)M^{-T}+BZCNM^{-T}\right\}<0, \tag{6.96}$$

where

$$X = \begin{bmatrix} X_1 & 0 \\ X_3 & X_4 \end{bmatrix}, \; Y = \begin{bmatrix} X_2 & 0 \\ 0 & 0 \end{bmatrix}. \tag{6.97}$$

The controller gain K is given by

$$K = ZUSV_1^T(\sin(\alpha\frac{\pi}{2})X_1 - \cos(\alpha\frac{\pi}{2})X_2)^{-1}V_1S^{-1}U^{-1}. \tag{6.98}$$

Remark 6.3.2 The proof of Theorems 6.3.6 and 6.3.7 are similar to those of Theorems 6.3.4 and 6.3.5 respectively , therefore, the proofs are omitted.

To check the regularity of the matrix $\sin(\alpha\frac{\pi}{2})X_1 - \cos(\alpha\frac{\pi}{2})X_2$ in Theorem 6.3.7, it suffices to see the proof of Theorem 3.2.3, since

$$\sin(\alpha\frac{\pi}{2})X_1 - \cos(\alpha\frac{\pi}{2})X_2 = r(X_1 + jX_2) + \overline{r(X_1 + jX_2)},$$

with $r = e^{j(1-\alpha)\frac{\pi}{2}}$, and

$$\begin{bmatrix} X_1 & X_2 \\ -X_2 & X_1 \end{bmatrix} > 0 \iff X_1 + jX_2 > .$$

Remark 6.3.3 Note that Theorem 6.3.4 only provides a sufficient condition for the admissibility of the closed-loop system, whereas Theorem 6.3.5 presents a necessary and sufficient condition based on the $\mathcal{LMI}$ formalism. However, the addition of Assumption 6.3.1 sometimes makes its application inconvenient.

The following theorem gives necessary and sufficient conditions for system (6.61) to be admissible.

Theorem 6.3.8 *[8] The closed-loop system* (6.61) *with order* $1 \le \alpha < 2$ *is admissible if and only if there exist matrices* $X > 0,\ Y$ *and scalar* $\gamma > 0$ *satisfying*

$$\text{sym}\left\{\Theta \otimes \left(A^T\left(XE + E_0Y^T\right)\right)\right\} - \frac{1}{\gamma}Q^TQ < 0. \tag{6.99}$$

and

$$KC = -\frac{\sin\theta}{\gamma}B^T\left(XE + E_0Y^T\right) \tag{6.100}$$

where $E_0 \in \mathbb{R}^{n\times(n-r)}$ *is any matrix of full column rank and satisfies* $E^TE_0 = 0$ *and*

$$Q = \begin{bmatrix} B^T\left(XE + E_0Y^T\right)\sin\theta & 0 \\ 0 & B^T\left(XE + E_0Y^T\right)\sin\theta \end{bmatrix},$$

$$\Theta = \begin{bmatrix} \sin\left(\alpha\frac{\pi}{2}\right) & -\cos\left(\alpha\frac{\pi}{2}\right) \\ \cos\left(\alpha\frac{\pi}{2}\right) & \sin\left(\alpha\frac{\pi}{2}\right) \end{bmatrix}, \text{ and } \theta = (1-\alpha)\frac{\pi}{2}.$$

Proof Necessity. By Theorem 5.4.1, we know that the system (6.61) is admissible if and only if there exist matrices $X > 0$ and Y satisfying

$$\text{sym}\left\{\Theta \otimes \left((A + BKC)^T \left(XE + E_0 Y^T\right)\right)\right\} < 0. \tag{6.101}$$

Inequality (6.101) is equivalent to

$$W + \text{sym}\left\{\Theta \otimes \left((KC)^T G\right)\right\} < 0 \tag{6.102}$$

with

$$W = sym\left\{\Theta \otimes (A^T \left(XE + E_0 Y^T\right))\right\}, G = B^T \left(XE + E_0 Y^T\right). \tag{6.103}$$

Inequality (6.102) is equivalent to

$$W + \text{sym}\left\{\left(I_2 \otimes (KC)^T\right)(\Theta \otimes G)\right\} < 0.$$

Then a sufficiently small scalar $\gamma > 0$ can always be found such that

$$W + \text{sym}\left\{F^T (\Theta \otimes G)\right\} + \gamma F^T F < 0, \tag{6.104}$$

with

$$F = I_2 \otimes (KC). \tag{6.105}$$

Rewriting the matrix $\Theta \otimes G$ by

$$\Theta \otimes G = Q + \widetilde{Q}, \tag{6.106}$$

where

$$Q = \begin{bmatrix} G \sin\theta & 0_{m\times n} \\ 0_{m\times n} & G \sin\theta \end{bmatrix}, \widetilde{Q} = \begin{bmatrix} 0_{m\times n} & G \cos\theta \\ -G \cos\theta & 0_{m\times n} \end{bmatrix}, \tag{6.107}$$

the inequality (6.104) can be rewritten as follows:

$$W + \gamma \left(F^T + \frac{1}{\gamma} Q^T\right)\left(F + \frac{1}{\gamma} Q\right) - \frac{1}{\gamma} Q^T Q + F^T \widetilde{Q} + \widetilde{Q}^T F < 0. \tag{6.108}$$

Choosing

$$KC = -\frac{\sin\theta}{\gamma} G, \tag{6.109}$$

we get $\left(F^T + \frac{1}{\gamma} Q^T\right)\left(F + \frac{1}{\gamma} Q\right) = 0_{2n}$ and $F^T \widetilde{Q} + \widetilde{Q}^T F = 0_{2n}$. Then, (6.101) leads to inequality (6.99).

Sufficiency - Suppose that (6.99) and (6.100) hold for some matrices $X > 0$, Y and some real $\gamma > 0$.

Indeed, for $KC = -\frac{\sin\theta}{\gamma}G$, (6.99) is equivalent to

$$W + \gamma\left(F^T + \frac{1}{\gamma}Q^T\right)\left(F + \frac{1}{\gamma}Q\right) - \frac{1}{\gamma}Q^T Q < 0, \tag{6.110}$$

or equivalently

$$W + F^T Q + Q^T F + \gamma F^T F < 0$$

Knowing that $\gamma F^T F \geq 0$, we obtain

$$W + F^T Q + Q^T F < 0$$

From (6.106), with $F^T \widetilde{Q} + \widetilde{Q}^T F = 0_{2n}$, we get

$$W + F^T (\Theta \otimes G) + (\Theta \otimes G)^T F < 0$$

which is equivalent to

$$\text{sym}\left\{\Theta \otimes \left((A + BKC)^T \left(XE + E_0 Y^T\right)\right)\right\} < 0$$

Using Theorem 5.4.1, we conclude that system (6.61) is admissible. This completes the proof of Theorem 6.3.8. □

Remark 6.3.4 Note that, in most practical applications, the state variables are not available. Then, an observer-based controller or output feedback controller are often considered. Here, the very challenging problem of static output feedback control is studied. Indeed, a necessary and sufficient condition to design a such controller is proposed to ensure the admissibility of the closed-loop system.

If all state variables are measurable which means that the output matrix C is equal to I_n, the problem treated becomes a stabilization by state feedback control of the SFOS (5.6), since the control (6.60) takes the form $u(t) = Kx(t)$. Therefore, the following result is deduced.

Theorem 6.3.9 *The SFOS* (6.61) *is admissible if and only if there exist matrices* $P \in \mathbb{R}^{n\times n} > 0$, $Q \in \mathbb{R}^{(n-r)\times n}$ *and scalar* $\varepsilon > 0$ *such that*

$$\begin{bmatrix} \Phi_1 & \Phi_2 \\ \Phi_2^T & \Phi_1 \end{bmatrix} < 0, \tag{6.111}$$

where

$$\Phi_1 = (A^T X + X^T A) \sin\left(\alpha\frac{\pi}{2}\right) - \frac{1}{\varepsilon}\sin^2\left(\alpha\frac{\pi}{2}\right) X^T B B^T X,$$
$$\Phi_2 = (A^T X - X^T A) \cos\left(\alpha\frac{\pi}{2}\right),$$
$$X = PE + SQ,$$

and S is any matrix of full column rank and satisfies $E^T S = 0$.

The controller gain K can be designed as

$$K = \frac{-\sin\left(\alpha\frac{\pi}{2}\right)}{\varepsilon} B^T (PE + SQ) . \quad (6.112)$$

In the following, two numerical examples are presented to demonstrate the effectiveness of the theoretical results.

Example 6.3.4 Consider the linear singular fractional-order system (5.6) with the following data:

$$E = \begin{bmatrix} 1 & 1 & 1 \\ 0 & 1 & 1 \\ 0 & 0 & 0 \end{bmatrix}, \quad A = \begin{bmatrix} 1 & 1 & -1 \\ 2 & -2 & -1 \\ 4 & 1 & -4 \end{bmatrix},$$
$$C = \begin{bmatrix} 1 & 1 & 2 \\ 1 & 1 & 1 \end{bmatrix}, \quad B = \begin{bmatrix} 1 \\ 1 \\ 1 \end{bmatrix}, \alpha = 1.2 \quad (6.113)$$

It can be seen that the system is regular and impulse free, since

$$\det(\lambda E - A) = 5\lambda^2 + 26\lambda - 3,$$

but it is not stable because its finite modes are in

$$\{0.1129, -5.3129\} .$$

Indeed, the time response of the open-loop system, i.e., with $u(t) = 0$, and initial condition $x(0) = [1, -1, 1.1328]^T$ is shown in Fig. 6.12. Figure 6.12 confirms the instability of the selected system.

The objective is to design an output feedback control law $u(t) = Ky(t)$ that stabilizes the closed-loop system (6.61). To this end, the inequality (6.99) in Theorem 6.3.8 is solved by using Matlab control toolbox. A feasible solution solution of (6.99) is given by

$$X = \begin{bmatrix} 0.0007 & -0.0013 & -0.0000 \\ -0.0013 & 0.0023 & 0.0000 \\ -0.0000 & 0.0000 & 444.4279 \end{bmatrix}, \quad Y = \begin{bmatrix} -0.0003 \\ -0.0013 \\ 0.0008 \end{bmatrix}, \gamma = 10^{-5}$$

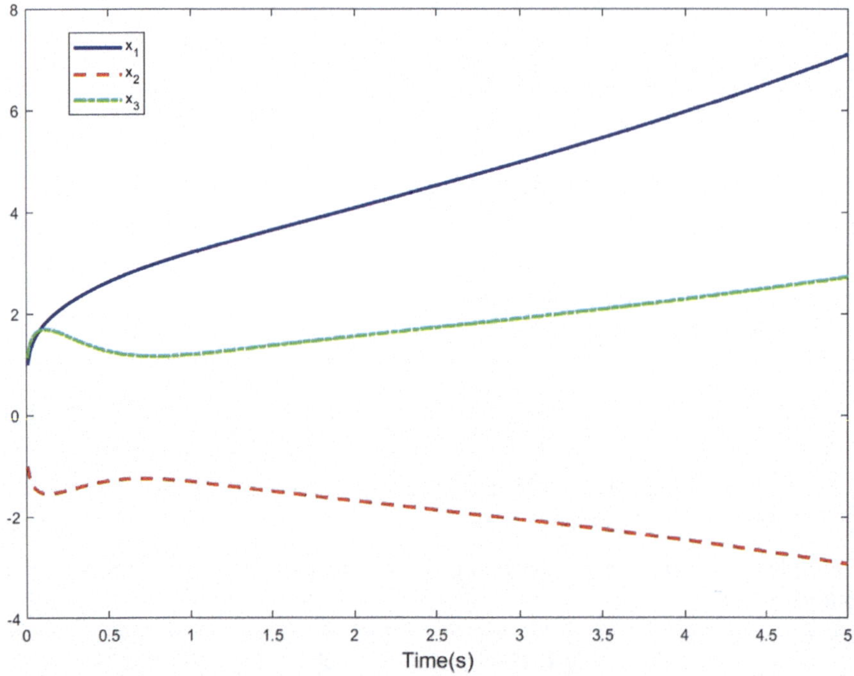

Fig. 6.12 Time response of the selected system

and the controller gain K is

$$K = \begin{bmatrix} -199.2581 \ 282.4540 \end{bmatrix}$$

We can now easily verify that the closed-loop system (6.61), i.e. $(E, A + BKC, \alpha)$, is admissible since $\det(\lambda E - A - BKC) = -\lambda - 26.4767$. Indeed, simulation results given in Figs. 6.13 and 6.14 confirm the admissibility of the closed-loop system (6.61) via the corresponding control input $u(t) = Ky(t)$.

6.4 Observer-Based Control for Singular Fractional-Order Systems

1. **Case** $0 < \alpha \leq 1$
   In the following, we deals with the design of observer-based controller for closed-loop fractional-order linear system (5.6), with $0 < \alpha \leq 1$, to be admissible. Indeed, we consider a Luenberger-type fractional-order linear observer of the form:

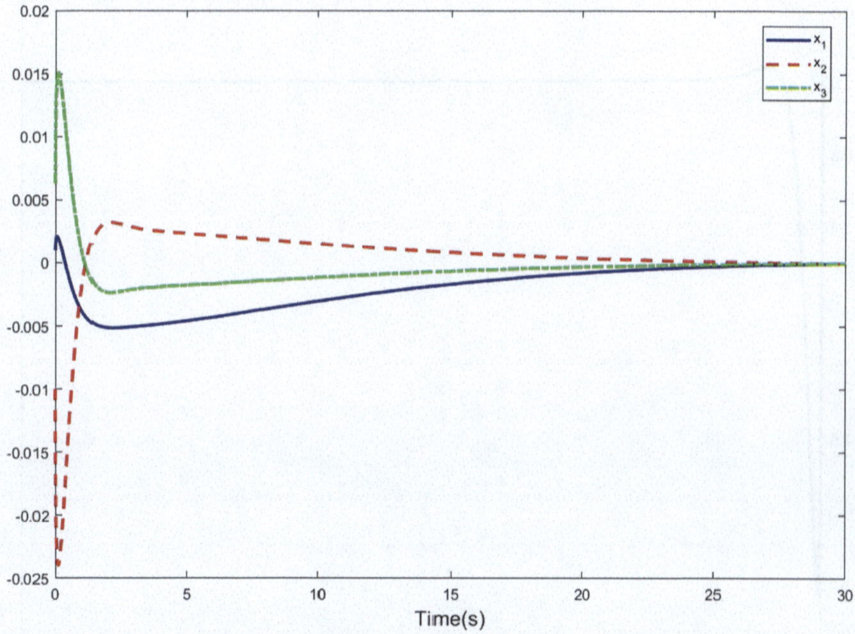

Fig. 6.13 Time response of the selected system with $u(t) = Ky(t)$

$$\begin{cases} ED^{\alpha}\hat{x}(t) = A\hat{x}(t) + Bu(t) + L\left(y(t) - \hat{y}(t)\right) \\ \hat{y}(t) \qquad\;\; = C\hat{x}(t) \end{cases} \tag{6.114}$$

coupled with the control law

$$u(t) = K\hat{x}(t) \tag{6.115}$$

where K and L are the parameter gains to determine. The closed-loop system is given by

$$\breve{E}D^{\alpha}v(t) = A_{cl}v(t) \tag{6.116}$$

with $e(t) = x(t) - \hat{x}(t)$ and

$$\breve{E} = \begin{bmatrix} E & 0 \\ 0 & E \end{bmatrix}, v(t) = \begin{bmatrix} \hat{x}(t) \\ e(t) \end{bmatrix}, A_{cl} = \begin{bmatrix} A + BK & LC \\ 0 & A - LC \end{bmatrix}$$

Conditions of the admissibility for the system (6.116), or the pair $\left(\breve{E}, A_{cl}\right)$, are given by the following results.

Lemma 6.4.1 *[9] The system* (6.116) *is admissible if and only if the pairs* $(E, A + BK)$ *and* $(E, A - LC)$ *are admissible.*

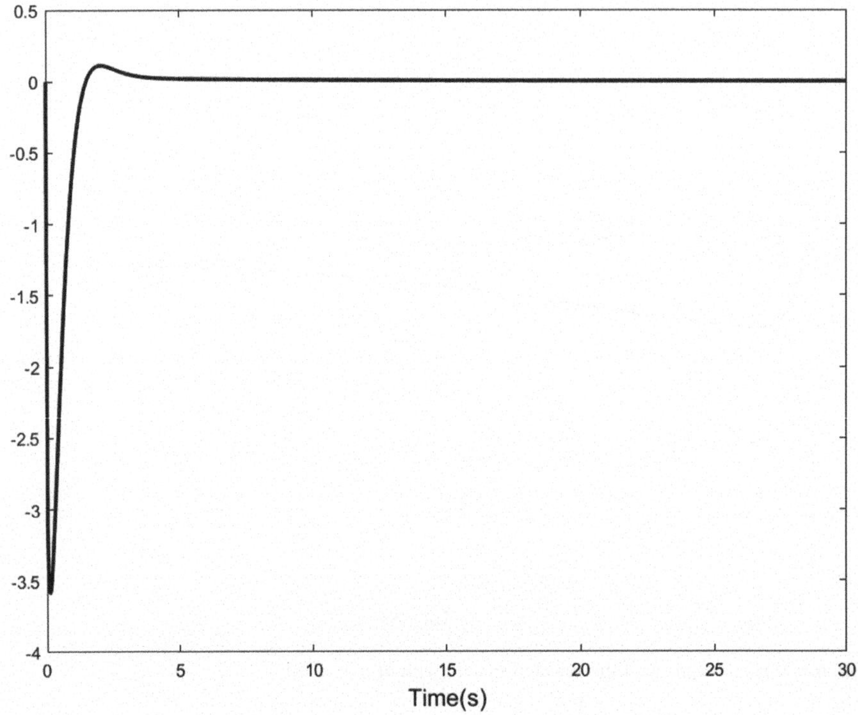

Fig. 6.14 Time response of the control input $u(t) = Ky(t)$

Proof We know that the system (6.116) is admissible if and only if the pair $\left(\check{E}, A_{cl}\right)$ is (i) regular, (ii) impulse free and (iii) stable.

Condition (i) means that $\det(s^\alpha \check{E} - A_{cl}) \neq 0$, for some $s \in \mathbb{C}$. Since

$$\det(s^\alpha \check{E} - A_{cl}) = \det(s^\alpha E - (A + BK).\det(s^\alpha E - (A - LC)) \tag{6.117}$$

we conclude that the condition (i) is equivalent to that the pairs $(E,\ A + BK)$ and $(E, A - LC)$ are simultaneously regular.

Condition (ii) means that for $\lambda \in \mathbb{C}$, $\deg\left(\det(\lambda\check{E} - A_{cl})\right) = rank\check{E} = 2r$. Note also that

$$\deg\left(\det(\lambda\check{E} - A_{cl})\right) = \deg\left(\det(\lambda E - (A + BK)\right) + \deg\left(\det(\lambda E - (A - LC)\right)$$

Since $\deg\left(\det(\lambda E - (A + BK)\right) \leq r$ and $\deg\left(\det(\lambda E - (A - LC)\right) \leq r$, we get

$$\deg\left(\det(\lambda E - (A + BK))\right) = \deg\left(\det(\lambda E - (A - LC)\right) = r$$

it follows that the pairs $(E, A + BK)$ and $(E, A - LC)$ are impulse free.

Finally from (6.117), we deduce that stability of the pair $\left(\breve{E}, A_{cl}\right)$ (condition (iii)) is equivalent to the stability of $(E, A + BK)$ and $(E, A - LC)$.

Consequently, the admissibility of the pair $\left(\breve{E}, A_{cl}\right)$ is equivalent to the admissibility of $(E, A + BK)$ and $(E, A - LC)$ separately. □

Based on the results of Lemma 6.4.1 and Corollary 5.4.2, the determination of gains K and L, such that the system (6.116) is admissible, are derived through the following theorem.

Theorem 6.4.1 *[9] The closed-loop system* (6.116) *is admissible if and only if there exist matrices* $X_1 = X_1^* \in \mathbb{C}^{n\times n} > 0$, $X_2 = X_2^* \in \mathbb{C}^{n\times n} > 0$, $Y_1 \in \mathbb{R}^{n\times(n-r)}$, $Y_2 \in \mathbb{R}^{n\times(n-r)}$, *scalars* $\gamma_1 > 0$ *and* $\gamma_2 > 0$ *satisfying*

$$\mathrm{sym}\left\{A^T\left((zX_1 + \overline{z}\overline{X}_1)E + E_0 Y_1^T\right)\right\} - \frac{1}{\gamma_1} Q_1^T Q_1 < 0 \tag{6.118}$$

$$\mathrm{sym}\left\{A\left((zX_2 + \overline{z}\overline{X}_2)E^T + \widetilde{E}_0 Y_2^T\right)\right\} - \frac{1}{\gamma_2} Q_2^T Q_2 < 0 \tag{6.119}$$

with

$$Q_1 = B^T\left((zX_1 + \overline{z}\overline{X}_1)E + E_0 Y_1^T\right) \tag{6.120}$$

$$Q_2 = -C\left((zX_2 + \overline{z}\overline{X}_2)E^T + \widetilde{E}_0 Y_2^T\right) \tag{6.121}$$

where $E_0 \in \mathbb{R}^{n\times(n-r)}$ *and* $\widetilde{E}_0 \in \mathbb{R}^{n\times(n-r)}$ *are full column rank satisfying* $E^T E_0 = 0$, $E\widetilde{E}_0 = 0$ *and* $z = e^{j(1-\alpha)\frac{\pi}{2}}$,

The gain matrices are given by

$$K = -\frac{1}{\gamma_1} B^T\left((zX_1 + \overline{z}\overline{X}_1)E + E_0 Y_1^T\right) \tag{6.122}$$

and

$$L = \frac{1}{\gamma_2}\left((zX_2 + \overline{z}\overline{X}_2)E^T + \widetilde{E}_0 Y_2^T\right)^T C^T \tag{6.123}$$

Proof Necessity. Assume that the system (6.116) is admissible, then by Lemma 6.4.1 the pairs $(E, A + BK)$ and $(E, A - LC)$ are admissible. By Corollary 5.4.2, the pair $(E, A + BK)$ is admissible if and only if there exist matrix $X_1 = X_1^* \in \mathbb{C}^{n\times n} > 0$, matrix $Y_1 \in \mathbb{R}^{n\times(n-r)}$ satisfying

$$\mathrm{sym}\left\{(A + BK)^T\left((zX_1 + \overline{z}\overline{X}_1)E + E_0 Y_1^T\right)\right\} < 0 \tag{6.124}$$

which is equivalent to

$$W + K^T Q_1 + Q_1^T K < 0 \tag{6.125}$$

with

$$W = \text{sym}\left\{A^T\left((zX_1 + \bar{z}\overline{X}_1)E + E_0 Y_1^T\right)\right\}$$

and Q_1 is given by (6.120). Then a sufficiently small scalar $\gamma_1 > 0$ can always be found such that

$$W + K^T Q_1 + Q_1^T K + \gamma_1 K^T K < 0 \tag{6.126}$$

or equivalently

$$W + \gamma_1\left(K + \frac{1}{\gamma_1}Q_1\right)^T\left(K + \frac{1}{\gamma_1}Q_1\right) - \frac{1}{\gamma_1}Q_1^T Q_1 < 0 \tag{6.127}$$

Then with the controller gain K given by (6.122), we get (6.118). By following the same approach, we get (6.119) with gain (6.123).

Sufficiency. Now suppose that condition (6.118) holds with the controller gain (6.122) for some matrices $X_1 = X_1^* \in \mathbb{C}^{n\times n} > 0$, $Y_1 \in \mathbb{R}^{n\times(n-r)}$ and a scalar $\gamma_1 > 0$.

Indeed, from (6.118) and (6.122) we get (6.126). Since $\gamma_1 K^T K > eq0$, we obtain (6.124). Using Theorem 5.4.1, we deduce that the pair $(E, A + BK)$ is admissible.

By the same reasoning, the proof of the admissibility of the pair $(E, A - LC)$ is derived. Which completes the proof of Theorem 6.4.1. □

Example 6.4.1 Consider the singular continuous fractional-order system described in (6.116) with the following data:

$$E = \begin{bmatrix} 0 & 1 & -3 \\ 1 & 1 & 3 \\ 0 & -1 & 3 \end{bmatrix}, \quad A = \begin{bmatrix} 1 & 1 & 2 \\ 1 & 1 & 3 \\ -0.5 & 1.5 & 2.7 \end{bmatrix}$$
$$B = \begin{bmatrix} 1 & 4 \\ -1 & 0 \\ -3 & -5 \end{bmatrix}, \quad C = \begin{bmatrix} 5 & 0 & 0 \\ 0 & 0 & 7 \end{bmatrix}, \alpha = 0.5 \tag{6.128}$$

The finite modes of the pair (E, A) are $\left\{\frac{5}{23}, 1\right\}$, then the unforced singular fractional order system (E, A) is unstable, which means that it is not admissible as can be seen in Fig. 6.15.

In the following, our objective is to design an observer-based controller such that the closed loop system defined in (6.116), with the control law (6.114)–(6.115), is admissible.

Solving the design conditions (6.118) and (6.119) of Theorem 6.4.1 with $\gamma_1 = 0.9$ and $\gamma_2 = 0.5$, we get the following results:

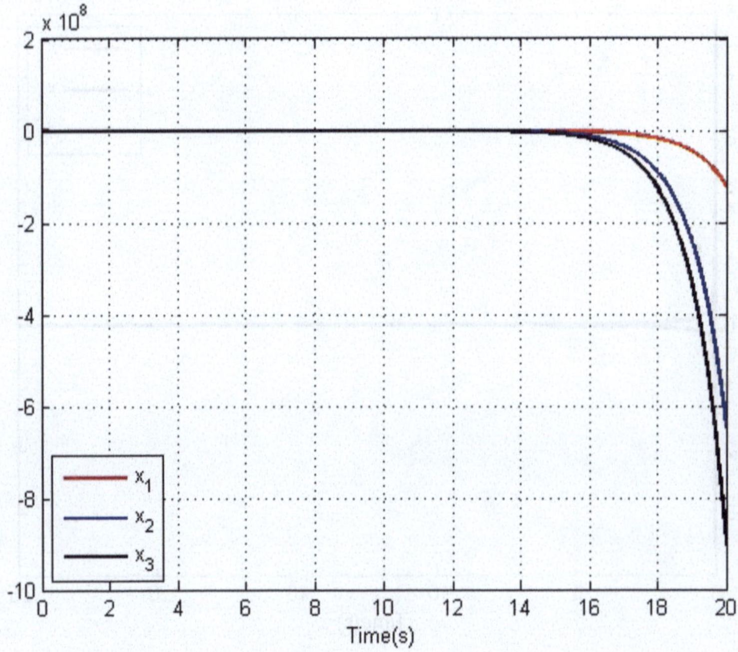

Fig. 6.15 State responses of the selected system in Example 6.4.1

$$X_1 = \begin{bmatrix} 1.5178 & -0.0445 & 1.2720 \\ -0.0445 & 4.0088 & 0.0445 \\ 1.2720 & 0.0445 & 1.5178 \end{bmatrix}, \ Y_1 = \begin{bmatrix} -4.3503 \\ -0.3467 \\ -11.5210 \end{bmatrix}$$

$$X_2 = \begin{bmatrix} 1.6308 & 0.2260 & 0.0215 \\ 0.2260 & 1.8035 & 0.4877 \\ 0.0215 & 0.4877 & 0.1797 \end{bmatrix}, \quad Y_2 = \begin{bmatrix} -0.3797 \\ 0.3401 \\ -0.0354 \end{bmatrix}.$$

The gain matrices can be designed as

$$K = \begin{bmatrix} -1.3897 & 1.8558 & -0.3648 \\ -1.8918 & 12.1136 & -2.8097 \\ 1.4474 & -1.4541 & -0.2180 \end{bmatrix}, L = \begin{bmatrix} 6.2675 & -1.5834 \\ 1.6911 & 6.3791 \\ -0.0399 & 0.1303 \end{bmatrix}.$$

We can easily verify that $(E, A + BK)$ and $(E, A - LC)$ are admissible. Indeed, in one hand we have

$$\det(\lambda E - A - BK) = -24.6521\lambda^2 - 83.2307\lambda - 37.2117$$

and

$$\det(\lambda E - A + LC) = 206.1997\lambda^2 + 1.454210^3\lambda + 2.0663 \times 10^3$$

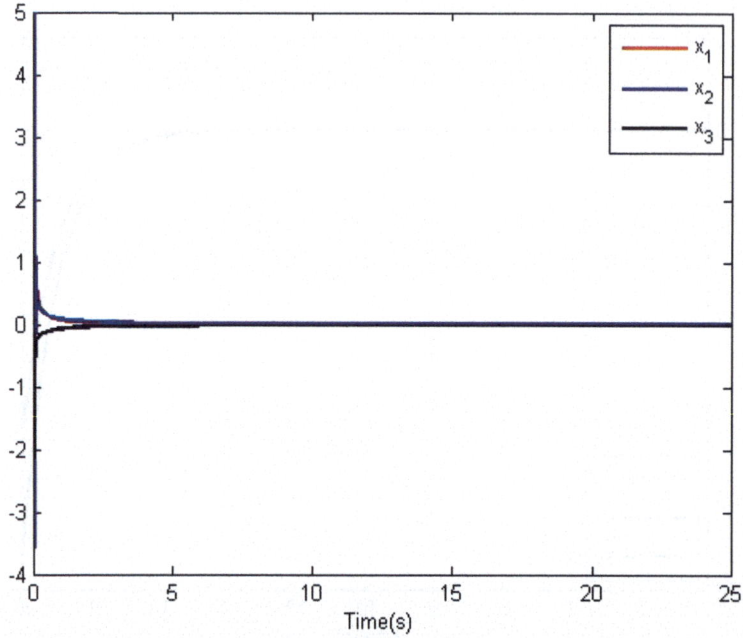

Fig. 6.16 State responses of the closed-loop system in Example 6.4.1

which means that the systems $(E, A + BK)$ and $(E, A - LC)$ are regular and impulse-free.

In the other hand, we have

$$\text{spec}(E, A + BK) = \{-2.8458, -0.5304\}$$

and

$$\text{spec}(E, A - LC) = \{-5.0797, -1.9728\}$$

meaning that the eigenvalues of the pairs $(E, A + BK)$ and $(E, A - LC)$ lie in the stable region. We conclude that the closed loop system of the singular fractional-order system (6.128) is now admissible. The histories of the system states and the observation errors are given in Figs. 6.16 and 6.17 respectively.

2. **Case** $1 \leq \alpha < 2$

   In the following, we deals with the design of observer-based controller for closed-loop fractional-order linear system (5.6), with $1 \leq \alpha < 2$, to be admissible. Indeed, we consider a Luenberger-type fractional-order linear observer of the form (6.114).

   Coupled with the control law of the form (6.115).

   The closed-loop system is given by

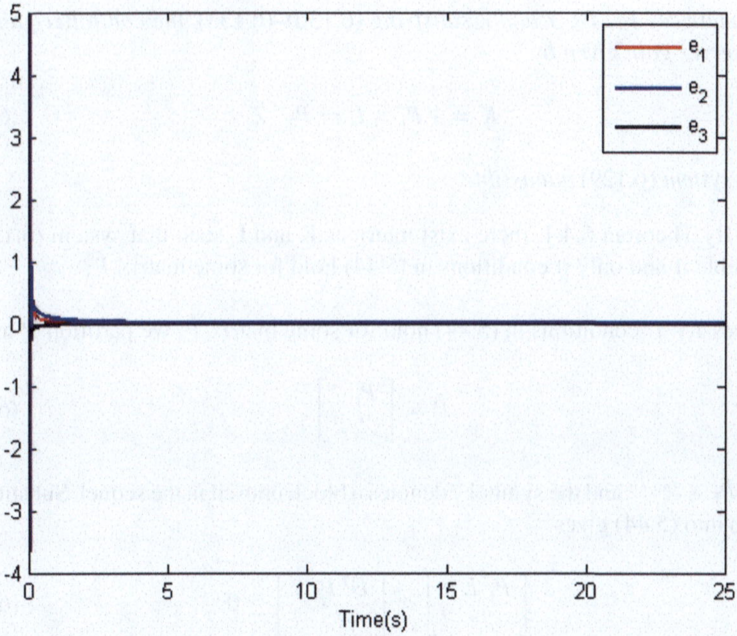

Fig. 6.17 The observation errors in Example 6.4.1

$$\bar{E}D^{\alpha}z(t) = \bar{A}z(t) \tag{6.129}$$

with $e(t) = x(t) - \hat{x}(t)$ and

$$\bar{E} = \begin{bmatrix} E & 0 \\ 0 & E \end{bmatrix}, z(t) = \begin{bmatrix} x(t) \\ e(t) \end{bmatrix}, \bar{A} = \begin{bmatrix} A + BK & -BK \\ 0 & A - LC \end{bmatrix}$$

The first result is as follows.

Theorem 6.4.2 *[10] There exist matrices K and L such that system* (6.129) *is admissible if and only if there exist matrices* P_i, $i = 1, 2$, *Y and Z, such that*

$$EP_1 = P_1^T E^T \geq 0, \tag{6.130}$$
$$\text{sym}\,\{\Theta \otimes (AP_1 + BY)\} < 0, \tag{6.131}$$
$$EP_2 = P_2^T E^T \geq 0, \tag{6.132}$$
$$\text{sym}\,\{\Theta \otimes (P_2 A - ZC)\} < 0, \tag{6.133}$$

where

$$\Theta = \begin{bmatrix} \sin(\pi - \alpha\frac{\pi}{2}) & \cos(\pi - \alpha\frac{\pi}{2}) \\ -\cos(\pi - \alpha\frac{\pi}{2}) & \sin(\pi - \alpha\frac{\pi}{2}) \end{bmatrix}, 1 \leq \alpha < 2. \tag{6.134}$$

Given matrices P_1, P_2, Y *and* Z *satisfying* (6.130)–(6.133), *the controller gain and the observer gain given by*

$$K = Y P_1^{-1}, L = P_2^{-1} Z \tag{6.135}$$

render system (6.129) *admissible.*

Proof By Theorem 5.4.1, there exist matrices K and L such that system (6.129) is admissible if and only if conditions in (5.44) hold for some matrix $\bar{P}$.

Necessity. If conditions in (5.44) hold for some matrix $\bar{P}$, we partition $\bar{P}$ as

$$\bar{P} = \begin{bmatrix} \bar{P}_1 & ? \\ ? & ? \end{bmatrix}, \tag{6.136}$$

where $\bar{P}_1 \in \mathbb{R}^{n \times n}$ and the symbol ? denotes a block unused in the sequel. Substituting (6.136) into (5.44) gives

$$\begin{bmatrix} \bar{P}_1^{\mathrm{T}} E & ? \\ ? & ? \end{bmatrix} = \begin{bmatrix} E^{\mathrm{T}} \bar{P}_1 & ? \\ ? & ? \end{bmatrix} \geq 0, \tag{6.137}$$

$$\operatorname{sym}\left\{\Theta \otimes \begin{bmatrix} \bar{P}_1^{\mathrm{T}}(A + BK) & ? \\ ? & ? \end{bmatrix}\right\} < 0. \tag{6.138}$$

From (6.138), we have

$$\sin\left(\pi - \alpha\frac{\pi}{2})\right) \operatorname{sym}\left\{\bar{P}_1^{\mathrm{T}}(A + BK)\right\} < 0, \tag{6.139}$$

which gives that $\bar{P}_1$ is invertible. Left- and right-multiplying (6.137) and (6.138) by $\begin{bmatrix} \bar{P}_1^{-\mathrm{T}} & 0 \end{bmatrix}$ and $I_2 \otimes \begin{bmatrix} \bar{P}_1^{-\mathrm{T}} & 0 \end{bmatrix}$ and their transposes, respectively, we obtain

$$E \bar{P}_1^{-\mathrm{T}} = \bar{P}_1^{-\mathrm{T}} E^{\mathrm{T}} \geq 0, \tag{6.140}$$

$$\operatorname{sym}\left\{\Theta \otimes \begin{bmatrix} (A + BK)\bar{P}_1^{-1} & ? \\ ? & ? \end{bmatrix}\right\} < 0. \tag{6.141}$$

By setting $P_1 = \bar{P}_1^{-1}$ and $Y = K \bar{P}_1^{-1}$, conditions (6.130) and (6.131) yield.

Next we have to prove (6.132) and (6.133). Left- and right-multiplying conditions in (5.44) by $\bar{P}^{-\mathrm{T}}$ and $I_2 \otimes \bar{P}^{-\mathrm{T}}$ and their transposes, respectively, we get

$$\bar{E} \bar{P}^{-1} = \bar{P}^{-\mathrm{T}} \bar{E}^{\mathrm{T}} \geq 0, \tag{6.142}$$

$$\operatorname{sym}\left\{\Theta \otimes (\bar{A} \bar{P}^{-1})\right\} < 0. \tag{6.143}$$

Letting

$$\bar{P}^{-1} = \begin{bmatrix} ? & ? \\ ? & \bar{P}_2 \end{bmatrix}, \tag{6.144}$$

where $\bar{P}_2 \in \mathbb{R}^{n\times n}$ and the symbol ? denotes a block unused in the sequel. Substituting (6.144) into (6.142) and (6.143) gives

$$\begin{bmatrix} ? & ? \\ ? & E\bar{P}_2 \end{bmatrix} = \begin{bmatrix} ? & ? \\ ? & \bar{P}_2^{\mathrm{T}} E^{\mathrm{T}} \end{bmatrix} \geq 0, \tag{6.145}$$

$$\mathrm{sym}\left\{\Theta \otimes \begin{bmatrix} ? & ? \\ ? & (A - LC)\bar{P}_2 \end{bmatrix}\right\} < 0. \tag{6.146}$$

From (6.146), we deduce that $\bar{P}_2$ is invertible.

Left- and right-multiplying (6.145) and (6.146) by $\begin{bmatrix} 0 & \bar{P}_2^{-\mathrm{T}} \end{bmatrix}$ and $I_2 \otimes \begin{bmatrix} 0 & \bar{P}_2^{-\mathrm{T}} \end{bmatrix}$ and their transposes, respectively, we have

$$\bar{P}_2^{-\mathrm{T}} E = E^{\mathrm{T}} \bar{P}_2^{-1} \geq 0, \tag{6.147}$$

$$\mathrm{sym}\left\{\Theta \otimes (\bar{P}_2^{-\mathrm{T}}(A - LC))\right\} < 0 \tag{6.148}$$

By setting $P_2 = \bar{P}_2^{-\mathrm{T}}$ and $Z = \bar{P}_2^{-\mathrm{T}} L$, conditions (6.132) and (6.133) are verified. This completes the proof of necessity.

Sufficiency. Suppose (6.130)–(6.133) hold for some matrices P_1 and P_2. Without loss of generality, assume that both P_1 and P_2 are invertible. Otherwise, P_1 and P_2 can be made invertible by adding small perturbations such that they retain to be solutions to (6.130)–(6.133).

Let $\bar{P}_1 = P_1^{-1}$, $\bar{P}_2 = P_2^{\mathrm{T}}$, $K = YP_1^{-1}$, $L = P_2^{-1}Z$. Then, (6.130)–(6.133) are equivalent to

$$\bar{P}_1^{\mathrm{T}} E = E^{\mathrm{T}} \bar{P}_1 \geq 0, \tag{6.149}$$

$$\mathrm{sym}\left\{\Theta \otimes (\bar{P}_1^{\mathrm{T}}(A + BK))\right\} < 0 \tag{6.150}$$

$$\bar{P}_2^{\mathrm{T}} E = E^{\mathrm{T}} \bar{P}_2 \geq 0, \tag{6.151}$$

$$\mathrm{sym}\left\{\Theta \otimes (\bar{P}_2^{\mathrm{T}}(A - LC))\right\} < 0 \tag{6.152}$$

Hence, there exists a scalar $\epsilon > 0$ such that

$$\mathrm{sym}\left\{\begin{bmatrix} \Theta \otimes (\bar{P}_1^{\mathrm{T}}(A + BK)) & \Theta \otimes (-\bar{P}_1^{\mathrm{T}} BK) \\ 0 & \epsilon\Theta \otimes (\bar{P}_2^{\mathrm{T}}(A - LC)) \end{bmatrix}\right\} < 0 \tag{6.153}$$

which is, by appropriate row and column permutations, equivalent to

$$\operatorname{sym}\left\{\Theta \otimes \begin{bmatrix} \bar{P}_1^{\mathrm{T}}(A+BK) & -\bar{P}_1^{\mathrm{T}}BK \\ 0 & \epsilon \bar{P}_2^{\mathrm{T}}(A-LC) \end{bmatrix}\right\} < 0 \tag{6.154}$$

By taking $\bar{P} = \operatorname{diag}(\bar{P}_1, \epsilon \bar{P}_2)$, the inequality (6.154) is exactly

$$\operatorname{sym}\left\{\Theta \otimes (\bar{P}\bar{A})\right\} < 0, \tag{6.155}$$

and both inequalities (6.149) and (6.151) imply

$$\bar{E}^{\mathrm{T}}\bar{P} = \bar{P}^{\mathrm{T}}\bar{E} \geq 0. \tag{6.156}$$

According to Theorem 5.4.1, (6.155) and (6.156) lead to the conclusion that the system (6.129) is admissible. This ends the proof of the theorem.

Remark 6.4.1 Theorem 6.4.2 gives necessary and sufficient conditions for the admissibility of the system (6.129) in terms of $\mathcal{LMI}$s. However, it is noted that the conditions in (6.130) and (6.132) are non-strict $\mathcal{LMI}$s, which contain equality constraints; this may result in numerical problems when checking such non-strict $\mathcal{LMI}$ conditions since equality constraints are fragile and usually not satisfied perfectly. Therefore, strict $\mathcal{LMI}$ conditions are more desirable than non-strict ones from the numerical point of view.

Considering Remark 6.4.1, the following theorem gives strict $\mathcal{LMI}$s conditions for admissibility.

Theorem 6.4.3 *[10] There exist matrices K and L such that system* (6.129) *is admissible if and only if there exist matrices $Q_i > 0$, M_i, $i = 1, 2$, Y and Z, such that*

$$\operatorname{sym}\left\{\Theta \otimes (A(Q_1E^T + S_1M_1) + BY)\right\} < 0 \tag{6.157}$$

$$\operatorname{sym}\left\{\Theta \otimes ((E^TQ_2 + M_2S_2)A - ZC)\right\} < 0 \tag{6.158}$$

where Θ is as in (6.134) *and $S_1 \in \mathbb{R}^{n\times(n-r)}$, $S_2 \in \mathbb{R}^{((n-r)\times n}$ are any matrices such that* $\operatorname{rank}(S_1) = \operatorname{rank}(S_2) = n - r$ *and satisfy $ES_1 = 0$ and $S_2E = 0$. Moreover, the controller gain and the observer gain are given by*

$$K = Y(Q_1E^T + S_1M_1)^{-1}, \ L = (E^TQ_2 + M_2S_2)^{-1}Z. \tag{6.159}$$

Proof The proof is fully based on the result of Theorem 6.4.2. We aim to remove the positive semi-definite terms of (6.130) and (6.132). It is seen from the proof of Theorem 6.4.2 that both $Q_1E^{\mathrm{T}} + S_1M_1$ and $E^{\mathrm{T}}Q_2 + M_2S_2$ could be made invertible if (6.157)–(6.158) are feasible. As in the literature, see for instance [11, 12], the rest of the proof follows from the well-used technique for removing the terms of (6.130) and (6.132) by setting $P_1 = Q_1E^{\mathrm{T}} + S_1M_1$ and $P_2 = E^{\mathrm{T}}Q_2 + M_2S_2$ in (6.130)–(6.133). The details are thus omitted. □

Example 6.4.2 Consider singular system (5.6) with the following parameters:

$$E = \begin{bmatrix} 1 & 0 & 0.5 \\ 2 & 1 & 1 \\ 0 & 0 & 0 \end{bmatrix}, \; A = \begin{bmatrix} 2.4 & 0.2 & 1.2 \\ 4 & 1.5 & 2 \\ 0 & 0 & 0 \end{bmatrix},$$
$$C = \begin{bmatrix} 1 & 1 & 0 \\ 0 & 0 & 1 \end{bmatrix}, \quad B = \begin{bmatrix} 4 & 1 \\ 1 & 1 \\ 1 & 2 \end{bmatrix}, \; \alpha = 1.23 \tag{6.160}$$

This example is borrowed from [2] by adding an output equation. This system is tested open-loop not regular and unstable. In [2] is normalized, then stabilized by state feedback controller. Here the results of Theorem 6.4.3 are applied to this example.

We choose $S_1 = \begin{bmatrix} -1 \\ 0 \\ 2 \end{bmatrix}$ and $S_2 \begin{bmatrix} 0 & 0 & 1 \end{bmatrix}$, which are with full column rank and row respectively and satisfy $ES_1 = 0$ and $S_2E = 0$. Then, the $\mathcal{LMI}$s of Theorem 6.4.3 were found feasible and the solutions, under non-singularity requirement of both $Q_1E^{\mathrm{T}} + S_1M_1$ and $E^{\mathrm{T}}Q_2 + M_2S_2$, are as follow:

$$Q_1 = \begin{bmatrix} 4.8235 & -8.3169 & -2.0862 \\ -8.3169 & 23.0151 & -4.1585 \\ -2.0862 & -4.1585 & 7.9528 \end{bmatrix}, \; Q_2 = \begin{bmatrix} 7.8723 & -7.1597 & 0 \\ -7.1597 & 6.5139 & 0 \\ 0 & 0 & 8.9959 \end{bmatrix},$$

$$M_1 = \begin{bmatrix} 5.9980 & 1.5000 & 4.4980 \end{bmatrix}, \; M_2 = \begin{bmatrix} 0 \\ 5.4980 \\ 11.4959 \end{bmatrix},$$

$$Y = \begin{bmatrix} -4.2323 & 0.2111 & 0.8301 \\ 1.9191 & -0.8717 & -2.4784 \end{bmatrix}, \; Z = \begin{bmatrix} 10.3820 & 4.0847 \\ 10.8492 & 4.3388 \\ 3.8718 & 6.7947 \end{bmatrix}.$$

The gain matrices are

$$K = \begin{bmatrix} -18.5795 & -7.9148 & -9.1975 \\ 16.5704 & 6.9793 & 8.0097 \end{bmatrix}, \; L = \begin{bmatrix} 14.7458 & 817.1811 \\ 17.9701 & 898.5172 \\ -0.1148 & 0.4134 \end{bmatrix}.$$

Let's note that the controller gain matrices obtained in Theorem 6.4.1 depend on the regularity of both $Q_1E^{\mathrm{T}} + S_1M_1$ and $E^{\mathrm{T}}Q_2 + M_2S_2$, which is not always guaranteed. Therefore, an alternative result is provided independently of these constraints.

Theorem 6.4.4 *[13] The closed-loop system* (6.116) *with* $1 \le \alpha < 2$ *is admissible if and only if there exist matrices* P_0, $P_1 \in \mathbb{R}^{n\times n} > 0$, Q_0, $Q_1 \in \mathbb{R}^{(n-r)\times n}$ *and scalars* $\varepsilon_0, \varepsilon_1 > 0$ *such that*

$$\begin{bmatrix} \Gamma_1 & \Gamma_2 \\ \Gamma_2^T & \Gamma_1 \end{bmatrix} < 0, \tag{6.161}$$

$$\begin{bmatrix} \Sigma_1 & \Sigma_2 \\ \Sigma_2^T & \Sigma_1 \end{bmatrix} < 0, \tag{6.162}$$

where

$$\begin{aligned}
\Gamma_1 &= \left(A^T\Phi + \Phi^T A\right)\sin\left(\alpha\frac{\pi}{2}\right) - \frac{1}{\varepsilon_0}\sin^2\left(\alpha\frac{\pi}{2}\right)\Phi^T B B^T \Phi, \\
\Gamma_2 &= \left(A^T\Phi - \Phi^T A\right)\cos\left(\alpha\frac{\pi}{2}\right), \\
\Sigma_1 &= \left(A\Psi + \Psi^T A^T\right)\sin\left(\alpha\frac{\pi}{2}\right) - \frac{1}{\varepsilon_1}\sin^2\left(\alpha\frac{\pi}{2}\right)\Psi^T C^T C \Psi, \\
\Sigma_2 &= \left(A\Psi - \Psi^T A^T\right)\cos\left(\alpha\frac{\pi}{2}\right), \\
\Phi &= P_0 E + S_0 Q_0, \\
\Psi &= P_1 E^T + S_1 Q_1,
\end{aligned}$$

and S_0, S_1 are any matrices of full column rank satisfying $E^T S_0 = 0$ and $E S_1 = 0$ respectively.

The controller gain matrices K, L can be designed by

$$K = -\frac{1}{\varepsilon_0}\sin\left(\alpha\frac{\pi}{2}\right) B^T \Phi, \tag{6.163}$$

and

$$L = \frac{1}{\varepsilon_1}\sin\left(\alpha\frac{\pi}{2}\right) \Psi^T C^T. \tag{6.164}$$

Proof The proof is the same as Theorem 6.4.1, so it is omitted. □

6.5 Dynamic Output Feedback Controller

This section deals with the design of a dynamic output feedback controllers for continuous-time singular fractional-order systems such that the closed-loop system is admissible. In the following, we will attempt to address the problem separately for both cases $\alpha \in (0, 1)$ and $\alpha \in (1, 2)$, and present as many existing results in the literature as possible.

Consider a singular fractional-order system given by

$$\begin{aligned} E D^\alpha x(t) &= A x(t) + B u(t), \\ y(t) &= C x(t). \end{aligned} \tag{6.165}$$

where α is the time fractional derivative order, $x(t) \in \mathbb{R}^n$ is the pseudo-state, $u(t) \in \mathbb{R}^m$ is the control input and $y(t) \in \mathbb{R}^p$ is the output. $A \in \mathbb{R}^{n\times n}$, $B \in \mathbb{R}^{n\times m}$, $C \in \mathbb{R}^{p\times n}$ with rank $C = p$, and matrix $E \in \mathbb{R}^{n\times n}$ is singular with rank $E = r < n$.

1. **Case one: $1 < \alpha < 2$** The following concerns a stabilization result by dynamic output feedback without normalization of the system (6.165). To this end, consider an dynamic output feedback controller

$$\begin{aligned} ED^{\alpha}x_c(t) &= A_c x_c(t) + B_c y(t), \\ u(t) &= C_c x_c(t) + D_c y(t), \end{aligned} \tag{6.166}$$

where $x_c(t) \in \mathbb{R}^n$, hence the resulting closed-loop control system can be rewritten as

$$\bar{E}D^{\alpha}\bar{x}(t) = (\bar{A} + \bar{B}F\bar{C})\bar{x}(t), \tag{6.167}$$

where

$$\bar{x}(t) = \begin{bmatrix} x(t) \\ x_c(t) \end{bmatrix}, \ \bar{E} = \begin{bmatrix} E & 0 \\ 0 & E \end{bmatrix}, \ \bar{A} = \begin{bmatrix} A & 0 \\ 0 & 0 \end{bmatrix},$$

$$\bar{B} = \begin{bmatrix} B & 0 \\ 0 & I \end{bmatrix}, \ F = \begin{bmatrix} D_c & C_c \\ B_c & A_c \end{bmatrix}, \ \bar{C} = \begin{bmatrix} C & 0 \\ 0 & I \end{bmatrix}.$$

Let $\bar{M}, \bar{N} \in \mathbb{R}^{2n\times 2n}$ be two arbitrary chosen non-singular matrices which make $\bar{M}\bar{E}\bar{N}$, $\bar{M}\bar{A}\bar{N}$ satisfying (5.24). An SVD of the matrix $\bar{C}\bar{N}$ with full row rank is obtained as

$$\bar{C}\bar{N} = \bar{U}\left[\,\bar{S}\ 0\,\right]\bar{V}^{\mathrm{T}}.$$

The concatenation of two state vectors of the system and the process into a single vector resulted in an augmented system (6.167), then the problem can be interpreted as a static output feedback on the augmented system. Thus, by using the same method as Theorem 6.3.4, the equivalent controller gain matrix F can be easily determined with the help of the following theorem.

Theorem 6.5.1 *[7] System* (6.167) *is admissible if there exist matrices* $X_1 \in \mathbb{R}^{2r\times 2r}$, $X_2 \in \mathbb{R}^{2(n-r)\times 2r}$ *and non-singular matrix* $X_3 \in \mathbb{R}^{2(n-r)\times 2(n-r)}$, $Z \in \mathbb{R}^{(n+m)\times(n+p)}$ *such that*

$$X_1 > 0, \tag{6.168}$$

$$\mathrm{sym}\left\{\Theta \otimes (\bar{A}\bar{N}X\bar{M}^{-T} + \bar{B}Z\bar{C}\bar{N}\bar{M}^{-T})\right\} < 0, \tag{6.169}$$

$$X = \begin{bmatrix} X_1 & 0 \\ X_2 & X_3 \end{bmatrix}, \tag{6.170}$$

$$\bar{V}X\bar{V}^T = \begin{bmatrix} Q_1 & 0 \\ Q_2 & Q_3 \end{bmatrix}, \tag{6.171}$$

with $Q_1 \in \mathbb{R}^{(n+p)\times(n+p)}$, $Q_2 \in \mathbb{R}^{(n-p)\times(n+p)}$, $Q_3 \in \mathbb{R}^{(n-p)\times(n-p)}$. *The controller gain* F *is given by*

$$F = Z\bar{U}\bar{S}Q_1^{-1}\bar{S}^{-1}\bar{U}^{-1}. \tag{6.172}$$

Proof The proof is similar to that of Theorem 6.3.4 and thus omitted. □

Example 6.5.1 [7] By using controller (6.166) for system (6.71), conditions (6.168), (6.169) in Theorem 6.5.1 are feasible, which implies that system (6.167) is stabilized by dynamic output feedback controller (6.166). Then, the feasible solutions are given by

$$X_1 = \begin{bmatrix} 0.1054 & -0.0085 & 0 & 0 \\ -0.0085 & 0.0271 & 0 & 0 \\ 0 & 0 & 1.0446 & 0 \\ 0 & 0 & 0 & 1.0446 \end{bmatrix},$$

$$X_2 = \begin{bmatrix} -0.0380 & 0.0085 & 0 & 0 \\ 0 & & 0 & 0 & 0 \end{bmatrix},$$

$$X_3 = \begin{bmatrix} 0.0674 & 0 \\ 0 & -1.0446 \end{bmatrix},$$

$$Z = \begin{bmatrix} -0.1672 & 0 & 0 & 0 \\ 0 & -0.2462 & 0 & 0 \\ 0 & 0.0492 & -0.1477 & 0 \\ 0 & 0 & 0 & -0.7386 \end{bmatrix}.$$

From (6.172), the gain matrix F is calculated as

$$F = \begin{bmatrix} -1.6274 & 0 & -0.7211 & 0 \\ 0 & -0.2357 & 0 & 0 \\ -0.3186 & 0.0471 & -5.5967 & 0 \\ 0 & 0 & 0 & -0.7071 \end{bmatrix}.$$

For given initial conditions

$$x(0) = \begin{bmatrix} 3.2 & 0 & -1 \end{bmatrix}^{\mathrm{T}},\ x_c(0) = \begin{bmatrix} 0 & 0 & 0 \end{bmatrix}^{\mathrm{T}},$$

the state responses of closed-loop system (6.167) and the corresponding dynamic output feedback (6.166) in Example 6.5.1 are shown in Figs. 6.18 and 6.19, respectively. The admissibility of the closed-loop (6.167) can be seen clearly from these two figures. Therefore, the stabilisation of system (6.167) is achieved by the dynamic output feedback technique.

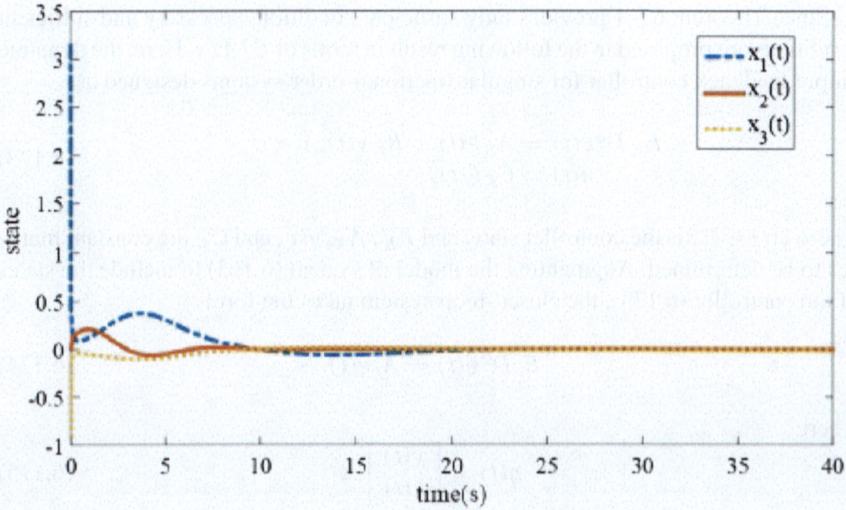

Fig. 6.18 State responses of the closed-loop system (6.167) in Example 6.5.1

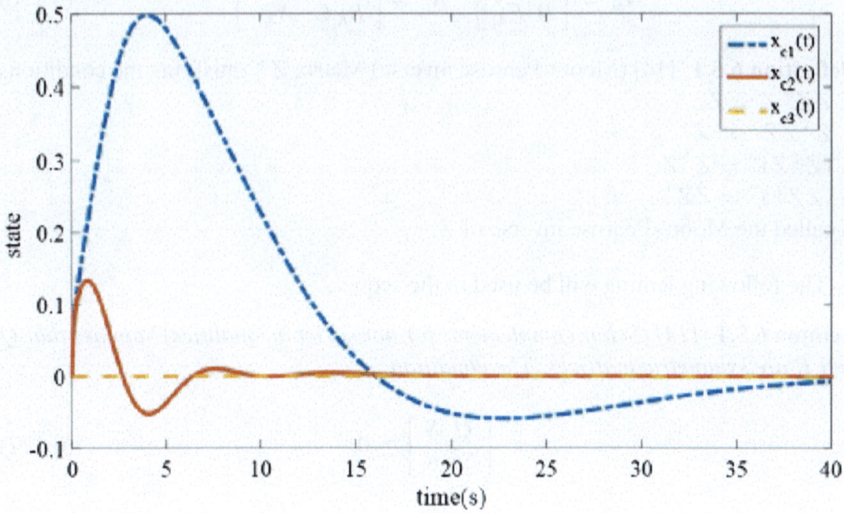

Fig. 6.19 State responses of the corresponding dynamic output feedback (6.166) in Example 6.5.1

Since Theorem 6.5.1 provides only sufficient condition, necessary and sufficient conditions are proposed in the following result in terms of $\mathcal{LMIs}$. Here, the dynamic output feedback controller for singular fractional-order systems designed as

$$\begin{aligned} E_K D^\alpha \xi(t) &= A_K \xi(t) + B_K y(t),\ 1 < \alpha < 2, \\ u(t) &= C_K \xi(t), \end{aligned} \tag{6.173}$$

where $\xi(t) \in \mathbb{R}^n$ is the controller state, and E_K, A_K, B_K, and C_K are constant matrices to be determined. Augmenting the model of system (6.165) to include the states of the controller (6.173), the closed-loop system takes the form

$$E_c D^\alpha \eta(t) = A_c \eta(t), \tag{6.174}$$

where

$$\eta(t) = \begin{bmatrix} x(t) \\ \xi(t) \end{bmatrix}^{\mathrm{T}}, \tag{6.175}$$

and

$$E_c = \begin{bmatrix} E & 0 \\ 0 & E_K \end{bmatrix},\ A_c = \begin{bmatrix} A & BC_K \\ B_K C & A_K \end{bmatrix}. \tag{6.176}$$

Definition 6.5.1 [14] (Moore-Penrose inverse) Matrix Z^+ satisfying the conditions
1. $ZZ^+Z = Z$
2. $Z^+ZZ^+ = Z^+$
3. $(Z^+Z)^{\mathrm{T}} = Z^+Z$
4. $(ZZ^+)^{\mathrm{T}} = ZZ^+$

is called the Moore-Penrose inverse of Z.

The following lemma will be used in the sequel.

Lemma 6.5.1 *[14] (Schur complements for non-strict inequalities) Suppose that Q and R are symmetric matrices. The condition*

$$\begin{bmatrix} Q & S \\ S^T & R \end{bmatrix} \geq 0 \tag{6.177}$$

is equivalent to

$$R \geq 0,\ Q - SR^+S^T \geq 0,\ S(I - RR^+) = 0, \tag{6.178}$$

where R^+ denotes the Moore-Penrose inverse of R.

The desired controller gains can be easily determined using the following theorem.

Theorem 6.5.2 *[15] Designing the controller* (6.173) *for the system* (6.165)*, the resulting closed-loop control system* (6.174) *is admissible if and only if there exist matrices* X, Y, Φ, *and* Ψ *of appropriate dimensions such that*

$$\begin{bmatrix} E^T & 0 \\ 0 & E \end{bmatrix}\begin{bmatrix} X & I \\ I & Y \end{bmatrix} = \begin{bmatrix} X^T & I \\ I & Y^T \end{bmatrix}\begin{bmatrix} E & 0 \\ 0 & E^T \end{bmatrix} \geq 0, \tag{6.179}$$

$$\begin{bmatrix} \text{Her}\left\{e^{j\theta}(A^TX + C^T\Phi^T\right\} & (e^{j\theta} - e^{-j\theta})A^T \\ -(e^{j\theta} - e^{-j\theta})A & \text{Her}\left\{e^{-j\theta}(AY - B\Psi)\right\} \end{bmatrix} < 0. \tag{6.180}$$

Furthermore, if the conditions (6.179) *and* (6.180) *hold, then there exist such matrices* X, Y, Φ *and* Ψ *satisfying* (6.179) *and* (6.180) *that both* Y *and* $X - Y^{-1}$ *are nonsingular and the parameters of the stabilizing dynamic output feedback controller* (6.173) *can be chosen as*

$$A_K = \Lambda(A^TY^{-1} + X^TA - X^TB\Psi Y^{-1} + \Phi C), \tag{6.181}$$

$$E_K = E, \tag{6.182}$$

$$B_K = -\Lambda\Phi, \tag{6.183}$$

$$C_K = -\Psi Y^{-1}, \tag{6.184}$$

where

$$\Lambda = (X - Y^{-1})^{-T}.$$

Proof (Sufficiency) Suppose that the conditions (6.179) and (6.180) are satisfied for some matrices X, Y, Φ, Ψ and both Y and $X - Y^{-1}$ are nonsingular. Note that the condition (6.179) implies

$$E^{\mathrm{T}}X = X^{\mathrm{T}}E \geq 0, \tag{6.185}$$

$$EY = Y^{\mathrm{T}}E^{\mathrm{T}} \geq 0. \tag{6.186}$$

If Y is singular, then we choose any nonsingular matrix $\tilde{Y}$satisfying $E\tilde{Y} = \tilde{Y}^{\mathrm{T}}E^{\mathrm{T}} \geq 0$. Let $\theta_1 \in (0, 1)$ not be an eigenvalue of $-Y^{-1}Y$ and be small enough to replace Y by $Y + \theta_1\tilde{Y}$, so that the inequalities (6.179) and (6.180) still hold. It is easy to verify that $Y + \theta_1\tilde{Y}$ is nonsingular. Indeed, suppose that $Y + \theta_1\tilde{Y}$ is singular, then there exists $x \neq 0$ to replace $(Y + \theta_1\tilde{Y})x = 0$. This implies $-\tilde{Y}^{-1}Yx = \theta_1 x$, which contradicts the fact that θ_1 is not an eigenvalue of $-\tilde{Y}^{-1}Y$. Let us prove that under (6.181)–(6.184) the closed-loop system (6.174) is admissible, which means that there exists $P_c > 0$ such that $E_c^{\mathrm{T}}P_c = P_c^{\mathrm{T}}E_c \geq 0$ and $\text{Her}\left\{e^{j\theta}A_c^{\mathrm{T}}P_c\right\} < 0$. Taking into account (6.186) and applying Lemma 6.5.1 to (6.179), we have

$$\begin{aligned} E^{\mathrm{T}}X - E^{\mathrm{T}}(EY)^{+}E &= E^{\mathrm{T}}X - Y^{-\mathrm{T}}Y^{\mathrm{T}}E^{\mathrm{T}}(EY)^{+}EYY^{-1} \geq 0, \\ &= E^{\mathrm{T}}X - Y^{-\mathrm{T}}(EY)^{\mathrm{T}}(EY)^{+}(EY)Y^{-1} \geq 0, \\ &= E^{\mathrm{T}}X - Y^{-\mathrm{T}}(EY)(EY)^{+}(EY)Y^{-1} \geq 0, \\ &= E^{\mathrm{T}}X - Y^{-\mathrm{T}}EYY^{-1} \geq 0, \\ &= E^{\mathrm{T}}X - Y^{-\mathrm{T}}E \geq 0, \\ &= X^{\mathrm{T}}E - Y^{-\mathrm{T}}E \geq 0, \\ &= (X^{\mathrm{T}} - Y^{-\mathrm{T}})E \geq 0, \\ &= (X - Y^{-1})^{\mathrm{T}}E \geq 0. \end{aligned}$$

Thus,

$$(X - Y^{-1})^{\mathrm{T}}E = E^{\mathrm{T}}(X - Y^{-1}) \geq 0. \tag{6.187}$$

If $X - Y^{-1}$ is singular, we choose any nonsingular matrix $\tilde{X}$ satisfying

$$E^{\mathrm{T}}\tilde{X} = \tilde{X}^{\mathrm{T}}E. \tag{6.188}$$

Let $\theta_2 \in (0, 1)$ not be an eigenvalue of $(X - Y^{-1})\tilde{X}^{-1}$ and be small enough to replace X by $X + \theta_2\tilde{X}$, so that the inequalities (6.179) and (6.180) still hold.Then, it is easy to see that $(X + \theta_2\tilde{X}) - Y^{-1}$ is nonsingular and satisfies (6.187). Considering the closed-loop system (6.174) with the parameters (6.181)–(6.184),

$$P_c = \begin{bmatrix} X & Y^{-1} - X \\ Y^{-1} - X & X - Y^{-1} \end{bmatrix}. \tag{6.189}$$

Let us prove that $E_c^{\mathrm{T}}P_c = P_c^{\mathrm{T}}E_c \geq 0$.

Taking into account (6.185) and (6.187), it can be deduced that

$$E_c^{\mathrm{T}}P_c = P_c^{\mathrm{T}}E_c, \tag{6.190}$$

and

$$\begin{aligned} &E^{\mathrm{T}}X - E^{\mathrm{T}}(Y^{-1} - X)(E^{\mathrm{T}}(X - Y^{-1}))^{+}E^{\mathrm{T}}(Y^{-1} - X) \\ &\quad = E^{\mathrm{T}}X - E^{\mathrm{T}}(X - Y^{-1})(E^{\mathrm{T}}(X - Y^{-1}))^{+}E^{\mathrm{T}}(X - Y^{-1}), \\ &\quad = E^{\mathrm{T}}X - E^{\mathrm{T}}(X - Y^{-1}), \\ &\quad = E^{\mathrm{T}}Y^{-1} \geq 0. \end{aligned} \tag{6.191}$$

Indeed, by pre- and post multiplying (6.186) by $Y^{-\mathrm{T}}$ and Y^{-1}, respectively, we get $Y^{-\mathrm{T}}E = E^{\mathrm{T}}Y^{-1} \geq 0$. By Definition 6.5.1 and (6.187), we obtain

$$
\begin{aligned}
&E^{\mathrm{T}}(Y^{-1}-X)\left[I-E^{\mathrm{T}}(X-Y^{-1})(E^{\mathrm{T}}(X-Y^{-1}))^{+}\right] \\
&\quad = E^{\mathrm{T}}(Y^{-1}-X)\left[I-(E^{\mathrm{T}}(X-Y^{-1})(E^{\mathrm{T}}(X-Y^{-1}))^{+})^{\mathrm{T}}\right], \\
&\quad = E^{\mathrm{T}}(Y^{-1}-X)-E^{\mathrm{T}}(Y^{-1}-X)((E^{\mathrm{T}}(X-Y^{-1}))^{+})^{\mathrm{T}}(E^{\mathrm{T}}(X-Y^{-1}))^{\mathrm{T}}, \\
&\quad = E^{\mathrm{T}}(Y^{-1}-X)+\left[E^{\mathrm{T}}(X-Y^{-1})\ (E^{\mathrm{T}}(X-Y^{-1}))^{+}E^{\mathrm{T}}(X-Y^{-1})\right]^{\mathrm{T}}, \\
&\quad = E^{\mathrm{T}}(Y^{-1}-X)+E^{\mathrm{T}}(X-Y^{-1})=0.
\end{aligned}
\tag{6.192}
$$

Taking into account (6.187), (6.191), and (6.192) and using Lemma 6.5.1, the condition

$$
E_c P_c = P_c^{\mathrm{T}} E_c^{\mathrm{T}} \geq 0 \tag{6.193}
$$

follows. Next, we have to prove $\mathrm{Her}\left\{e^{j\theta} A_c^{\mathrm{T}} P_c\right\} < 0$.

Pre- and post-multiplying (6.180) by $\begin{bmatrix} I & 0 \\ -I & Y^{-\mathrm{T}} \end{bmatrix}$ and $\begin{bmatrix} I & -I \\ 0 & Y^{-1} \end{bmatrix}$, respectively, we obtain that (6.180) is equivalent to

$$
\begin{bmatrix} \Sigma_{11} & \Sigma_{12} \\ \Sigma_{12}^{*} & \Sigma_{22} \end{bmatrix} < 0, \tag{6.194}
$$

where

$$
\begin{aligned}
\Sigma_{11} &= \mathrm{Her}\left\{e^{j\theta}(A^{T}X+C^{\mathrm{T}}\Phi^{\mathrm{T}})\right\}, \\
\Sigma_{12} &= -\,\mathrm{Her}\left\{e^{j\theta}(A^{T}X+C^{\mathrm{T}}\Phi^{\mathrm{T}})\right\}+(e^{j\theta}-e^{-j\theta})A^{\mathrm{T}}Y^{-1}, \\
\Sigma_{22} &= \mathrm{Her}\left\{e^{j\theta}(A^{\mathrm{T}}X+C^{\mathrm{T}}\Phi^{\mathrm{T}}+Y^{-\mathrm{T}}A-Y^{-\mathrm{T}}\Psi^{\mathrm{T}}B^{\mathrm{T}}Y^{-1})\right\}.
\end{aligned}
$$

That is,

$$
\mathrm{Her}\left\{e^{j\theta} A_c^{\mathrm{T}} P_c\right\} < 0. \tag{6.195}
$$

Therefore, by (6.193), (6.195) and Lemma 5.4.2, the closed-loop system (6.174) is admissible. This completes the proof of sufficiency.

(Necessity) Assume that there exists a dynamic output feedback controller in (6.181)–(6.184) such that the closed-loop system (6.174) is admissible. Then, it follows that there exists a real matrix P_c such that

$$
E_c^{\mathrm{T}} P_c = P_c^{\mathrm{T}} E_c \geq 0, \tag{6.196}
$$

$$
\mathrm{Her}\left\{e^{j\theta} A_c^{\mathrm{T}} P_c\right\} < 0. \tag{6.197}
$$

The matrix P_c can be written as

$$
P_c = \begin{bmatrix} P_1 & P_2 \\ P_3 & P_4 \end{bmatrix},
$$

where the partition is compatible with that for A_c and E_c in (6.176). Then, it follows from (6.196) that

$$E^{\mathrm{T}} P_1 = P_1^{\mathrm{T}} E \geq 0, \tag{6.198}$$
$$E^{\mathrm{T}} P_2 = P_3^{\mathrm{T}} E_K, \tag{6.199}$$
$$E_K^{\mathrm{T}} P_4 = P_4^{\mathrm{T}} E_K \geq 0. \tag{6.200}$$

Without loss of generality, one can assume P_4 is nonsingular. If not, by (6.200), one can choose any nonsingular matrix $\tilde{P}_4$ satisfying

$$E_K^{\mathrm{T}} \tilde{P}_4 = \tilde{P}_4^{\mathrm{T}} E_K \geq 0.$$

Let $\alpha_1 \in (0, 1)$ not be an eigenvalue of $-P_4 \tilde{P}_4^{-1}$ and be small enough to replace P_c by

$$P_c + \alpha_1 \begin{bmatrix} 0 & 0 \\ 0 & \tilde{P}_4 \end{bmatrix},$$

so that the inequality in (6.197) still holds. Then, it is easy to see that $P_4 + \alpha_1 \tilde{P}_4$ is nonsingular. Similarly, without loss of generality, one can assume $P_1 - P_2 P_4^{-1} P_3$ is nonsingular. Now, by the block 1-1 of (6.197), we have

$$\operatorname{Her}\left\{e^{j\theta}(A^T P_1 + C^T B_K^{\mathrm{T}} P_3)\right\} < 0. \tag{6.201}$$

Set

$$X = P_1, \ \Phi = P_3^{\mathrm{T}} B_K. \tag{6.202}$$

Then, (6.201) implies

$$\operatorname{Her}\left\{e^{j\theta}(A^T P_1 + C^{\mathrm{T}} \Phi^{\mathrm{T}}\right\} < 0, \tag{6.203}$$

and (6.198) yields

$$E^{\mathrm{T}} X = X^{\mathrm{T}} E \geq 0. \tag{6.204}$$

Now, let

$$N = \begin{bmatrix} I & 0 \\ -P_4^{-1} P_3 & I \end{bmatrix}. \tag{6.205}$$

If (6.197) is pre- and post-multiplied by N^T and N, respectively, considering the block 1-1, one obtains

$$\operatorname{Her}\left\{e^{j\theta} A^{\mathrm{T}}(P_1 - P_2 P_4^{-1} P_3)\right\} - \operatorname{Her}\left\{e^{j\theta} P_3^{\mathrm{T}} P_4^{-\mathrm{T}} C_K^{\mathrm{T}} B^{\mathrm{T}}(P_1 - P_2 P_4^{-1} P_3)\right\} < 0. \tag{6.206}$$

If (6.196) is pre- and post-multiplied by N^T and N, respectively, considering the block 1-1, one obtains

$$E^{\mathrm{T}}(P_1 - P_2 P_4^{-1} P_3) = (P_1 - P_2 P_4^{-1} P_3)^{\mathrm{T}} E \geq 0. \tag{6.207}$$

By setting

$$\tilde{Y} = (P_1 - P_2 P_4^{-1} P_3)^{-1}, \tag{6.208}$$

the relation (6.207) becomes

$$E^{\mathrm{T}} \tilde{Y}^{-1} = \tilde{Y}^{-\mathrm{T}} E \geq 0. \tag{6.209}$$

If (6.209) is pre- and post-multiplied by $\tilde{Y}^{\mathrm{T}}$ and $\tilde{Y}$, respectively, the inequality (6.209) can be rewritten as

$$\tilde{Y}^{\mathrm{T}} E^{\mathrm{T}} = E \tilde{Y} \geq 0. \tag{6.210}$$

Pre- and post-multiplying (6.206) by $\tilde{Y}^{\mathrm{T}}$ and $\tilde{Y}$, respectively, and setting

$$\tilde{\Psi} = C_K P_4^{-1} P_3 \tilde{Y}, \tag{6.211}$$

the inequality (6.206) can be rewritten as

$$\mathrm{Her}\left\{e^{j\theta} \tilde{Y}^{\mathrm{T}} A^{\mathrm{T}}\right\} - \mathrm{Her}\left\{e^{j\theta} \tilde{\Psi}^{\mathrm{T}} B\mathrm{T}\right\} < 0. \tag{6.212}$$

By setting Ξ_{11} and Ξ_{22} to the left-hand sides of (6.203) and (6.212), respectively, we have

$$\Xi_{11} < 0, \ \Xi_{22} < 0. \tag{6.213}$$

Using properties of negative definite matrices, the inequality (6.203) implies

$$-(e^{j\theta} - e^{-j\theta}) A \Xi_{11}^{-1} (e^{j\theta} - e^{-j\theta}) A^{\mathrm{T}} \leq 0. \tag{6.214}$$

Therefore, there exists $\epsilon_1 > 0$ such that

$$\epsilon_1 \Xi_{22} - \left[-(e^{j\theta} - e^{-j\theta}) A \Xi_{11}^{-1} (e^{j\theta} - e^{-j\theta}) A^{\mathrm{T}}\right] < 0, \tag{6.215}$$

which is equivalent to

$$\epsilon_1 \Xi_{22} < -(e^{j\theta} - e^{-j\theta}) A \Xi_{11}^{-1} (e^{j\theta} - e^{-j\theta}) A^{\mathrm{T}}. \tag{6.216}$$

Taking in to account (6.204) and (6.209), there exists $\epsilon_2 > 0$ such that

$$\epsilon_2 X^{\mathrm{T}} E \geq \tilde{Y}^{-\mathrm{T}} E. \tag{6.217}$$

For

$$\epsilon = \max(\epsilon_1, \epsilon_2), \tag{6.218}$$

we have

$$\epsilon\,\Xi_{22} \leq \epsilon_1\,\Xi_{22}, \tag{6.219}$$

and, using (6.216), we obtain

$$\epsilon\,\Xi_{22} - \left[-(e^{j\theta} - e^{-j\theta})A\,\Xi_{11}^{-1}(e^{j\theta} - e^{-j\theta})A^{\mathrm{T}}\right] < 0. \tag{6.220}$$

By Shur complement, (6.220) is equivalent to

$$\begin{bmatrix} \Xi_{11} & (e^{j\theta} - e^{-j\theta})A^{\mathrm{T}} \\ -(e^{j\theta} - e^{-j\theta})A & \epsilon\,\Xi_{22} \end{bmatrix} < 0. \tag{6.221}$$

Defining

$$Y = \epsilon\tilde{Y},\ \Psi = \epsilon\tilde{\Psi}, \tag{6.222}$$

the condition (6.180) is satisfied.

The equality (6.210) yields

$$EY = Y^{\mathrm{T}}E^{\mathrm{T}}. \tag{6.223}$$

Taking into account (6.204), (6.217) and (6.218) and applying Definition 6.5.1, we have

$$\begin{aligned} E^{\mathrm{T}}X - E^{\mathrm{T}}(EY)^{+}E &= E^{\mathrm{T}}X - E^{\mathrm{T}}(\epsilon E\tilde{Y})^{+}E \\ &= E^{\mathrm{T}}X - \frac{1}{\epsilon}\tilde{Y}^{-\mathrm{T}}\epsilon\tilde{Y}^{\mathrm{T}}E^{\mathrm{T}}(\epsilon E\tilde{Y})^{+}E\epsilon\tilde{Y}\frac{1}{\epsilon}\tilde{Y}^{-1} \\ &= E^{\mathrm{T}}X - \frac{1}{\epsilon}\tilde{Y}^{-\mathrm{T}}(\epsilon E\tilde{Y})(\epsilon E\tilde{Y})^{+}(\epsilon E\tilde{Y})\frac{1}{\epsilon}\tilde{Y}^{-1} \\ &= E^{\mathrm{T}}X - \frac{1}{\epsilon}\tilde{Y}^{-\mathrm{T}}(\epsilon E\tilde{Y})\frac{1}{\epsilon}\tilde{Y}^{-1} \\ &= E^{\mathrm{T}}X - \frac{1}{\epsilon}\tilde{Y}^{-\mathrm{T}}E \geq 0 \end{aligned} \tag{6.224}$$

and

$$\begin{aligned} E^{\mathrm{T}}\left(I - (EY)(EY)^{+}\right) &= E^{\mathrm{T}} - E^{\mathrm{T}}\left[(EY)(EY)^{+}\right]^{\mathrm{T}} \\ &= E^{\mathrm{T}} - E^{\mathrm{T}}((EY)^{+})^{\mathrm{T}}(EY)^{\mathrm{T}} \\ &= E^{\mathrm{T}} - Y^{-\mathrm{T}}Y^{\mathrm{T}}E^{\mathrm{T}}((EY)^{+})^{\mathrm{T}}(EY)^{\mathrm{T}} \\ &= E^{\mathrm{T}} - Y^{-\mathrm{T}}(EY)^{\mathrm{T}}((EY)^{+})^{\mathrm{T}}(EY)^{\mathrm{T}} \\ &= E^{\mathrm{T}} - Y^{-\mathrm{T}}\left[(EY)(EY)^{+}(EY)\right]^{\mathrm{T}} \\ &= E^{\mathrm{T}} - Y^{-\mathrm{T}}(EY)^{\mathrm{T}} \\ &= E^{\mathrm{T}} - Y^{-\mathrm{T}}Y^{\mathrm{T}}E^{\mathrm{T}} \\ &= 0. \end{aligned} \tag{6.225}$$

Taking into account (6.204), (6.223), (6.224), and (6.225) and applying Lemma 6.5.1, the condition (6.179) is satisfied. The necessity is proven. □

Remark 6.5.1 (a) Note that if $\alpha = 1$, we recover necessary and sufficient conditions for the existence of a dynamic output feedback controller for integer singular systems. Indeed, the conditions (6.179) and (6.180) are equivalent to conditions (3.17)–(3.19) in Theorem 3.2 in [11].

(b) The proposed result of Theorem 15 in [16] yields only sufficient conditions for admissibility of SFOS with a dynamic output feedback controller, whereas Theorem 6.5.2 is more insightful, since it provides both necessary and sufficient conditions in terms of $\mathcal{LMIs}$.

Example 6.5.2 [15] Consider a singular fractional-order linear time invariant system described in (6.165) with fractional-order derivative $\alpha = 1.2$ as follows:

$$E = \begin{bmatrix} 1 & 1 & 1 \\ 1 & 1 & 1 \\ 0 & 0 & 1 \end{bmatrix}, \; A = \begin{bmatrix} 0.1 & -0.3 & -2 \\ 0.5 & -3 & -1 \\ 0.2 & 0.4 & -0.1 \end{bmatrix},$$

$$B = \begin{bmatrix} -1 & 1 \\ 1 & -1 \\ 1 & 1 \end{bmatrix}, \; C = \begin{bmatrix} 1 & 0 & -1 \\ 0 & 1 & 0 \end{bmatrix}. \tag{6.226}$$

It can be verified that the (E, A) is regular and impulse-free but unstable since its finite modes are -1.0157 and 0.5447, so they are not all located in the stability region. Indeed, the time histories of the system (6.226) are shown in Fig. 6.20, which demonstrate that it is not asymptotically stable and its states are not convergent. Therefore, the pair (E, A) is not admissible.

Note that this example is borrowed from [3], where a derivative controller is used to make its resultant system normalized and then an additional dynamic output feedback is used to achieve the stabilization. However, the dynamic output feedback stabilization method of [3] provides only a sufficient condition. By applying result of Theorem 6.5.2, this system is stabilized by a dynamic output feedback control input, where the condition is necessary and sufficient and without normalization. Solving the $\mathcal{LMIs}$ (6.179)–(6.180) by using Matlab toolbox with the help of Yalmip, a set of solutions can be obtained as follows:

$$X = \begin{bmatrix} 67.3157 & 81.7229 & -57.5542 \\ 81.7229 & 440.1209 & 226.8670 \\ -57.5542 & 226.8670 & 400.5739 \end{bmatrix}, \; Y = \begin{bmatrix} 254.3216 & 74.1253 & 104.9356 \\ 74.1253 & 171.8979 & 111.7615 \\ 104.9356 & 111.7615 & 304.2964 \end{bmatrix},$$

$$\Phi = 10^3 \begin{bmatrix} -0.1486 & 0.3045 \\ -0.3004 & 1.0879 \\ -0.2397 & 0.5071 \end{bmatrix}, \quad \Psi = \begin{bmatrix} -156.5427 & -148.0563 & -186.4033 \\ 227.9592 & 221.0982 & 400.5955 \end{bmatrix}.$$

According to Theorem 6.5.2, we conclude that the singular fractional-order system (6.226) is asymptotically stabilizable using the obtained dynamic output feedback controllers parameters:

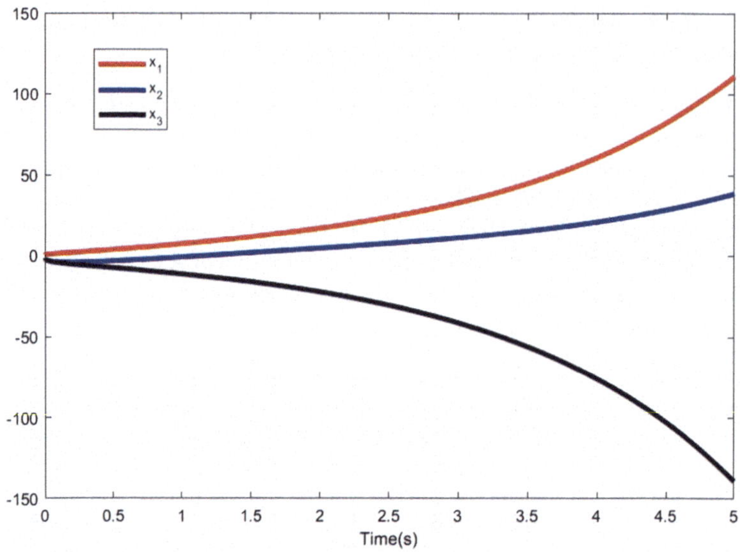

Fig. 6.20 Time histories of system (6.226) with $u(t) = 0$

$$E_K = E,\ A_K = \begin{bmatrix} -7.7473 & 3.5991 & 7.8405 \\ 3.2768 & -1.2339 & -1.7703 \\ -2.6162 & 1.9710 & 1.9996 \end{bmatrix},$$

$$B_K = \begin{bmatrix} -6.5607 & 4.9395 \\ 2.6860 & -1.3724 \\ 0.1110 & 0.8696 \end{bmatrix},\ C_K = \begin{bmatrix} 0.3388 & 0.5162 & 0.3062 \\ -0.3344 & -0.4744 & -1.0269 \end{bmatrix}. \tag{6.227}$$

The time histories of the resultant closed-loop SFOS in the form (6.174) via the obtained controllers are shown in Fig. 6.21 where all states asymptotically converge to zero.

Therefore, by Theorem 6.5.2, we conclude that the closed-loop singular fractional-order system (6.226)–(6.227) is admissible.

Example 6.5.3 [15] Consider a singular fractional-order linear time invariant system (6.165) with fractional-order derivative $\alpha = 1.25$ as follows:

$$E = \begin{bmatrix} 1 & 1 & 1 \\ 1 & 1 & 1 \\ 0 & 0 & 0 \end{bmatrix}, A = \begin{bmatrix} 2 & 0 & 1.2 \\ 0 & 0 & 1.5 \\ 0 & -3 & -2 \end{bmatrix},$$

$$B = \begin{bmatrix} -1 & 1 \\ 1 & -1 \\ 0 & 0 \end{bmatrix}, C = \begin{bmatrix} 1 & 0 & 1.3 \\ -2 & 0 & 2 \end{bmatrix}. \tag{6.228}$$

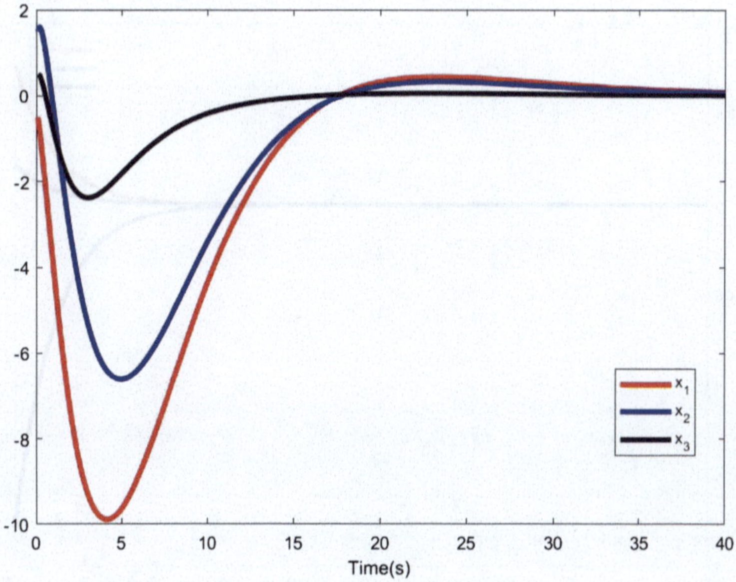

Fig. 6.21 Time histories of closed-loop system (6.226)–(6.227)

Then, it can be verified that the (E, A) is regular and impulse-free but unstable since its finite mode is $\frac{90}{29}$, so it is not located in the stability region. Indeed, the time histories of the system (6.228) are shown in Fig. 6.22, which demonstrate that it is not asymptotically stable and its states are not convergent. Therefore, the pair (E, A) is not admissible. Note that the dynamic output feedback control design method for SFOS with fractional-order derivative α belonging to $1 < \alpha < 2$ in [16] and [3] provides only sufficient conditions. In addition, the method of [3] requires the normalization of the system, imposing the constraint rank $\left(\begin{bmatrix} E & B \end{bmatrix}\right) = n$. According to Lemma 6.2.1, the system cannot be normalized since rank $\left(\begin{bmatrix} E & B \end{bmatrix}\right) = 2 \neq 3$, so the $\mathcal{LMI}$ (8) of Theorem 1 in [3] cannot be verified. Thus, the approach proposed in [3] is not always feasible.

In contrast to [3], the results presented in this paper are not conservative. Note that by introducing free variables and there by more relaxation, the proposed new formulations are more suitable for the design problem. Solving the $\mathcal{LMI}s$ (6.179)–(6.180) in Theorem 6.5.2 by using Matlab toolbox with the help of Yalmip, a set of feasible solutions can be obtained as follows:

$$X = 10^5 \begin{bmatrix} 6.9052 & 0.1671 & 1.2020 \\ 0.1671 & 7.2748 & 1.8989 \\ 1.2020 & 1.8989 & 2.1414 \end{bmatrix}, \; Y = 10^5 \begin{bmatrix} 3.9554 & 2.6386 & -4.3233 \\ 2.6386 & 5.8612 & -3.0950 \\ -4.3233 & -3.0950 & 9.1277 \end{bmatrix},$$

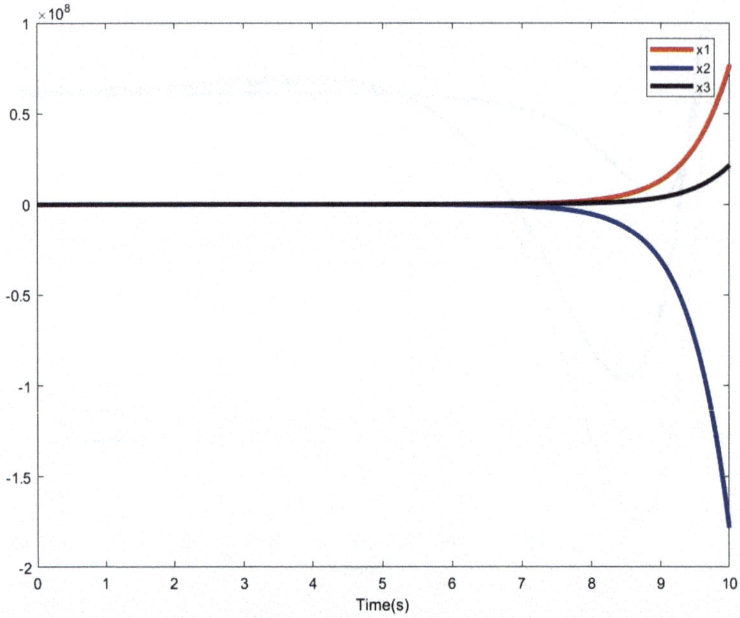

Fig. 6.22 Time histories of system (6.228) with $u(t) = 0$

$$\Phi = 10^6 \begin{bmatrix} -1.0658 & 0.3575 \\ -0.0251 & -0.1220 \\ -0.3385 & -0.0174 \end{bmatrix}, \quad \Psi = 10^5 \begin{bmatrix} -2.7523 & 0.4886 & 0.2330 \\ 2.7523 & -0.4886 & -0.2330 \end{bmatrix}.$$

According to Theorem 6.5.2, we conclude that the singular fractional-order system (6.228) is asymptotically stabilizable using the obtained dynamic output feedback controllers parameters:

$$E_K = E, \; A_K = \begin{bmatrix} -3.9457 & 1.0715 & -0.6351 \\ 3.8969 & -1.0715 & 2.8677 \\ -0.4203 & -3.0000 & -4.0071 \end{bmatrix},$$

$$B_K = \begin{bmatrix} 1.3664 & -0.5754 \\ 0.2724 & 0.0980 \\ 1.0554 & 0.3176 \end{bmatrix}, \; C_K = \begin{bmatrix} 1.7143 & -0.5358 & 0.6048 \\ -1.7143 & 0.5358 & -0.6048 \end{bmatrix}. \tag{6.229}$$

The time histories of the resultant closed-loop SFOS in the form (6.174) via the obtained controllers are shown in Fig. 6.23, where all states asymptotically converge to zero.

Therefore, by Theorem 6.5.2, we conclude that the closed-loop singular fractional-order system (6.228)–(6.229) is admissible.

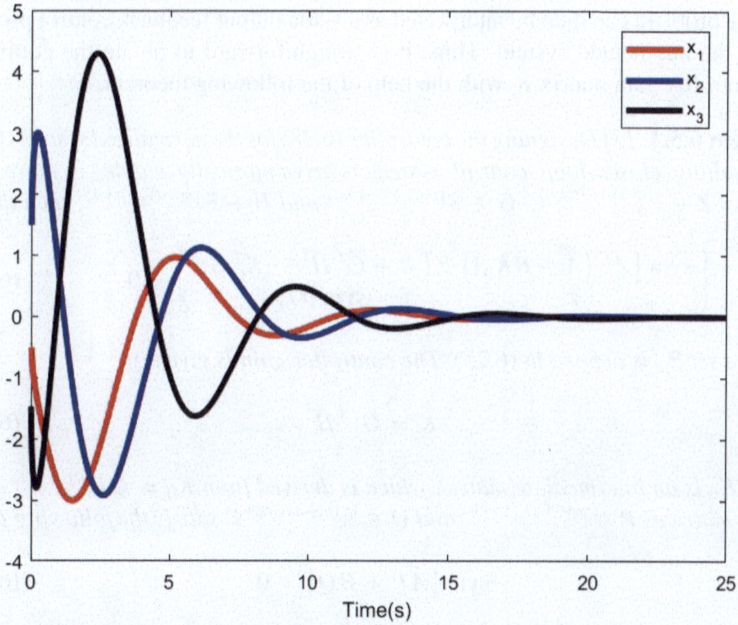

Fig. 6.23 Time histories of closed-loop system (6.228)–(6.229)

2. **Case two: 0 < α < 1**
   In a first step, we focus on a normalized singular fractional-order system. Assume that the controller (6.28) is designed for the system (6.165). Our objective is to estimate for the normalized system (6.40) a dynamic output feedback controller

$$\begin{cases} D^{\alpha} z(t) = A_k z(t) + B_k y(t) \\ v(t) \quad = C_k z(t) + D_k y(t) \end{cases} \tag{6.230}$$

where $z(t) \in \mathbb{R}^q$. We can concatenate the two state vectors of the system and the process into a single vector, resulting in an augmented system as

$$D^{\alpha} \bar{x}(t) = (\bar{A} + \bar{B}\bar{K}\bar{C})\bar{x}(t), \tag{6.231}$$

with

$$\bar{A} = \begin{bmatrix} A_1 & 0_{n\times q} \\ 0_{q\times n} & 0_q \end{bmatrix}, \ \bar{B} = \begin{bmatrix} 0_{n\times q} & B_1 \\ I_q & 0_{q\times m} \end{bmatrix}, \ \bar{C} = \begin{bmatrix} 0_{q\times n} & I_q \\ C & 0_{p\times q} \end{bmatrix},$$

$$\bar{K} = \begin{bmatrix} A_k & B_k \\ C_k & D_k \end{bmatrix}, \ \bar{x}(t) = \begin{bmatrix} x(t) \\ z(t) \end{bmatrix}.$$

The problem can then be interpreted as a static output feedback control problem on the augmented system. Thus, it is straightforward to obtain the equivalent controller gain matrix $\bar{K}$ with the help of the following theorem.

Theorem 6.5.3 *[3] Designing the controller* (6.28) *for the normalized system* (6.40), *the resulting closed-loop control system is asymptotically stable, if there exist matrices* $Z \in \mathbb{P}_{\alpha}^{(n+q)\times(n+q)}$, $G \in \mathbb{R}^{(m+q)\times(m+q)}$ *and* $H \in \mathbb{R}^{(m+q)\times(p+q)}$, *such that*

$$\begin{bmatrix} \mathrm{sym}\left\{Z^T\left(\bar{A}+\bar{B}K_0\right)\right\} & Z^T\bar{B}+\bar{C}^TH^{T}-K_0^TG^T \\ * & \mathrm{sym}\left\{G\right\} \end{bmatrix} < 0, \tag{6.232}$$

with the set $\mathbb{P}_{\alpha}$ *is defined in* (6.25). *The controller gain is given by*

$$\bar{K} = G^{-1}H \tag{6.233}$$

where K_0 *is an intermediate matrix, which is derived from* $K_0 = QP^{-1}$.

The matrices $P \in \mathbb{P}_{\alpha}^{(n+q)\times(n+q)}$ *and* $Q \in \mathbb{R}^{(m+q)\times(n+q)}$ *satisfy the following* $\mathcal{LMI}$

$$\mathrm{sym}\left\{\bar{A}P + BQ\right\} < 0 \tag{6.234}$$

Proof The proof is a similar procedure as that of Theorem 6.3.2, hence it is omitted

Example 6.5.4 We take the system (6.54), its normalized system is given in (6.56). If we consider the exact plant controlled by the dynamic output feedback controller (6.230), by solving the $\mathcal{LMI}s$ in Theorem 6.5.3, one has the feasible controller parameters

$$\bar{K} = \begin{bmatrix} A_k & B_k \\ C_k & D_k \end{bmatrix}$$

as

$$\bar{K} = \begin{bmatrix} -1.5990 & 0.1316 & -0.0710 \\ 0.1360 & -0.5269 & 1.0564 \\ 0.8553 & -2.4134 & 2.1184 \end{bmatrix}, \tag{6.235}$$

$$\bar{K} = \begin{bmatrix} -1.4316 & -0.0669 & 0.0753 & -0.0263 \\ -0.0669 & -1.4316 & 0.0753 & -0.0263 \\ 0.1368 & 0.1368 & -0.4288 & 0.9405 \\ 0.8517 & 0.8517 & -2.2979 & 1.9446 \end{bmatrix}, \tag{6.236}$$

$$\bar{K} = \begin{bmatrix} -1.4203 & -0.0778 & -0.0778 & 0.0772 & 0.0030 \\ -0.0778 & -1.4203 & -0.0778 & 0.0772 & 0.0030 \\ -0.0778 & -0.0778 & -1.4203 & 0.0772 & 0.0030 \\ 0.1397 & 0.1397 & 0.1397 & -0.4126 & 0.8501 \\ 0.7972 & 0.7972 & 0.7972 & 2.5687 & 1.8854 \end{bmatrix}, \tag{6.237}$$

for $q = 1,\ q = 2$ and $q = 3$ respectively.

Now, consider the system (6.165), with $0 < \alpha < 1$. Applying the controller (6.166), we can obtain the closed-loop system (6.167).
Then, the following theorem is presented.

Theorem 6.5.4 *[7] System (6.167), with $0 < \alpha < 1$, is admissible if there exist matrices $X_1 \in \mathbb{R}^{2r\times 2r}$, $X_2 \in \mathbb{R}^{2(n-r)\times 2r}$ and non-singular matrix $X_3 \in \mathbb{R}^{2(n-r)\times 2(n-r)}$, $Z \in \mathbb{R}^{(n+m)\times(n+p)}$ such that*

$$\begin{bmatrix} X_1 & X_2 \\ -X_2 & X_1 \end{bmatrix} > 0, \tag{6.238}$$

$$\operatorname{sym}\left\{\bar{A}\bar{N}(\sin(\alpha\frac{\pi}{2})X - \cos(\alpha\frac{\pi}{2})Y)\bar{M}^{-T} + \bar{B}Z\bar{C}\bar{N}\bar{M}^{-T})\right\} < 0, \tag{6.239}$$

$$X = \begin{bmatrix} X_1 & 0 \\ X_3 & X_4 \end{bmatrix}, \ Y = \begin{bmatrix} X_2 & 0 \\ 0 & 0 \end{bmatrix}, \tag{6.240}$$

$$\bar{V}(\sin(\alpha\frac{\pi}{2})X - \cos(\alpha\frac{\pi}{2})Y)\bar{V}^T = \begin{bmatrix} \tilde{Q}_1 & 0 \\ \tilde{Q}_2 & \tilde{Q}_3 \end{bmatrix}, \tag{6.241}$$

with $\tilde{Q}_1 \in \mathbb{R}^{(n+p)\times(n+p)}$, $\tilde{Q}_2 \in \mathbb{R}^{(n-p)\times(n+p)}$, $\tilde{Q}_3 \in \mathbb{R}^{(n-p)\times(n-p)}$.
The controller gain F is given by

$$F = Z\bar{U}\bar{S}\tilde{Q}_1^{-1}\bar{S}^{-1}\bar{U}^{-1}. \tag{6.242}$$

Since the proofs of Theorem 6.5.4 is similar to that of Theorem 6.3.4, therefore, the proof is omitted.

6.6 Robust Control of Uncertain SFOS

This section focuses on the robust stabilization of fractional-order singular systems for the two cases $(0, 1)$ and $(1, 2)$. Robust stabilization is a critical aspect of control design, ensuring that the system remains stable and performs as desired despite the presence of uncertainties and disturbances. Fractional-order singular systems, with their unique dynamics and non-integer derivatives, pose additional challenges in achieving robust stabilization compared to classical systems. The analysis builds upon the admissibility and stability results established earlier, extending them to develop sufficient conditions for the design of stabilizing controllers. A Linear Matrix Inequality (LMI)-based framework is proposed, providing a systematic and efficient approach to achieve robust stabilization. This approach ensures not only the stabilization of the system but also its robustness against parameter uncertainties, making it highly applicable in practical scenarios.

6.6.1 Stabilization of Normalized Uncertain SFOS

Consider the following uncertain descriptor fractional-order systems

$$\begin{array}{ll} ED^{\alpha}x(t) = (A + \Delta A)\,x(t) + Bu(t), & \\ y(t) \quad = Cx(t), & 0 < \alpha < 2 \end{array} \tag{6.243}$$

ΔA is time-invariant matrix representing a norm-bounded parameter uncertainty, and is assumed be of the following form

$$\Delta A = M_A \Delta N_A, \tag{6.244}$$

where M_A and N_A are known real constant matrices of appropriate dimensions, and the uncertain matrix Δ satisfies

$$\Delta\Delta^{\mathrm{T}} \leq I. \tag{6.245}$$

Consider the feedback control for system (6.243) in the following form

$$u(t) = -LD^{\alpha}x(t) + Kx(t), \tag{6.246}$$

where K and L are gain matrices that need to be determined to ensure that the system (6.243) is normalized and the resulting normalized system is asymptotically stable.

Based on the result of Lemma 6.2.1 and Theorem 6.2.1, the following normalized system is obtained

$$D^{\alpha}x(t) = (A_1 + B_1 K + E_1 \Delta A)\,x(t), \tag{6.247}$$

where $E_1 = (E + BL)^{-1}$, $A_1 = E_1 A$ and $B_1 = E_1 B$ represent the newly obtained nominal matrices. The design of the gain matrix K to robustly stabilize the singular fractional-order system (6.243) is thoroughly analyzed for the fractional order α belonging to $1 \leq \alpha < 2$ and $0 < \alpha < 1$.

Case one: $1 \leq \alpha < 2$

In this context, the robust asymptotic stabilization of the uncertain singular fractional-order system (6.243) is achieved by utilizing Theorem 2.3.1 and the condition (2.14) outlined in Remark 2.3.1.

The following theorem provides a sufficient condition for designing the gain matrix K.

Theorem 6.6.1 *[2] Assume that* (6.243) *is normalizable, then there exists a gain matrix K such that the uncertain singular fractional-order system* (6.243) *with fractional order $1 \leq \alpha < 2$ controlled by the control* (6.246) *is asymptotically stable, if there exist matrices $X \in \mathbb{R}^{m\times n}$, $P_0 = P_0^{\mathrm{T}} > 0 \in \mathbb{R}^{n\times n}$ and a real scalar $\delta > 0$, such that*

$$\begin{bmatrix} \Omega_{11} & * & * & * \\ \Omega_{21} & \Omega_{22} & * & * \\ N_A P_0 & 0 & -\delta I & * \\ 0 & N_A P_0 & 0 & -\delta I \end{bmatrix} < 0, \tag{6.248}$$

where

$$\begin{aligned} \Omega_{11} &= \Omega_{22} \\ &= (A_1 P_0 + P_0 A_1^T + B_1 X + X^{\mathrm{T}} B_1^{\mathrm{T}}) \sin\theta + \delta E_1 M_A (E_1 M_A)^{\mathrm{T}}, \\ \Omega_{21} &= (P_0 A_1^{\mathrm{T}} - A_1 P_0 + X^{\mathrm{T}} B_1^{\mathrm{T}} - B_1 X) \cos\theta, \end{aligned}$$

where $\theta = \pi - \alpha\frac{\pi}{2}$ *and matrices P and Y are given by* $\mathcal{LMI}$ (6.13) *in Theorem 6.2.1. Moreover, the gain matrix K is given by*

$$K = X P_0^{-1}. \tag{6.249}$$

Case two: 0 < α < 1
In this context, a sufficient condition for the robust asymptotic stabilization of the uncertain singular fractional-order system (6.243) is formulated using Linear Matrix Inequalities (LMIs).

Theorem 6.6.2 *[2] Assume that* (6.243) *is normalizable, then there exists a gain matrix K such that the uncertain singular fractional-order system* (6.243) *with fractional order* $0 < \alpha < 1$ *controlled by the control* (6.246) *is asymptotically stable, if there exist matrices* $X \in \mathbb{R}^{m\times n}$, $P_0 = P_0^{\mathrm{T}} > 0 \in \mathbb{R}^{n\times n}$ *and two real scalars* $\delta_i > 0$ $(i = 1, 2)$, *such tha such that*

$$\begin{bmatrix} W_{11} & W_{12} \\ W_{12}^{\mathrm{T}} & W_{22} \end{bmatrix} < 0, \tag{6.250}$$

where

$$\begin{aligned} W_{11} &= \sum_{i=1}^{2} \mathrm{sym}\,\{\Gamma_{i1} \otimes (A_1 P_0 + B_1 k)\} + \sum_{i=1}^{2} \delta_i \left\{ I_2 \otimes (E_1 M_A)(E_1 M_A)^{\mathrm{T}} \right\}, \\ W_{12} &= \left[I_2 \otimes (N_A P_0)^{\mathrm{T}} \;\; I_2 \otimes (N_A P_0)^{\mathrm{T}} \right], \\ W_{22} &= -\,\mathrm{diag}(\delta_1, \delta_2) \otimes I_{2n}, \\ \Gamma_{11} &= \begin{bmatrix} \sin\left(\alpha\frac{\pi}{2}\right) & -\cos\left(\alpha\frac{\pi}{2}\right) \\ \cos\left(\alpha\frac{\pi}{2}\right) & \sin\left(\alpha\frac{\pi}{2}\right) \end{bmatrix}, \end{aligned}$$

$$\Gamma_{21} = \begin{bmatrix} \sin\left(\alpha\frac{\pi}{2}\right) & \cos\left(\alpha\frac{\pi}{2}\right) \\ -\cos\left(\alpha\frac{\pi}{2}\right) & \sin\left(\alpha\frac{\pi}{2}\right) \end{bmatrix},$$

and P and Y are determined by $\mathcal{LMI}$ (6.13) in Theorem 6.2.1.

Moreover, the gain matrix K is given by

$$K = XP_0^{-1}. \tag{6.251}$$

Example 6.6.1 [2] Consider the robust stabilization of the following uncertain descriptor fractional-order system described in (6.243) with parameters as follows

$$E = \begin{bmatrix} 1 & 0 & 0.5 \\ 2 & 1 & 1 \\ 0 & 0 & 0 \end{bmatrix}, \quad A = \begin{bmatrix} 2.4 & 0.2 & 1.2 \\ 4 & 1.5 & 2 \\ 0 & 0 & 0 \end{bmatrix}, \quad B = \begin{bmatrix} 4 & 1 \\ 1 & 1 \\ 1 & 2 \end{bmatrix},$$

$$M_A = \begin{bmatrix} 0.1 & 0 & 0 \\ 0.1 & 0.3 & 0.4 \\ 0 & 0.2 & 0 \end{bmatrix}, \quad N_A = \begin{bmatrix} 0 & 0 & 0 \\ 0.2 & 0 & 0.4 \\ 0 & 0.1 & 0 \end{bmatrix}, \quad \alpha = 1.23.$$

The nominal descriptor fractional-order system is not admissible because

$$\det(\lambda E - A) = 0.$$

The purpose of this example is to design a state feedback control law that ensures the resulting normalized fractional-order system is asymptotically stable.

The solution of the $\mathcal{LMI}$ (6.13) with respect to P and Y gives

$$P = 10^8 \begin{bmatrix} 0.2613 & -1.5587 & 1.0635 \\ -1.5587 & -1.2010 & -0.7272 \\ 1.0635 & -0.7272 & 0.4664 \end{bmatrix}, \quad Y = 10^8 \begin{bmatrix} -1.4992 & 0.7485 & 0.3345 \\ 0.6535 & -0.2534 & -2.4425 \end{bmatrix},$$

and the gain matrix L is given by

$$L = YP^{-1} = \begin{bmatrix} 1.0224 & -0.5004 & -2.3943 \\ -2.8015 & 1.6202 & 3.6775 \end{bmatrix}.$$

This means that the system can be normalized by the full state controller. Using the full state derivative controller, we obtain the new matrices

$$A_1 = \begin{bmatrix} 0.9963 & 0.4454 & 0.4981 \\ 1.9559 & 0.5388 & 0.9780 \\ -0.1605 & 0.1136 & -0.0802 \end{bmatrix}, \quad B_1 = \begin{bmatrix} -0.3541 & -0.3010 \\ 1.5889 & 0.9091 \\ -1.0030 & -0.3770 \end{bmatrix}.$$

A feasible solution of $\mathcal{LMI}$ (6.248) is as follows

$$P_0 = \begin{bmatrix} 27.6967 & -6.7669 & -14.8319 \\ -6.7669 & 9.7767 & 1.0669 \\ -14.8319 & 1.0669 & 11.3889 \end{bmatrix}, \quad X = \begin{bmatrix} 0.0118 & 0.0004 & -0.0023 \\ -37.6733 & -2.1008 & 18.7534 \end{bmatrix},$$

and

$$\delta = 1.784.$$

Finally, the asymptotically stabilizing state-feedback gain is obtained as

$$K = XP_0^{-1} = \begin{bmatrix} 0.0017 & 0.0010 & 0.0019 \\ -2.8547 & -1.9850 & -1.8851 \end{bmatrix}.$$

The state responses of the selected system when

$$\Delta_A = \begin{bmatrix} \cos(0.8) & 0 & 0 \\ 0 & \mathrm{e}^{-0.8} & 0 \\ 0 & 0 & \sin(0.1) \end{bmatrix}$$

is shown in Fig. 6.24 which shows that is asymptotically stable and its states converge to zero.

Example 6.6.2 Consider the robust stabilization of the following uncertain descriptor fractional-order system described in (6.243) for the fractional order α belonging $0 < \alpha < 1$ and with parameters as follows

$$E = \begin{bmatrix} 1 & 0 \\ 0 & 0 \end{bmatrix}, \quad A = \begin{bmatrix} 2 & 1 \\ 1 & -1 \end{bmatrix}, \quad B = \begin{bmatrix} 2 \\ 1 \end{bmatrix},$$

$$M_A = \begin{bmatrix} 0.5 & 0.3 \\ 0.1 & 0.2 \end{bmatrix}, \quad N_A = \begin{bmatrix} 0.2 & 0.1 \\ 0.1 & 0.4 \end{bmatrix}, \quad \alpha = 0.5.$$

It is easy to see that the nominal descriptor fractional-order system is regular, impulse free but not asymptotically stable since

$$\det(\lambda E - A) = \lambda - 3.$$

The purpose of this example is to design a state feedback control law that ensures the resulting normalized fractional-order system is asymptotically stable.

The solution of the $\mathcal{LMI}$ (6.13) with respect to P and Y gives

$$P = \begin{bmatrix} 0.6453 & 0.3928 \\ 0.3928 & 1.0381 \end{bmatrix}, \quad Y = \begin{bmatrix} -0.5331 & -0.1263 \end{bmatrix},$$

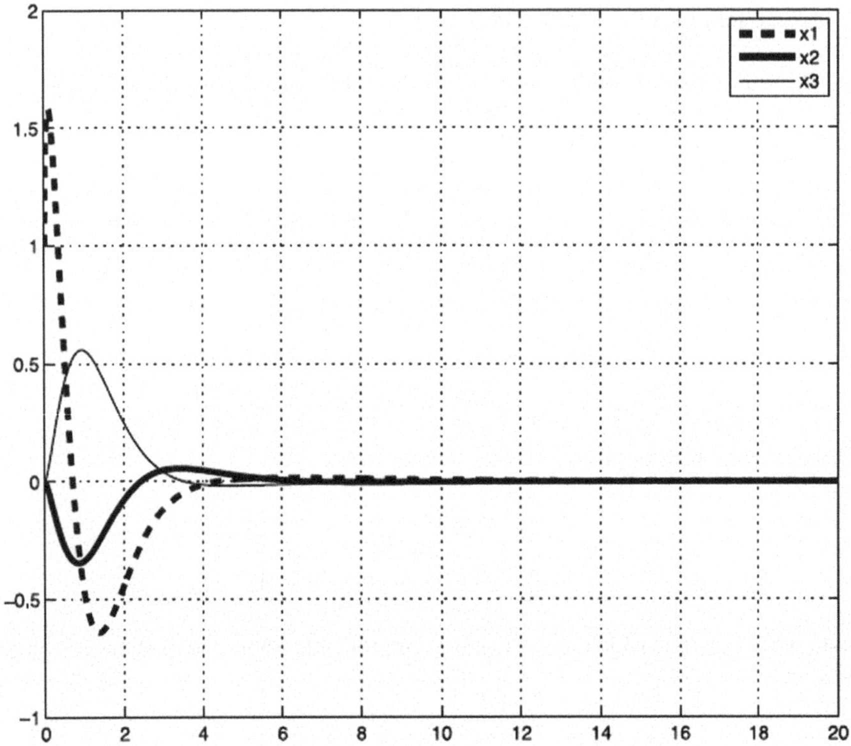

Fig. 6.24 Time response of the selected system in Example 6.6.1 with $\alpha = 1.23$

and the gain matrix L is given by

$$L = YP^{-1} = \begin{bmatrix} -0.9771 & 0.2481 \end{bmatrix}.$$

This means that the system can be normalized by the full state controller. Using the full state derivative controller, we obtain the new matrices

$$E_1 = \begin{bmatrix} 1.0000 & -2.0000 \\ 3.9385 & -3.8462 \end{bmatrix}, \ A_1 = \begin{bmatrix} 0 & 3.0000 \\ 4.0308 & 7.7846 \end{bmatrix}, \ B_1 = \begin{bmatrix} 0 \\ 4.0308 \end{bmatrix}.$$

A feasible solution of $\mathcal{LMI}$ (6.250) is as follows

$$P_0 = \begin{bmatrix} 41.4716 & -6.5223 \\ -6.5223 & 24.6018 \end{bmatrix}, \ X = \begin{bmatrix} -47.1856 & -45.8083 \end{bmatrix},$$

and

$$\delta_1 = 54.9195, \ \delta_2 = 54.9195.$$

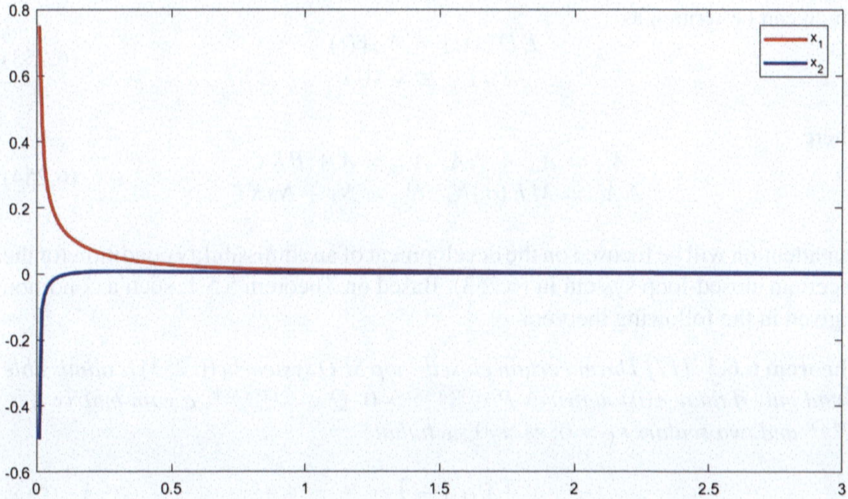

Fig. 6.25 Time response of the selected system in Example 6.6.2 with $\alpha = 0.5$

Finally,the asymptotically stabilizing state-feedback gain is obtained as

$$K = XP_0^{-1} = \begin{bmatrix} -1.4929 & -2.2578 \end{bmatrix}.$$

The state responses of the selected system when

$$\Delta_A = \begin{bmatrix} \cos(0.8) & 0 \\ 0 & \sin(0.1) \end{bmatrix}$$

is shown in Fig. 6.25 which shows that is asymptotically stable and its states converge to zero.

6.6.2 Stabilization of Uncertain SFOS

In the following, we will examine the stabilization of a fractional-order singular system with uncertainties without resorting to normalization.

Case one: $1 \leq \alpha < 2$
Considering the uncertain singular fractional-order system in (5.151) under the static output feedback control $u(t) = Ky(t)$, we obtain the following closed-loop system

$$\begin{aligned} ED^{\alpha}x(t) &= ((A + \Delta A) + (B + \Delta B)KC)x(t) \\ y(t) &= Cx(t) \end{aligned} \tag{6.252}$$

which can be written as

$$\begin{aligned} ED^{\alpha}x(t) &= A_{cl}x(t) \\ y(t) &= Cx(t) \end{aligned} \tag{6.253}$$

where

$$\begin{aligned} A_{cl} &= A_{oc} + \Delta A_c \; A_{oc} = A + BKC \\ \Delta A_c &= MF(\sigma)N_c \; N_c = N_1 + N_2KC \end{aligned} \tag{6.254}$$

Our attention will be focused on the development of an admissibility condition for the uncertain closed-loop system in (6.253). Based on Theorem 5.5.1, such a condition is given in the following theorem.

Theorem 6.6.3 *[17] The uncertain closed-loop SFO system in* (6.253) *is admissible if and only if there exist matrices $P \in \mathbb{R}^{n\times n} > 0$, $Q \in \mathbb{R}^{(n-r)\times n}$, a gain matrix $K \in \mathbb{R}^{m\times p}$ and two scalars $\varepsilon_1 > 0$, $\varepsilon_2 > 0$ such that*

$$\begin{bmatrix} \Gamma_{11} & \Gamma_{12} \\ \Gamma_{12}^T & \Gamma_{22} \end{bmatrix} < 0 \tag{6.255}$$

where

$$\begin{aligned} \Gamma_{11} &= \mathrm{sym}\{\Theta \otimes A_{oc}G\} + (\varepsilon_1 + \varepsilon_2)(I_2 \otimes MM^T) \\ \Gamma_{12} &= \left[I_2 \otimes (N_1G)^T \; I_2 \otimes (N_2KCG)^T \right] \\ \Gamma_{22} &= -\,\mathrm{diag}(\varepsilon_1, \varepsilon_1, \varepsilon_2, \varepsilon_2) \otimes I_q \\ A_{oc} &= A + BKC, \; G = PE^T + SQ \end{aligned}$$

Proof Sufficiency: Suppose that the inequality (6.255) holds for some symmetric matrix $P > 0$, matrix Q together with gain matrix K and scalars $\varepsilon_1 > 0$, $\varepsilon_2 > 0$. Note that from (6.254), we have

$$\begin{aligned} \Phi &= \mathrm{sym}\left\{\Theta \otimes (A_{cl}(PE^{\mathrm{T}} + SQ))\right\} \\ &= \Phi_1 + \Phi_2 + \Phi_3 \end{aligned} \tag{6.256}$$

with

$$\begin{aligned} \Phi_1 &= \mathrm{sym}\left\{\Theta \otimes (A_{oc}(PE^{\mathrm{T}} + SQ))\right\} \\ \Phi_2 &= \mathrm{sym}\left\{\Theta \otimes (\Delta A(PE^{\mathrm{T}} + SQ))\right\} \\ \Phi_3 &= \mathrm{sym}\left\{\Theta \otimes (\Delta BKC(PE^{\mathrm{T}} + SQ))\right\} \end{aligned}$$

It is easy to see that

$$\Phi_2 = \mathrm{sym}\left\{(\Theta \otimes MF(\sigma))(I_2 \otimes N_1(PE^{\mathrm{T}} + SQ))\right\} \tag{6.257}$$

Applying the Lemma 3.4.1 to (6.257), we obtain for any scalar $\varepsilon_1 > 0$

$$\begin{aligned}\Phi_2 &\le \varepsilon_1((\Theta \otimes M)(I_2 \otimes F(\sigma))(I_2 \otimes F(\sigma))^{\mathrm{T}}(\Theta \otimes M)^{\mathrm{T}}) \\ &+ \frac{1}{\varepsilon_1}(I_2 \otimes N_1(PE^{\mathrm{T}} + SQ))^{\mathrm{T}}(I_2 \otimes N_1(PE^{\mathrm{T}} + SQ)) \\ &\le \varepsilon_1(\Theta \otimes M)(I_2 \otimes F(\sigma)F(\sigma)^{\mathrm{T}})(\Theta \otimes M)^{\mathrm{T}} \\ &+ \frac{1}{\varepsilon_1}(I_2 \otimes N_1(PE^{\mathrm{T}} + SQ))^{\mathrm{T}}(I_2 \otimes N_1(PE^{\mathrm{T}} + SQ))\end{aligned} \tag{6.258}$$

It follows from (5.160) that (6.258) implies

$$\begin{aligned}\Phi_2 &\le \varepsilon_1(\Theta \otimes M)(\Theta \otimes M)^{\mathrm{T}} \\ &+ \frac{1}{\varepsilon_1}(I_2 \otimes N_1(PE^{\mathrm{T}} + SQ))^{\mathrm{T}}(I_2 \otimes N_1(PE^{\mathrm{T}} + SQ))\end{aligned} \tag{6.259}$$

In the same way, we have for any scalar $\varepsilon_2 > 0$

$$\begin{aligned}\Phi_3 &= \mathrm{sym}\left\{\Theta \otimes MF(\sigma)N_2KC(PE^{\mathrm{T}} + SQ)\right\} \\ &\le \varepsilon_2(\Theta \otimes M)(\Theta \otimes M)^{\mathrm{T}} + \frac{1}{\varepsilon_2}(I_2 \otimes N_2KC(PE^{\mathrm{T}} \\ &+ SQ))^{\mathrm{T}}(I_2 \otimes N_2KC(PE^{\mathrm{T}} + SQ))\end{aligned} \tag{6.260}$$

Substituting both (6.259) and (6.260) into (6.256), we get

$$\begin{aligned}\Phi \ &\le \ \Phi_1 + \varepsilon_1(\Theta \otimes M)(\Theta \otimes M)^{\mathrm{T}} + \frac{1}{\varepsilon_1}(I_2 \otimes N_1(PE^{\mathrm{T}} + SQ))^{\mathrm{T}} \\ &\times (I_2 \otimes N_1(PE^{\mathrm{T}} + SQ)) + \varepsilon_2(\Theta \otimes M)(\Theta \otimes M)^{\mathrm{T}} \\ &+ \frac{1}{\varepsilon_2}(I_2 \otimes N_2KC(PE^{\mathrm{T}} + SQ))^{\mathrm{T}}(I_2 \otimes N_2KC(PE^{\mathrm{T}} + SQ))\end{aligned} \tag{6.261}$$

Note that

$$\begin{aligned}(\Theta \otimes M)(\Theta \otimes M)^{\mathrm{T}} &= \begin{bmatrix} MM^{\mathrm{T}} & 0 \\ 0 & MM^{\mathrm{T}} \end{bmatrix} \\ &= I_2 \otimes MM^{\mathrm{T}}\end{aligned} \tag{6.262}$$

Using (6.262), the inequality (6.261) becomes

$$\begin{aligned}\Phi &\le \Phi_1 + (\varepsilon_1 + \varepsilon_2)(I_2 \otimes MM^{\mathrm{T}}) \\ &+ \frac{1}{\varepsilon_1}(I_2 \otimes N_1(PE^{\mathrm{T}} + SQ))^{\mathrm{T}} \times (I_2 \otimes N_1(PE^{\mathrm{T}} + SQ)) \\ &+ \frac{1}{\varepsilon_2}(I_2 \otimes N_2KC(PE^{\mathrm{T}} + SQ))^{\mathrm{T}}(I_2 \otimes N_2KC(PE^{\mathrm{T}} + SQ))\end{aligned} \tag{6.263}$$

Applying the Schur complement to (6.255) implies that the right side of the inequality (6.263) is negative definite. Then according to (6.256) we get

$$\operatorname{sym}\left\{\Theta \otimes (A_{cl}(PE^{\mathrm{T}} + SQ))\right\} < 0 \tag{6.264}$$

which means that the uncertain closed-loop system (6.253) is admissible.

Necessity: Suppose that the closed-loop system (6.253) is admissible then there exist matrices $P > 0$ and Q such that for all $F(\sigma)$ satisfying (5.152) and (5.153) we have

$$\Phi = \operatorname{sym}\left\{\Theta \otimes (A_{cl}(PE^{\mathrm{T}} + SQ))\right\} < 0 \tag{6.265}$$

By using the same decomposition as in (6.256), inequality (6.265) implies that for any $\xi \in \mathbb{R}^{2n}$ and $\xi \neq 0$

$$\xi^{\mathrm{T}}(\Phi_1 + \Phi_3)\xi + \xi^{\mathrm{T}}\Phi_2\xi < 0 \tag{6.266}$$

which is equivalent to

$$\begin{aligned}\xi^{\mathrm{T}}(\Phi_1 + \Phi_3)\xi &< -2\xi^{\mathrm{T}}(\Theta \otimes MF(\sigma)N_1(PE^{\mathrm{T}} + SQ))\xi \\ &= -2\xi^{\mathrm{T}}(\Theta \otimes M)(I_2 \otimes F(\sigma))(I_2 \otimes N_1(PE^{\mathrm{T}} + SQ))\xi\end{aligned} \tag{6.267}$$

By defining $\tilde{F}(\sigma) = I_2 \otimes F(\sigma)$ and according to (5.166), hence

$$\xi^{\mathrm{T}}(\Phi_1 + \Phi_3)\xi < -2 \times \max\left\{\xi^{\mathrm{T}}(\Theta \otimes M)\tilde{F}(\sigma)(I_2 \otimes N_1(PE^{\mathrm{T}} + SQ))\xi\right\} \tag{6.268}$$

which implies

$$(\xi^{\mathrm{T}}(\Phi_1 + \Phi_3)\xi)^2 > 4\left(\max\left\{\xi^{\mathrm{T}}(\Theta \otimes M)\tilde{F}(\sigma)(I_2 \otimes N_1(PE^{\mathrm{T}} + SQ))\xi\right\}\right)^2 \tag{6.269}$$

Using (5.174), we get

$$(\xi^{\mathrm{T}}(\Phi_1 + \Phi_3)\xi)^2 > 4 \times \left\{\max\left(\xi^{\mathrm{T}}(\Theta \otimes M)\tilde{F}(\sigma)(I_2 \otimes N_1(PE^{\mathrm{T}} + SQ))\xi\right)^2\right\} \tag{6.270}$$

Applying the Lemma 3.4.3, the inequality (6.270) becomes

$$\begin{aligned}&(\xi^{\mathrm{T}}(\Phi_1 + \Phi_3)\xi)^2 > 4\left(\xi^{\mathrm{T}}(\Theta \otimes M)(\Theta \otimes M)^{\mathrm{T}}\xi\right) \\ &\times \left(\xi^{\mathrm{T}}(I_2 \otimes N_1(PE^{\mathrm{T}} + SQ))^{\mathrm{T}}(I_2 \otimes N_1(PE^{\mathrm{T}} + SQ))\xi\right)\end{aligned} \tag{6.271}$$

Applying Lemma 3.4.4 to (6.271), it follows that there exists a scalar $\varepsilon_1 > 0$ such that

$$\varepsilon_1^2(\Theta \otimes M)(\Theta \otimes M)^{\mathrm{T}} + \varepsilon_1(\Phi_1 + \Phi_3) \\ + (I_2 \otimes N_1(PE^{\mathrm{T}} + SQ))^{\mathrm{T}}(I_2 \otimes N_1(PE^{\mathrm{T}} + SQ)) < 0 \tag{6.272}$$

Using the equality (6.262), we can write (6.272) as follows

$$\Phi_1 + \Phi_3 + \varepsilon_1(I_2 \otimes MM^{\mathrm{T}}) \\ + \frac{1}{\varepsilon_1}(I_2 \otimes (N_1(PE^{\mathrm{T}} + SQ))^{\mathrm{T}}(N_1(PE^{\mathrm{T}} + SQ))) < 0 \tag{6.273}$$

By the same argument, we can conclude that there exists another scalar $\varepsilon_2 > 0$ such that

$$\Phi_1 + (\varepsilon_1 + \varepsilon_2)(I_2 \otimes MM^{\mathrm{T}}) \\ + \frac{1}{\varepsilon_1}(I_2 \otimes (N_1(PE^{\mathrm{T}} + SQ))^{\mathrm{T}}(N_1(PE^{\mathrm{T}} + SQ))) \\ + \frac{1}{\varepsilon_2}(I_2 \otimes (N_2KC(PE^{\mathrm{T}} + SQ))^{\mathrm{T}}(N_2KC(PE^{\mathrm{T}} + SQ))) < 0 \tag{6.274}$$

To check that (6.255) is verified, it is enough to apply the Shur complement to the inequality (6.274). This ends the proof. □

Remark 6.6.1 Due to the existence of the bilinear term $KC(PE^{\mathrm{T}} + SQ)$, the inequality (6.255) is not an $\mathcal{LMI}$ condition. In addition the matrix $PE^{\mathrm{T}} + SQ$ is not always invertible which further increases the difficulty in determining the matrix gain K [11].

In order to develop $\mathcal{LMI}$ design condition, let us consider the following system

$$\bar{E}\dot{\bar{x}}(t) = \bar{A}\bar{x}(t) \tag{6.275a}$$
$$y(t) = \bar{C}\bar{x}(t) \tag{6.275b}$$

where

$$\bar{E} = \begin{bmatrix} E & 0 \\ 0 & 0 \end{bmatrix}, \ \bar{A} = \begin{bmatrix} A + \Delta A & B + \Delta B \\ KC & -I_m \end{bmatrix} \\ \bar{C} = \begin{bmatrix} C & 0 \end{bmatrix}, \ \bar{x}(t) = \begin{bmatrix} x(t)^{\mathrm{T}} & u(t)^{\mathrm{T}} \end{bmatrix}^{\mathrm{T}} \tag{6.276}$$

Then we have the following result.

Lemma 6.6.1 *[17] System* (6.253) *is admissible if and only if the system* (6.275) *is admissible.*

Proof From the system (6.275) we have

$$\begin{aligned}\lambda\bar{E}-\bar{A} &= \begin{bmatrix}\lambda E-(A+\Delta A) & -(B+\Delta B)\\ -KC & I_m\end{bmatrix}\\ &= \begin{bmatrix}I_n & -(B+\Delta B)\\ 0 & I_m\end{bmatrix}\\ &\times\begin{bmatrix}\lambda E-(A+\Delta A)-(B+\Delta B)KC & 0\\ 0 & I_m\end{bmatrix}\\ &\times\begin{bmatrix}I_n & 0\\ -KC & I_m\end{bmatrix}\end{aligned}$$

Then it is easy to check that the regularity and the stability of the system (6.253) are equivalent to those of the system (6.275) with reference to the properties of determinants. For the impulsiveness, it suffices to see if $\text{rank}(\bar{E})=\text{rank}(E)$, which proves the lemma. □

In the other hand, the system (6.275) is uncertain since the matrix $\bar{A}$ can be written as

$$\begin{aligned}\bar{A} &= \begin{bmatrix}A & B\\ KC & -I_m\end{bmatrix}+\begin{bmatrix}\Delta A & \Delta B\\ 0 & 0\end{bmatrix}\\ &= \bar{A}_0+\Delta\bar{A}_0\end{aligned} \tag{6.277}$$

and

$$\begin{aligned}\Delta\bar{A}_0 &= \begin{bmatrix}M & 0\\ 0 & 0\end{bmatrix}\begin{bmatrix}F(\sigma) & 0\\ 0 & F(\sigma)\end{bmatrix}\begin{bmatrix}N_1 & N_2\\ 0 & 0\end{bmatrix}\\ &= \bar{M}\bar{F}(\sigma)\bar{N}\end{aligned} \tag{6.278}$$

Now, by using the results of the Lemma 6.6.1 and Theorem 5.5.1, $\mathcal{LMI}$ design condition is proposed.

Theorem 6.6.4 *[17] The system* (6.275) *is admissible if there exist matrices* $P_1\in\mathbb{R}^{n\times n}>0$, $P_3\in\mathbb{R}^{m\times m}>0$, $Q_1\in\mathbb{R}^{(n-r)\times n}$, $Q_2\in\mathbb{R}^{(n-r)\times m}$, $Y\in\mathbb{R}^{m\times p}$, *two nonsingular matrices* $Q_3\in\mathbb{R}^{m\times m}$, $X\in\mathbb{R}^{m\times m}$ *and a scalar* $\xi>0$ *such that the following* $\mathcal{LMI}$ *holds*

$$\begin{bmatrix}\Psi & \Sigma^T\\ \Sigma & -\xi I_{4h}\end{bmatrix}<0 \tag{6.279}$$

where the matrix

$$\Psi=\begin{bmatrix}\Psi_{11} & * & * & *\\ \Psi_{21} & \Psi_{22} & * & *\\ \Psi_{31} & \Psi_{32} & \Psi_{33} & *\\ \Psi_{41} & \Psi_{42} & \Psi_{43} & \Psi_{44}\end{bmatrix}$$

is symmetric defined as

$$\begin{aligned}
\Psi_{11} &= \text{sym}\left\{A^T(P_1E + S_1Q_1) + A^TS_2Q_3J + C^TY^TJ\right\} \\
&\quad \times \sin(\alpha\frac{\pi}{2}) + \xi N_1^TN_1 \\
\Psi_{21} &= (B^T(P1E + S_1Q_1) + B^TS_2Q_3J - XJ)\sin(\alpha\frac{\pi}{2}) \\
&\quad + (Q_2^TS_1^TA + Q_3^TS_2^TA + YC)\sin(\alpha\frac{\pi}{2}) + \xi N_2^TN_1 \\
\Psi_{22} &= \text{sym}\left\{B^TS_1Q_2 + B^TS_2Q_3 - X\right\}\sin(\alpha\frac{\pi}{2}) + \xi N_2^TN_2 \\
\Psi_{31} &= (A^T(P_1E + S_1Q_1) + A^TS_2Q_3J + C^TY^TJ)\cos(\alpha\frac{\pi}{2}) \\
&\quad - ((P_1E + S_1Q_1)^TA - J^TQ_3^TS_2^TA - J^TYC)\cos(\alpha\frac{\pi}{2}) \\
\Psi_{32} &= (A^TS_1Q_2 + A^TS_2Q_3 + C^TY^T + J^TX^T)\cos(\alpha\frac{\pi}{2}) \\
&\quad - ((P_1E + S_1Q_1)^TB + J^TQ_3^TS_2^TB)\cos(\alpha\frac{\pi}{2}) \\
\Psi_{42} &= (B^T(S_1Q_2 + S_2Q_3) + X^T)\cos(\alpha\frac{\pi}{2}) \\
&\quad - (B^T(S_1Q_2 + S_2Q_3) + X^T)^T\cos(\alpha\frac{\pi}{2}) \\
\Psi_{33} &= \Psi_{11},\ \Psi_{41} = -\Psi_{32}^T,\ \Psi_{43} = \Psi_{21},\ \Psi_{44} = \Psi_{22},
\end{aligned}$$

The matrix $\Sigma = \left[\Sigma_{ij}\right]_{1\le i,j\le 4}$ *is as follows*

$$\begin{aligned}
\Sigma_{11} &= (M^T(P_1E + S_1Q_1) + M^TS_2Q_3J)\sin(\alpha\tfrac{\pi}{2}) \\
\Sigma_{12} &= (M^TS_1Q_2 + M^TS_2Q_3)\sin(\alpha\tfrac{\pi}{2}) \\
\Sigma_{13} &= -(M^T(P_1E + S_1Q_1) + M^TS_2Q_3J)\cos(\alpha\tfrac{\pi}{2}) \\
\Sigma_{14} &= -(M^TS_1Q_2 + M^TS_2Q_3)\cos(\alpha\tfrac{\pi}{2}) \\
\Sigma_{2j} &= \Sigma_{4j} = 0,\ j = 1, 2, 3, 4 \\
\Sigma_{31} &= -\Sigma_{13},\ \Sigma_{32} = -\Sigma_{14},\ \Sigma_{33} = \Sigma_{11},\ \Sigma_{34} = \Sigma_{12}
\end{aligned}$$

The matrix $S_1 \in \mathbb{R}^{n\times(n-r)}$ *is any matrix of full column rank satisfying* $E^TS_1 = 0$, $S_2 \in \mathbb{R}^{n\times m}$ *such that* $E^TS_2 = 0$. *J is a known matrix.*

The controller gain matrix K is given by

$$K = X^{-T}Y \tag{6.280}$$

Proof Suppose that the $\mathcal{LMI}$ (6.279) for some matrices P_1, P_3, Q_1, Q_2, Q_3, X, Y and some scalar ξ. Consider the matrices

$$\bar{P} = \begin{bmatrix} P_1 & 0 \\ 0 & P_3 \end{bmatrix}, \bar{S} = \begin{bmatrix} S_1 & S_2 \\ 0 & P_3 \end{bmatrix}, \bar{Q} = \begin{bmatrix} Q_1 & Q_2 \\ Q_3 J & Q_3 \end{bmatrix}, \text{ then we have}$$

$$\bar{E}^{\mathrm{T}}\bar{S} = \begin{bmatrix} E^{\mathrm{T}} & 0 \\ 0 & 0 \end{bmatrix}\begin{bmatrix} S_1 & S_2 \\ 0 & Q_3 \end{bmatrix} = 0_{(n+m)\times(n+m)}$$

and it is clear that $\bar{P} > 0$. By computation of the matrix

$$\Xi = \begin{bmatrix} \mathrm{sym}\left\{\Theta \otimes \bar{A}_0^{\mathrm{T}}(\bar{P}\bar{E} + \bar{S}\bar{Q})\right\} + \xi I_2 \otimes \bar{N}^{\mathrm{T}}\bar{N} & * \\ \Theta \otimes (\bar{M}^{\mathrm{T}}(\bar{P}\bar{E} + \bar{S}\bar{Q}))) & -\xi I_{4h} \end{bmatrix}$$

and by defining the matrices $X = P_3 Q_3$ and $Y^{\mathrm{T}} = K^{\mathrm{T}} P_3 Q_3$ and taking into account the $\mathcal{LMI}$ (6.279), we get that

$$\Xi < 0$$

which means that the system (6.275) is admissible. In this case, the gain matrix K is deduced by $K = X^{-\mathrm{T}}Y$. This ends the proof of the theorem. □

Example 6.6.3 Consider the robust stabilization of the following uncertain singular fractional-order linear time invariant system described in (5.151) with fractional-order derivative $\alpha = 1.5$ and the parameters:

$$E = \begin{bmatrix} 1 & 2 \\ 1 & 2 \end{bmatrix}, \ A = \begin{bmatrix} 5 & -3 \\ -2 & 3 \end{bmatrix}, \ B = \begin{bmatrix} 2 \\ 3 \end{bmatrix}$$

$$C = \begin{bmatrix} 1 & 0 \\ 0 & 1 \end{bmatrix}, \ N_1 = \begin{bmatrix} 1 & 0.1 \end{bmatrix}, \ N_2 = 0.2$$

$$M = \begin{bmatrix} 0.1 \\ 0.22 \end{bmatrix}$$

Note that this unforced system ($u(t) = 0$) is regular, impulse-free but unstable because $\det(\lambda E - A) = -20\lambda + 9$, which means that the finite mode of the system is outside the stable area. The time response of the system is shown in Fig. 6.26 where all its variables are not convergent.

Note that this system is normalizable since rank $\begin{bmatrix} E & B \end{bmatrix} = 2$ [2]. However the corresponding method uses the derivative $D^{\alpha}x(t)$ which is not easy to compute in practice. In order to avoid this difficulty, note Theorem 6.6.4 is used. Indeed, the corresponding strict $\mathcal{LMI}$ condition is feasible and leads to the following solution:

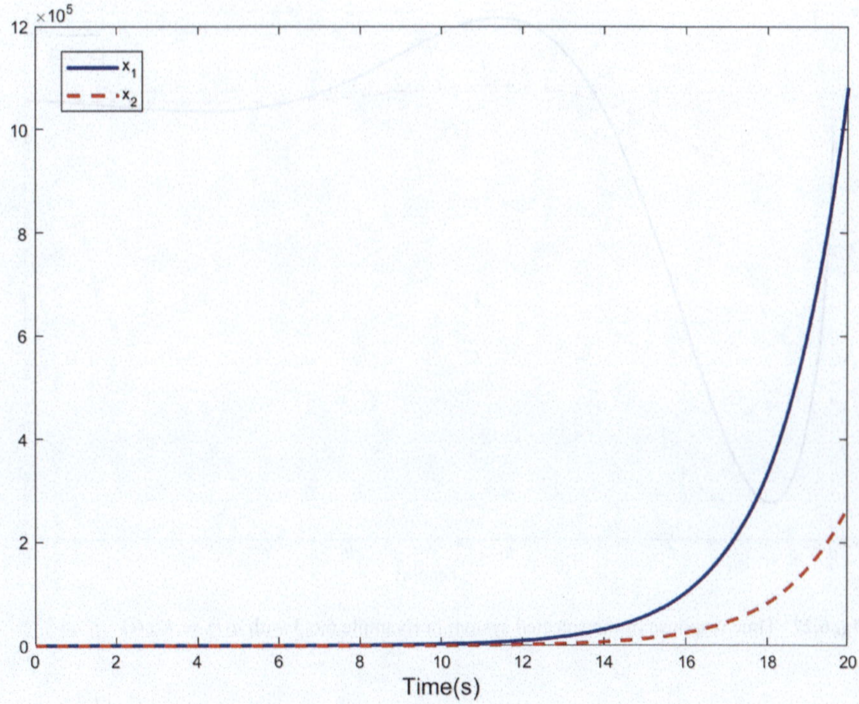

Fig. 6.26 Time response of the selected system in Example 6.6.3 with $u(t) = 0$

$$P_1 = \begin{bmatrix} 1.9071 & -1.7945 \\ -1.7945 & 1.9071 \end{bmatrix}, \; P_3 = 3.7016$$
$$Q_1 = 10^3 \times \begin{bmatrix} -1.7829 & -0.0002 \end{bmatrix}, \; Q_2 = -1.7832 \times 10^3$$
$$Y = \begin{bmatrix} -11.9673 & 5.6941 \end{bmatrix}, \; X = 4.2967$$
$$Q_3 = 1.1608, \; \xi = 0.4878$$

and there exists a state-feedback robust admissible controller for the system (5.151) and it can be designed as

$$u(t) = \begin{bmatrix} -2.7852 & 1.3252 \end{bmatrix} x(t)$$

By applying this control law, the simulation results in Fig. 6.27 show that x_1, x_2 converge to zero.

Case two: $0 < \alpha < 1$

For a singular fractional-order system, it is important to develop conditions which guarantee that the given closed-loop singular fractional-order system is not only stable but also regular and impulse-free. In the following, robust stabilization of closed-loop system (5.151) is studied by considering the control law:

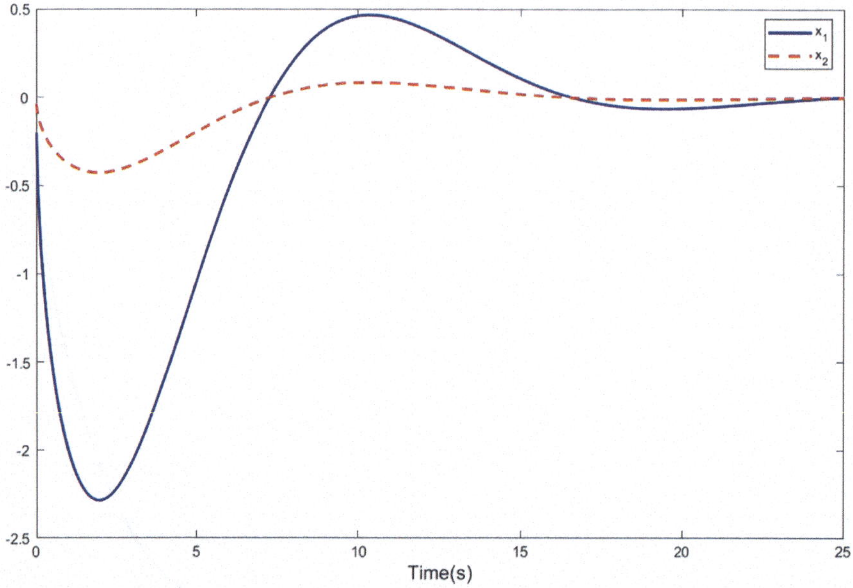

Fig. 6.27 Time response of the selected system in Example 6.6.3 with $u(t) = Kx(t)$

$$u(t) = Kx(t), \tag{6.281}$$

where K is the gain parameter to be determined. Then, the closed-loop system (5.151) becomes:

$$ED^{\alpha}x(t) = (A + \Delta A + (B + \Delta B)K)x(t). \tag{6.282}$$

Applying Definition 5.5.4 to the closed-loop system (6.282), denoting $W = K(\mathcal{P}E^{\mathrm{T}} + SQ)$ with $\mathcal{P} = \sin(\alpha\frac{\pi}{2})X - \cos(\alpha\frac{\pi}{2})Y$ and using Lemma 3.4.2, we have the following result.

Theorem 6.6.5 *[18] The uncertain singular closed-loop system* (6.282) *is quadratically admissible if and only if there exist appropriate dimensional matrices X, Y, Q and W such that* (5.185) *holds and*

$$\begin{bmatrix} \Sigma & (\mathcal{P}E^T + SQ)^T N_1^T + W^T N_2^T \\ N_1(\mathcal{P}E^T + SQ) + N_2 W & -\epsilon I \end{bmatrix} < 0 \tag{6.283}$$

where
$\mathcal{P} = \sin(\alpha\frac{\pi}{2})X - \cos(\alpha\frac{\pi}{2})Y$
$\Sigma = \mathrm{sym}\left\{A(\mathcal{P}E^T + SQ) + BW\right\} + \epsilon MM^T$ *and $S \in \mathbb{R}^{n\times(n-r)}$ is any matrix of full column rank and satisfies $ES = 0$.*

Then, a stabilizing state-feedback controller gain matrix can be calculated by

$$K = W(\mathcal{P}E^T + SQ)^{-1} \tag{6.284}$$

Example 6.6.4 Consider the robust stabilization of the following uncertain singular fractional-order system described in (5.151) with parameters as follows

$$E = \begin{bmatrix} -1 & -1 & -1 & -2 \\ 4 & 3 & 2 & 5 \\ -6 & 0 & 2 & 2 \\ 1 & 1 & 1 & 2 \end{bmatrix}, \; A = \begin{bmatrix} -2 & -4 & -2 & -6 \\ 9 & 15 & 5 & 18 \\ -18 & -4 & 2 & -4 \\ 2 & 5 & 2 & 6 \end{bmatrix}, \; B = \begin{bmatrix} 10 & 1 \\ -1 & 0 \\ 0 & 1 \\ 2 & -1 \end{bmatrix}$$

$$M = \begin{bmatrix} 1 & 0 & 0 & 2 \\ 0 & 1 & 0 & 0 \\ 0 & 0 & 1 & 1 \\ 0 & 0 & 0 & 1 \end{bmatrix}, \; N_1 = \begin{bmatrix} 1 & 0 & 0 & 0 \\ 1 & 1 & 0 & 1 \\ 1 & 1 & 0 & 0 \\ 0 & 0 & 1 & 0 \end{bmatrix}, \; N_2 = \begin{bmatrix} 1 & 1 \\ 0 & 1 \\ 2 & 1 \\ -1 & 1 \end{bmatrix}, \; \alpha = 0.5$$

The nominal singular fractional-order system is not admissible. Indeed,

$$\det(\lambda E - A) = -4\lambda^3 + 28\lambda^2 - 64\lambda + 40,$$

so, the system is regular, impulse-free but unstable because its finite poles $\lambda \in \{1, 3+i, 3-i\}$, don't satisfy $|\arg(\lambda)| > \alpha\frac{\pi}{2}$.

Solving conditions (5.185) and (6.283) in Theorem 6.6.5, we obtain the feasible solution as follows

$$X = 10^7 \begin{bmatrix} 0.7133 & -1.3020 & 1.4685 & 0.1664 \\ -1.3020 & 4.5053 & -1.1043 & -2.4056 \\ 1.4685 & -1.1043 & 5.8480 & -1.0629 \\ 0.1664 & -2.4056 & -1.0629 & 2.3382 \end{bmatrix},$$

$$Y = 10^7 \begin{bmatrix} 0 & 0.1220 & -0.5682 & -0.4463 \\ -0.1220 & 0 & 0.7430 & 0.7430 \\ 0.5682 & -0.7430 & 0 & -0.7430 \\ 0.4463 & -0.7430 & 0.7430 & 0 \end{bmatrix},$$

$$Q = 10^7 \begin{bmatrix} 3.2830 & 1.4671 & 1.7740 & 3.0269 \end{bmatrix},$$

$$W = 10^7 \begin{bmatrix} -2.2121 & 1.3048 & -0.5902 & -0.0323 \\ 3.0002 & -2.1730 & -0.9574 & 1.7394 \end{bmatrix},$$

$$\epsilon = 4.4819 \times 10^7.$$

Finally, for $S = \begin{bmatrix} 0 \\ -1 \\ -1 \\ 1 \end{bmatrix}$, the asymptotically stabilizing state-feedback gain is obtained as

$$K = W((\sin(\alpha\frac{\pi}{2})X - \cos(\alpha\frac{\pi}{2})Y)E^{\mathrm{T}} + SQ)^{-1}$$
$$= \begin{bmatrix} -10.7860 & -5.2611 & 1.3266 & -4.2902 \\ -46.6692 & -22.3633 & 4.5844 & -17.0277 \end{bmatrix}.$$

The singular system system (5.151) is now asymptotically stable since its finite modes are $\lambda = 1.0111 \pm 2.5104i$ and satisfy the condition $|\arg(\lambda)| > \alpha\frac{\pi}{2}$.

6.7 Summary

This chapter focused on the admissibility conditions of continuous closed-loop singular fractional-order systems. For the case (1, 2),an $\mathcal{LMI}$-based approach was proposed to design both state feedback and static output feedback controllers. For the case (0, 1), the design was extended to output feedback controllers using an observer-based framework. Additionally, robust stabilization methods were addressed to handle uncertain singular fractional-order systems effectively.

References

1. Dai, L.: Singular Control Systems. Springer, Berlin (1989)
2. N'Doye, I., Darouach, M., Zasadzinski, M., Radhy, N.: Robust stabilisation of uncertain descriptor fractional-order systems. Automatica **49**, 1907–1913 (2013). June
3. Wei, Y., Tse, P.W., Yao, Z., Wang, Y.: The output feedback control synthesis for a class of singular fractional order systems. ISA Trans. **69**, 1–9 (2017)
4. Wei, Y.H., Chen, Y.Q., Cheng, S.S., Wang, Y.: Completeness on the stability criterion of fractional order LTI systems. Fract. Calc. Appl. Anal. **20**(1), 17 (2017). https://doi.org/10.1515/fca-2017-0008
5. Khargonakar, P.P., Petersen, I.R., Zhou, K.: Robust stabilization of uncertain linear systems: quadratic stability and h_∞ control theory. IEEE Trans. Automat. Contr. **35**, 356–361 (1990)
6. Saadni, S.M., Chaabane, M., Mehdi, D.: Robust stability and stabilization of a class of singular systems with multiple time-varying delays. Asian J. Control **1**, 1–11 (2006)
7. Zhang, X., Zhao, Z.N., Wang, Q.: Static and dynamic output feedback stabilisation of descriptor fractional order systems. IET Control Theory Appl. **14** (2019). https://doi.org/10.1049/iet-cta.2019.0694
8. Marir, S., Chadli, M., Bouagada, D.: New admissibility conditions for singular linear continuous-time fractional-order systems. J. Franklin Inst. **354**, 752–766 (2017)
9. Marir, S., Chadli, M., Bouagada, D.: A novel approach of admissibility for singular linear continuous-time fractional-order systems. Int. J. Control Autom. Syst. (2017). https://doi.org/10.1007/s12555-016-0003-0
10. Lin, C., Chen, B., Shi, P., Yu, J.P.: Necessary and sufficient conditions of observer-based stabilization for a class of fractional-order descriptor systems. Syst. Control Lett. **112**, 31–35 (2018)
11. Xu, S.Y., Lam, J.: Robust Control and Filtering of Singular Systems. Springer, Berlin (2006)
12. Uezato, E., Ikeda, M.: Strict LMI conditions for stability, robust stability, and $\mathcal{H}_\infty$ control of descriptor systems. In: Proceedings of the 37th IEEE Conference on Decision and Control, pp. 4092–4097 (1999)

13. Marir, S., Chadli, M., Basin, M.V.: Stabilization and $\mathcal{H}_\infty$ control for descriptor fractional-order systems. In: 2023 IEEE International Conference on Networking, Sensing and Control (ICNSC), Marseille, France, pp. 1–6, Oct 2023. https://doi.org/10.1109/ICNSC58704.2023.10319004
14. Boyd, S., Elghaoui, L., Féron, E., Balakrishnan, V.: Linear Matrix Inequality in Systems and Control Theory. SIAM, Philadelphia (1994)
15. Marir, S., Chadli, M., Basin, M.V.: Necessary and sufficient admissibility conditions of dynamic output feedback for singular linear fractional–order systems. Asian J. Control **25**, 2439–2450 (2022). https://doi.org/10.1002/asjc.3001
16. Liu, Y., Cui, L., Duan, D.: Dynamic output feedback stabilization of singular fractional-order systems. Math. Problems Eng. **2016**, 1–8 (2016). https://doi.org/10.1155/2016/9694780
17. Marir, S., Chadli, M.: Robust admissibility and stabilization of uncertain singular fractional-order linear time-invariant systems. IEEE/CAA J. Automatica Sinica **6**(3), 685–692 (2019). https://doi.org/10.1109/JAS.2019.1911480
18. Xuefeng, Z., Chen, Y.Q.: Admissibility and robust stabilization of continuous linear singular fractional order systems with the fractional order α: the $0 < \alpha < 1$ case. ISA Trans. **82**, 42–50 (2018)

Chapter 7
$\mathcal{H}_\infty$ Control for Singular Fractional-Order Systems

7.1 Introduction

As pointed out in Chap. 4, considerable efforts have been devoted to stabilization for fractional-order systems. Over the last years singular values have been developed as a tool to deal with this problem. In general terms, the controller should be chosen so that the closed-loop transfer function matrix has certain characteristics that are derived from the specifications. An optimal design minimizes the maximum singular value of the desired loop shape, subject to a closed-loop stability constraint. This is an $\mathcal{H}^\infty$ optimization problem, for which a considerable mathematical theory is available [1, 2].

7.2 Bounded Real Lemma for SFOS

Singular fractional linear continuous systems performance can be evaluated through different norms of transfer function. In this section, we consider the $\mathcal{H}_\infty$ control problem. The corresponding versions of the bounded real lemma are proposed in terms of $\mathcal{LMI}$s. Note that for FOS, the bounded real lemma characterizes only the $\mathcal{H}_\infty$ norm and has nothing to do with stability. This is different from the $SFOS$ case, where the transfer function is well defined if and only if the system is regular. For these reasons, the admissibility conditions are added to the statement of the bounded real lemma for $SFOS$. Consider the following fractional order system admitting a pseudo state representation of the form

$$\begin{cases} ED^\alpha x(t) = Ax(t) + Fw(t), \ 1 \leq \alpha < 2, \\ \quad y(t) = Cx(t), \end{cases} \tag{7.1}$$

S. Marir and M. Chadli, *Control of Singular Fractional Order Systems: LMI Approach*, Studies in Systems, Decision and Control 292,
https://doi.org/10.1007/978-3-031-87382-9_7

where α is the fractional commensurate order, $x(t) \in \mathbb{R}^n$ is the pseudo-state vector, $w(t) \in \mathbb{R}^q$ is the exogenous input vector and $y(t) \in \mathbb{R}^p$ is the output vector. The real-valued matrices A, F, C are assumed to be known and constant with appropriate dimensions.The matrix $E \in \mathbb{R}^{n\times n}$is singular, $rank(E) = r < n$. The system (7.1) is assumed to be of minimal realization. If the system (7.1) is regular, the transfer function matrix between $w(t)$ to $y(t)$ is

$$G_{wy}(s) = C(s^\alpha E - A)^{-1} F. \tag{7.2}$$

Theorem 7.2.1 *[3] Let the SFOS* (7.1) *be admissible and* $\gamma > 0$. *For* $\theta = (1-\alpha)\frac{\pi}{2}$, *the following statements are equivalent.*

(i) The SFOS (7.1) *with its transfer function* $G_{wy}(s)$ *in* (7.2) *verifies*

$$\left\| G_{wy}(s) \right\|_{\mathcal{H}_\infty} < \gamma.$$

(ii) There exist matrices $P \in \mathbb{C}^{n\times n}$ *such that*

$$EP = P^* E^T \geq 0 \tag{7.3a}$$

$$\begin{bmatrix} \mathrm{Her}\{e^{-j\theta} AP\} & * & * \\ e^{-j\theta} CP & -\gamma^2 I_p & * \\ F^T & 0 & -I_q \end{bmatrix} < 0. \tag{7.3b}$$

(iii) There exist matrices $P \in \mathbb{C}^{n\times n} > 0$ *and* $Q \in \mathbb{C}^{(n-r)\times n}$ *such that the following* $\mathcal{LMI}$ *holds:*

$$\begin{bmatrix} \mathrm{Her}\{e^{-j\theta} A\left(PE^T + SQ\right)\} & * & * \\ e^{-j\theta} C\left(PE^T + SQ\right) & -\gamma^2 I_p & * \\ F^T & 0 & -I_q \end{bmatrix} < 0, \tag{7.4}$$

where $S \in \mathbb{R}^{n\times(n-r)}$ *is any matrix with full column rank and satisfies* $ES = 0$.

(iv) There exist matrices $P \in \mathbb{C}^{n\times n} > 0$, $Q \in \mathbb{C}^{(n-r)\times n}$, $\mathcal{V} \in \mathbb{R}^{(n+p)\times(n+p)}$, *and* $\mathcal{U} \in \mathbb{C}^{(n+p)\times(n+p)}$ *such that*

$$\begin{bmatrix} \mathrm{Her}\left\{e^{-j\theta} \mathcal{A}_c \mathcal{U}\right\} + \mathcal{F}\mathcal{F}^T & * \\ \mathcal{P}_c \mathcal{E}^T + \mathcal{S}\mathcal{Q}_c + e^{j\theta} \mathcal{V}^T \mathcal{A}_c^T - \mathcal{U} & -\mathcal{V} - \mathcal{V}^T \end{bmatrix} < 0, \tag{7.5}$$

with $\mathcal{E}_c \mathcal{S} = 0$, *and*

$$\mathcal{E}_c = \begin{bmatrix} E & 0_{n\times p} \\ 0_{p\times n} & \frac{\gamma^2}{2} I_p \end{bmatrix}, \ \mathcal{A}_c = \begin{bmatrix} A & 0_{n\times p} \\ C & -I_p \end{bmatrix}, \ \mathcal{F} = \begin{bmatrix} F \\ 0_{p\times q} \end{bmatrix}, \tag{7.6a}$$

$$\mathcal{P}_c = \begin{bmatrix} P & 0_{n\times p} \\ 0_{p\times n} & e^{j\theta} I_p \end{bmatrix}, \ \mathcal{Q}_c = \begin{bmatrix} Q & 0_{(n-r)\times p} \end{bmatrix}. \tag{7.6b}$$

Proof We start demonstrating the equivalence between (i) and (ii).
Necessity. Since the system (7.1) is admissible, then, by Lemma 5.4.1, nonsingular matrices M and N can always be found such that

$$E = M\begin{bmatrix} I_r & 0 \\ 0 & 0 \end{bmatrix}N, \ A = M\begin{bmatrix} A_1 & 0 \\ 0 & I_{n-r} \end{bmatrix}N. \tag{7.7}$$

We also write

$$F = M\begin{bmatrix} F_1 \\ F_2 \end{bmatrix}, C = \begin{bmatrix} C_1 \ C_2 \end{bmatrix}N, \tag{7.8}$$

where the partition is compatible with that of A in (7.7). Then, it is easy to see that the transfer function (7.2) is represented as

$$G_{wy}(s) = C_1(s^\alpha I_r - A_1)^{-1}F_1 - C_2F_2. \tag{7.9}$$

Then, from (7.9) and Theorem 4.3.1, the inequality $\|G_{wy}(s)\|_{\mathcal{H}_\infty} < \gamma$ implies that there exists a matrix $P_1 \in H_r > 0$ such that

$$\begin{bmatrix} \mathrm{Her}(e^{-j\theta}A_1P_1) & * & * \\ e^{-j\theta}C_1P_1 & -\gamma^2 I_p & * \\ F_1^{\mathrm{T}} & -F_2^{\mathrm{T}}C_2^{\mathrm{T}} & -I_q \end{bmatrix} < 0. \tag{7.10}$$

Applying Schur complement to (7.10), we obtain

$$\begin{bmatrix} \mathrm{Her}(e^{-j\theta}A_1P_1) & * \\ e^{-j\theta}C_1P_1 & -\gamma^2 I_p \end{bmatrix} + \begin{bmatrix} F_1F_1^{\mathrm{T}} & * \\ -C_2F_2F_1^{\mathrm{T}} & C_2F_2F_2^{\mathrm{T}}C_2^T \end{bmatrix} < 0,$$

which is equivalent to

$$\begin{bmatrix} \mathrm{Her}(e^{-j\theta}A_1P_1) + F_1F_1^{\mathrm{T}} & * \\ e^{-j\theta}C_1P_1 - C_2F_2F_1^{\mathrm{T}} & C_2F_2F_2^{\mathrm{T}}C_2^{\mathrm{T}} - \gamma^2 I_p \end{bmatrix} < 0.$$

Then a sufficiently small scalar $\delta > 0$ can always be found such that

$$\begin{bmatrix} \mathrm{Her}(e^{-j\theta}A_1P_1) + F_1F_1^{\mathrm{T}} & * \\ e^{-j\theta}C_1P_1 - C_2F_2F_1^{\mathrm{T}} & C_2(F_2F_2^{\mathrm{T}} + \delta I)C_2^T - \gamma^2 I_p \end{bmatrix} < 0. \tag{7.11}$$

Let $\Psi = F_2F_2^{\mathrm{T}} + \delta I$, then $\Psi > 0$, and by (7.11) we have

$$\begin{bmatrix} \mathrm{Her}(e^{-j\theta}A_1P_1) + F_1F_1^{\mathrm{T}} & * \\ e^{-j\theta}C_1P_1 - C_2F_2F_1^{\mathrm{T}} & C_2\Psi\Psi^{-1}\Psi C_2^{\mathrm{T}} - \gamma^2 I_p \end{bmatrix} < 0. \tag{7.12}$$

Using the Shur complement, we get

$$\begin{bmatrix} \mathrm{Her}(e^{-j\theta}A_1P_1) + F_1F_1^{\mathrm{T}} & * & * \\ e^{-j\theta}C_1P_1 - C_2F_2F_1^{\mathrm{T}} & -\gamma^2 I_p & * \\ 0 & \Psi C_2^{\mathrm{T}} & -\Psi \end{bmatrix} < 0. \tag{7.13}$$

Since $-\delta I < 0$, then

$$\begin{bmatrix} 0 & 0 & 0 \\ 0 & 0 & 0 \\ 0 & 0 & -\delta I \end{bmatrix} \leq 0,$$

the inequality

$$\begin{bmatrix} \mathrm{Her}(e^{-j\theta}A_1P_1) + F_1F_1^{\mathrm{T}} & * & * \\ e^{-j\theta}C_1P_1 - C_2F_2F_1^{T} & -\gamma^2 I_p & * \\ 0 & \Psi C_2^{\mathrm{T}} & -\Psi - \delta I \end{bmatrix} < 0 \tag{7.14}$$

is also valid. Replacing $\delta I = \Psi + F_2F_2^{\mathrm{T}}$, (7.14) becomes

$$\begin{bmatrix} \mathrm{Her}(e^{-j\theta}A_1P_1) + F_1F_1^{\mathrm{T}} & * & * \\ e^{-j\theta}C_1P_1 - C_2F_2F_1^{\mathrm{T}} & -\gamma^2 I_p & * \\ 0 & \Psi C_2^{\mathrm{T}} & F_2F_2^{\mathrm{T}} - 2\Psi \end{bmatrix} < 0, \tag{7.15}$$

which is equivalent to

$$\begin{bmatrix} \mathrm{Her}(e^{-j\theta}A_1P_1) + F_1F_1^{\mathrm{T}} & * & * \\ 0 & F_2F_2^{\mathrm{T}} - 2\Psi & * \\ e^{-j\theta}C_1P_1 - C_2F_2F_1^{\mathrm{T}} & -C_2\Psi & -\gamma^2 I_p \end{bmatrix} < 0. \tag{7.16}$$

Taking into account

$$P = N^{-1}\begin{bmatrix} P_1 & 0 \\ -e^{j\theta}F_2F_1^{\mathrm{T}} & -e^{j\theta}\Psi \end{bmatrix} M^{\mathrm{T}}, \tag{7.17}$$

we get

$$\begin{aligned} EP &= M\begin{bmatrix} I & 0 \\ 0 & 0 \end{bmatrix} NN^{-1}\begin{bmatrix} P_1 & 0 \\ -e^{j\theta}B_2B_1^{\mathrm{T}} & -e^{j\theta}\Psi \end{bmatrix} M^{\mathrm{T}} \\ &= M\begin{bmatrix} P_1 & 0 \\ 0 & 0 \end{bmatrix} M^{\mathrm{T}} \geq 0. \end{aligned}$$

Likewise,

$$P^*E^{\mathrm{T}} = M \begin{bmatrix} P_1 & -e^{-j\theta} B_1 B_2^{\mathrm{T}} \\ 0 & -e^{-j\theta} \Psi \end{bmatrix} N^{-\mathrm{T}} N^{\mathrm{T}} \begin{bmatrix} I & 0 \\ 0 & 0 \end{bmatrix} M^{\mathrm{T}}$$
$$= M \begin{bmatrix} P_1 & 0 \\ 0 & 0 \end{bmatrix} M^{\mathrm{T}},$$

then (7.3a) holds. Now, to verify that (7.3b) holds, it suffices to show

$$\begin{bmatrix} \mathrm{Her}(e^{-j\theta} AP) + FF^{\mathrm{T}} & * \\ e^{-j\theta} CP & -\gamma^2 I_p \end{bmatrix} < 0. \tag{7.18}$$

Noting that

$$\Sigma = \begin{bmatrix} \mathrm{Her}(e^{-j\theta} AP) + FF^{\mathrm{T}} & * \\ e^{-j\theta} CP & -\gamma^2 I_p \end{bmatrix}$$

and using (7.7), (7.8), and (7.17), we obtain

$$\Sigma = \begin{bmatrix} \Sigma_{11} & \Sigma_{12} \\ \Sigma_{21} & \Sigma_{22} \end{bmatrix}$$

with

$$\Sigma_{12} = \Sigma_{21}^*$$
$$\Sigma_{22} = -\gamma^2 I_p$$

and

$$\begin{aligned}
\Sigma_{11} &= \mathrm{Her}\left\{ \mathrm{e}^{-j\theta} M \begin{bmatrix} A_1 & 0 \\ 0 & I \end{bmatrix} N N^{-1} \begin{bmatrix} P_1 & 0 \\ -\mathrm{e}^{j\theta} F_2 F_1^{\mathrm{T}} & -\mathrm{e}^{j\theta} \Psi \end{bmatrix} M^{\mathrm{T}} \right\} \\
&\quad + M \begin{bmatrix} F_1 \\ F_2 \end{bmatrix} \begin{bmatrix} F_1^T & F_2^{\mathrm{T}} \end{bmatrix} M^{\mathrm{T}} \\
&= \mathrm{Her}\left\{ \mathrm{e}^{-j\theta} M \begin{bmatrix} A_1 P_1 & 0 \\ -\mathrm{e}^{j\theta} F_2 F_1^{\mathrm{T}} & -\mathrm{e}^{j\theta} \Psi \end{bmatrix} M^{\mathrm{T}} \right\} + M \begin{bmatrix} F_1 F_1^{\mathrm{T}} & F_1 F_2^{\mathrm{T}} \\ F_2 F_1^{\mathrm{T}} & F_2 F_2^{\mathrm{T}} \end{bmatrix} M^{\mathrm{T}} \\
&= M \left[\mathrm{Her} \begin{pmatrix} \mathrm{e}^{-j\theta} A_1 P_1 & 0 \\ -F_2 F_1^{\mathrm{T}} & -\Psi \end{pmatrix} + \begin{pmatrix} F_1 F_1^{\mathrm{T}} & F_1 F_2^{\mathrm{T}} \\ F_2 F_1^{\mathrm{T}} & F_2 F_2^{\mathrm{T}} \end{pmatrix} \right] M^{\mathrm{T}} \\
&= M \begin{bmatrix} \mathrm{Her}\left(\mathrm{e}^{-j\theta} A_1 P_1\right) + F_1 F_1^{\mathrm{T}} & -F_1 F_2^{\mathrm{T}} + F_1 F_2^{\mathrm{T}} \\ -F_2 F_1^{\mathrm{T}} + F_2 F_1^{\mathrm{T}} & -2\Psi + F_2 F_2^{\mathrm{T}} \end{bmatrix} M^{\mathrm{T}} \\
\Sigma_{21} &= \mathrm{e}^{-j\theta} \begin{bmatrix} C_1 & C_2 \end{bmatrix} N N^{-1} \begin{bmatrix} P_1 & 0 \\ -\mathrm{e}^{j\theta} F_2 F_1^{\mathrm{T}} & -\mathrm{e}^{j\theta} \Psi \end{bmatrix} M^{\mathrm{T}} \\
&= \mathrm{e}^{-j\theta} \begin{bmatrix} C_1 & C_2 \end{bmatrix} \begin{bmatrix} P_1 & 0 \\ -\mathrm{e}^{j\theta} F_2 F_1^{\mathrm{T}} & -\mathrm{e}^{j\theta} \Psi \end{bmatrix} M^{\mathrm{T}} \\
&= \mathrm{e}^{-j\theta} \begin{bmatrix} C_1 P_1 - \mathrm{e}^{j\theta} C_2 F_2 F_1^{\mathrm{T}} & -\mathrm{e}^{j\theta} C_2 \Psi \end{bmatrix} M^{\mathrm{T}} \\
&= \begin{bmatrix} \mathrm{e}^{-j\theta} C_1 P_1 - C_2 F_2 F_1^{\mathrm{T}} & -C_2 \Psi \end{bmatrix} M^{\mathrm{T}}.
\end{aligned}$$

After a few transformations, we get

$$\Sigma = \begin{bmatrix} M & 0 \\ 0 & I_p \end{bmatrix} \begin{bmatrix} \mathrm{Her}\left(e^{-j\theta} A_1 P_1\right) + F_1 F_1^{\mathrm{T}} & * & * \\ 0 & F_2 F_2^{\mathrm{T}} - 2\Psi & * \\ e^{-j\theta} C_1 P_1 - C_2 F_2 F_1^{\mathrm{T}} & -C_2 \Psi & -\gamma^2 I_p \end{bmatrix} \begin{bmatrix} M^{\mathrm{T}} & 0 \\ 0 & I_p \end{bmatrix}.$$

Using (7.16), it follows that (7.18) is true and the condition (7.3b) holds for this P.

Sufficiency. Assume that there exists a matrix $P \in \mathbb{C}^{n\times n}$ such that conditions in (ii) are satisfied. In the following, we show that the transfer function $G_{wy}(s)$ in (7.2) satisfies $\left\|G_{wy}(s)\right\|_{\mathcal{H}_\infty} < \gamma$. To this end, we apply Schur complement to (7.3b) and obtain

$$\begin{bmatrix} \mathrm{Her}(e^{-j\theta} A P) & * \\ e^{-j\theta} C P & -\gamma^2 I_p \end{bmatrix} + \begin{bmatrix} F \\ 0 \end{bmatrix} \begin{bmatrix} F^{\mathrm{T}} & 0 \end{bmatrix} < 0,$$

i.e.,

$$\begin{bmatrix} \mathrm{Her}(e^{-j\theta} A P) + F F^{\mathrm{T}} & * \\ e^{-j\theta} C P & -\gamma^2 I_p \end{bmatrix} < 0.$$

Applying the Schur complement again, we get

$$\mathrm{Her}(e^{-j\theta} A P) + F F^{\mathrm{T}} + \frac{1}{\gamma^2}(e^{-j\theta} C P)^* e^{-j\theta} C P < 0,$$

which is the same as

$$\mathrm{Her}(e^{-j\theta} A P) + F F^{\mathrm{T}} + \frac{1}{\gamma^2} P^* C^{\mathrm{T}} C P < 0.$$

In this case, a matrix $\omega > 0$ can always be found such that

$$\mathrm{Her}(e^{-j\theta} A P) + F F^{\mathrm{T}} + \frac{1}{\gamma^2} P^* C^{\mathrm{T}} C P + \omega < 0. \tag{7.19}$$

Recall, it is a fact that for any subregion Ω in the complex plane satisfying

$$\Omega \cup \bar{\Omega} = \left\{ s^\alpha / s \in \mathbb{C}, \mathrm{Re}(s) \geq 0 \right\},$$

where $\bar{\Omega}$ is the symmetrical region of Ω with respect to the real axis, there must hold

$$\left\|G_{wy}(s)\right\|_{\mathcal{H}_\infty} = \sup_{\mathrm{Re}(s)\geq 0} \overline{\sigma}(G_{wy}(s)) = \sup_{s\in\Omega} \overline{\sigma}(G_{wy}(s))$$

This just follows from the maximum modulus principle and the complex conjugate symmetry of $G_{wy}(s)$.

Now, we can specify a proper region corresponding to $\mathcal{H}_\infty$ norm.

By taking

$$\Phi = \begin{bmatrix} 0 & e^{-j\theta)} \\ e^{j\theta} & 0 \end{bmatrix}, \quad \Psi = 0$$

with $\theta = (1-\alpha)\frac{\pi}{2}$, we have

$$\begin{aligned} \Upsilon(\Phi, 0) &= \left\{\lambda/\lambda \in \mathbb{C}, \lambda e^{j\theta} + \lambda^* e^{-j\theta} \geq 0\right\}, \\ \overline{\Upsilon}(\Phi, 0) &= \left\{\lambda/\lambda \in \mathbb{C}, \lambda^* e^{j\theta} + \lambda e^{-j\theta} \geq 0\right\}, \end{aligned} \tag{7.20}$$

with $\Upsilon(\Phi, \Psi)$ is defined in Lemma 4.2.3 by the formula (4.7). It is easy to see that

$$\{s^\alpha/s \in \mathbb{C}, \Re e(s) \geq 0\} = \Upsilon(\Phi, 0) \cup \overline{\Upsilon}(\Phi, 0), \tag{7.21}$$

where $\overline{\Upsilon}(\Phi, 0)$ is the symmetric region of $\Upsilon(\Phi, 0)$ with respect to the real axis. Then, we can write

$$\begin{aligned} \left\|G_{wy}(s)\right\|_{\mathcal{H}_\infty} &= \sup_{\Re e(s)\geq 0} \overline{\sigma}\left(G_{wy}(s)\right) = \sup_{s\in\Upsilon(\Phi,0)} \overline{\sigma}\left(G_{wy}(s)\right) \\ &= \sup_{s\in\overline{\Upsilon}(\Phi,0)} \overline{\sigma}\left(G_{wy}(s)\right), \end{aligned} \tag{7.22}$$

which follows from

$$\sigma\left(G_{wy}(s)\right) = \sigma\left(G_{wy}(\overline{s})\right).$$

From (7.20), (7.3a), and (7.19), it can be seen that for all $\lambda \in \Upsilon(\Phi, 0)$

$$\left(\lambda e^{j\theta} + \lambda^* e^{-j\theta}\right) EP - \text{Her}(e^{-j\theta} AP) - FF^{\mathrm{T}} - \mathrm{W} > 0, \tag{7.23}$$

where

$$\mathrm{W} = \frac{1}{\gamma^2} P^* C^{\mathrm{T}} C P + \omega. \tag{7.24}$$

The inequality (7.23) can be written as

$$e^{j\theta} P^*((\lambda^*)^* E^{\mathrm{T}} - A^{\mathrm{T}}) + e^{-j\theta}\left(\lambda^* E - A\right) P - FF^{\mathrm{T}} - \mathrm{W} > 0. \tag{7.25}$$

Since $\lambda \in \Upsilon(\Phi, 0)$, then $\lambda^* \in \overline{\Upsilon}(\Phi, 0)$ and, taking (7.21) into account, we put $\lambda^* = s^\alpha$. Therefore, inequality (7.25) becomes

$$e^{j\theta} P^*\left((s^\alpha)^* E^{\mathrm{T}} - A^{\mathrm{T}}\right) + e^{-j\theta}\left(s^\alpha E - A\right) P - FF^{\mathrm{T}} - \mathrm{W} > 0,$$

which is the same as

$$e^{j\theta} P^*\left(s^\alpha E - A\right)^* + e^{-j\theta}\left(s^\alpha E - A\right) P - FF^{\mathrm{T}} - \mathrm{W} > 0. \tag{7.26}$$

The matrix $(s^\alpha E - A)^{-1}$ is well defined, since the system (7.1) is admissible. Thus, pre- and post-multiplying the inequality (7.26) by $C(s^\alpha E - A)^{-1}$ and $\left[C(s^\alpha E - A)^{-1}\right]^*$, respectively, we obtain

$$\begin{gathered} e^{j\theta}C(s^\alpha E - A)^{-1}P^*(s^\alpha E - A)^*\left[C(s^\alpha E - A)^{-1}\right]^* \\ +e^{-j\theta}C(s^\alpha E - A)^{-1}(s^\alpha E - A)P\left[C(s^\alpha E - A)^{-1}\right]^* \\ -C(s^\alpha E - A)^{-1}FF^{\mathrm{T}}\left[C(s^\alpha E - A)^{-1}\right]^* \\ -C(s^\alpha E - A)^{-1}\mathrm{W}\left[C(s^\alpha E - A)^{-1}\right]^* \geq 0. \end{gathered}$$

i.e.,

$$\begin{gathered} e^{j\theta}C(s^\alpha E - A)^{-1}P^*C^{\mathrm{T}} + e^{-j\theta}CP\left[C(s^\alpha E - A)^{-1}\right]^* \\ -C(s^\alpha E - A)^{-1}F\left[C(s^\alpha E - A)^{-1}F\right]^* \\ -C(s^\alpha E - A)^{-1}\mathrm{W}\left[C(s^\alpha E - A)^{-1}\right]^* \geq 0 \end{gathered}$$

Then,

$$\begin{gathered} -C(s^\alpha E - A)^{-1}F\left[C(s^\alpha E - A)^{-1}F\right]^* \\ \geq C(s^\alpha E - A)^{-1}\mathrm{W}\left[C(s^\alpha E - A)^{-1}\right]^* \\ -e^{j\theta}C(s^\alpha E - A)^{-1}P^*C^T - e^{-j\theta}CP\left[C(s^\alpha E - A)^{-1}\right]^*. \end{gathered}$$

By adding $\gamma^2 I_p$, we get

$$\begin{gathered} \gamma^2 I_p - C(s^\alpha E - A)^{-1}F\left[C(s^\alpha E - A)^{-1}F\right]^* \\ \geq \gamma^2 I_p + C(s^\alpha E - A)^{-1}\mathrm{W}\left[C(s^\alpha E - A)^{-1}\right]^* \\ -e^{j\theta}C(s^\alpha E - A)^{-1}P^*C^T - e^{-j\theta}CP\left[C(s^\alpha E - A)^{-1}\right]^*, \end{gathered} \tag{7.27}$$

since

$$\begin{aligned} &\gamma^2 I_p + C(s^\alpha E - A)^{-1}\mathrm{W}\left[C(s^\alpha E - A)^{-1}\right]^* \\ &-e^{j\theta}C(s^\alpha E - A)^{-1}P^*C^{\mathrm{T}} - e^{-j\theta}CP\left[C(s^\alpha E - A)^{-1}\right]^* \\ &= \gamma^2 I_p + \left[C(s^\alpha E - A)^{-1} - e^{-j\theta}CP\mathrm{W}^{-1}\right] \\ &\times\mathrm{W}\left[\left(C(s^\alpha E - A)^{-1}\right)^* - e^{j\theta}\mathrm{W}^{-1}P^*C^T\right] - e^{-j\theta}CP\mathrm{W}^{-1}e^{j\theta}P^*C^T \\ &= \gamma^2 I_p + \left[C(s^\alpha E - A)^{-1} - e^{-j\theta}CP\mathrm{W}^{-1}\right] \\ &\times\mathrm{W}\left[\left(C(s^\alpha E - A)^{-1}\right)^* - e^{j\theta}\mathrm{W}^{-1}P^*C^{\mathrm{T}}\right] - CP\mathrm{W}^{-1}P^*C^{\mathrm{T}}. \end{aligned}$$

and $\mathrm{W} = \frac{1}{\gamma^2}P^*C^{\mathrm{T}}CP + \omega > 0$, therefore

$$\left[C(s^\alpha E - A)^{-1} - e^{-j\theta}CP\mathrm{W}^{-1}\right] \times \mathrm{W}\left[\left(C(s^\alpha E - A)^{-1}\right)^* - e^{j\theta}\mathrm{W}^{-1}P^*C^{\mathrm{T}}\right] \geq 0.$$

Then from (7.27), the following inequality

$$\gamma^2 I_p - G_{wy}(s)G^*_{wy}(s) \geq \gamma^2 I_p - CP\mathrm{W}^{-1}P^*C^{\mathrm{T}} \tag{7.28}$$

is valid. Note that $\omega = \mathrm{W} - \frac{1}{\gamma^2} P^* C^{\mathrm{T}} C P > 0$ is equivalent to

$$\begin{bmatrix} \mathrm{W} & P^* C^{\mathrm{T}} \\ C P & \gamma^2 I_p \end{bmatrix} > 0$$

or equivalently

$$\gamma^2 I_p - C P \mathrm{W}^{-1} P^* C^{\mathrm{T}} > 0$$

Then, from (7.28) we get $\gamma^2 I_p - G_{wy}(s) G_{wy}^*(s) > 0$, which means that for all elements of the set $\{s^\alpha / s \in \mathbb{C}, \mathrm{Re}(s) \geq 0\}$,

$$\|G_{wy}(s)\|_{\mathcal{H}_\infty} < \gamma.$$

Now, we demonstrate the equivalence between (ii) and (iii).

Sufficiency. Assume that the $\mathcal{LMI}$ (7.4) in (iii) is satisfied for some matrices $P \in \mathbb{C}^{n \times n} > 0$ and $Q \in \mathbb{C}^{(n-r) \times n}$. Let $X = P E^{\mathrm{T}} + S Q$, then we have

$$\begin{aligned} E X &= E P E^{\mathrm{T}} + E S Q = E P E^{\mathrm{T}} \geq 0, \\ X^* E^{\mathrm{T}} &= E P^* E^{\mathrm{T}} + Q^{\mathrm{T}} S^{\mathrm{T}} E^{\mathrm{T}} = E P E^{\mathrm{T}} \geq 0, \end{aligned}$$

which means that the conditions (7.3a) in (ii) are satisfied.

By replacing $P E^{\mathrm{T}} + S Q$ by X in (7.4), that the condition (7.3b) in (ii) is also satisfied. We conclude then that the transfer function of the system (7.1) verifies $\|G_{uy}(s)\|_{\mathcal{H}_\infty} < \gamma$.

Necessity. Assume that the system (7.1) is admissible and its transfer function (7.2) verifies $\|G_{uy}(s)\|_{\mathcal{H}_\infty} < \gamma$ for some scalar $\gamma > 0$. Since the system (7.1) is admissible, then, without loss the generality, the matrix E is assumed to have the form

$$E = \begin{bmatrix} I_r & 0 \\ 0 & 0 \end{bmatrix}.$$

Assume that there exists matrix $X \in \mathbb{C}^{n \times n}$ that verifies (ii). From (7.3a), we deduce that the matrix X takes the following form:

$$X = \begin{bmatrix} P_1 & 0 \\ Q_3 & Q_4 \end{bmatrix}, \quad P_1 > 0.$$

We can then write X as follows:

$$X = \begin{bmatrix} P_1 & 0 \\ 0 & I_{n-r} \end{bmatrix} \begin{bmatrix} I_r & 0 \\ 0 & 0 \end{bmatrix} + \begin{bmatrix} 0 \\ I_{n-r} \end{bmatrix} \begin{bmatrix} Q_3 & Q_4 \end{bmatrix},$$

i.e.,

$$X = PE^{\mathrm{T}} + SQ,$$

with

$$P = \begin{bmatrix} P_1 & 0 \\ 0 & I_{n-r} \end{bmatrix}, \; S = \begin{bmatrix} 0 \\ I_{n-r} \end{bmatrix}, \; Q = \begin{bmatrix} Q_3 & Q_4 \end{bmatrix}.$$

It is clear that $ES = 0$ and the matrix P is positive definite.

By (7.3b), the condition (7.4) in (iii) is satisfied, so the equivalence between (ii) and (iii) is proven.

In the following, we demonstrate that the statements (iii) and (vi) are equivalent.

Sufficiency. By pre-multiplying and post-multiplying (7.5) by $\begin{bmatrix} I & e^{-j\theta}\mathcal{A}_c \end{bmatrix}$ and $\begin{bmatrix} I & e^{-j\theta}\mathcal{A}_c \end{bmatrix}^*$, respectively, we obtain

$$e^{-j\theta}\mathcal{A}_c\left(\mathcal{P}_c\mathcal{E}_c^{\mathrm{T}} + \mathcal{S}\mathcal{Q}_c\right) + e^{j\theta}\left(\mathcal{P}_c\mathcal{E}_c^{\mathrm{T}} + \mathcal{S}\mathcal{Q}_c\right)^*\mathcal{A}_c^* + \mathcal{F}\mathcal{F}^{\mathrm{T}} < 0. \tag{7.29}$$

Setting $\mathcal{S} = \begin{bmatrix} S \\ 0 \end{bmatrix}$, where S is any matrix of full column rank and verifies $ES = 0$, implies $\mathcal{E}_c\mathcal{S} = 0$. Using (7.6), (7.29), the development of

$$\mathrm{Her}\left\{e^{-j\theta}\begin{bmatrix} A & 0 \\ C & -I_p \end{bmatrix}\left(\begin{bmatrix} P & 0 \\ 0 & e^{j\theta}I_p \end{bmatrix}\begin{bmatrix} E^T & 0 \\ 0 & \frac{\gamma^2}{2}I_p \end{bmatrix} + \begin{bmatrix} S \\ 0 \end{bmatrix}\begin{bmatrix} Q & 0 \end{bmatrix}\right)\right\}$$

$$+\begin{bmatrix} F \\ 0 \end{bmatrix}\begin{bmatrix} F^T & 0 \end{bmatrix} \prec 0$$

yields (7.4).

Necessity. The condition (7.4) can be written as

$$e^{-j\theta}\mathcal{A}_c\Omega + e^{j\theta}\Omega^*\mathcal{A}_c^* + \mathcal{F}\mathcal{F}^{\mathrm{T}} < 0, \tag{7.30}$$

where $\mathcal{A}_c$ and $\mathcal{F}$ are defined in (7.6) and

$$\Omega = \begin{bmatrix} PE^T + SQ & 0 \\ 0 & e^{j\theta}\frac{\gamma^2}{2}I_p \end{bmatrix}. \tag{7.31}$$

Therefore, there exists a scalar $\epsilon > 0$ such that:

$$e^{-j\theta}\mathcal{A}_c\Omega + e^{j\theta}\Omega^*\mathcal{A}_c^{\mathrm{T}} + \mathcal{F}\mathcal{F}^{\mathrm{T}} < -\epsilon\mathcal{A}_c\mathcal{A}_c^{\mathrm{T}}. \tag{7.32}$$

By Shur complement, the last inequality is equivalent to

$$\begin{bmatrix} e^{-j\theta}\mathcal{A}_c\Omega + e^{j\theta}\Omega^*\mathcal{A}_c^{\mathrm{T}} + \mathcal{F}\mathcal{F}^{\mathrm{T}} & \epsilon e^{-j\theta}\mathcal{A}_c \\ \epsilon e^{j\theta}\mathcal{A}_c^{\mathrm{T}} & -\epsilon I_{n+p} \end{bmatrix} < 0. \tag{7.33}$$

Since $\epsilon e^{-j\theta}\mathcal{A}_c$ can be written as

$$\epsilon e^{-j\theta}\mathcal{A}_c = e^{-j\theta}\mathcal{A}_c\mathcal{V}^{\mathrm{T}} + e^{-j\theta}\mathcal{A}_c\mathcal{V}$$

with

$$\mathcal{V} = \frac{\epsilon}{2}I_{n+p},$$

taking into account

$$\mathcal{U} = \Omega - e^{j\theta}\mathcal{V}\mathcal{A}_c^{\mathrm{T}},$$

the inequality (7.33) is equivalent to

$$\begin{bmatrix} e^{-j\theta}\mathcal{A}_c\mathcal{U} + e^{j\theta}\mathcal{U}^*\mathcal{A}_c^* + \mathcal{F}\mathcal{F}^{\mathrm{T}} + 2\mathcal{A}_c\mathcal{V}\mathcal{A}_c^{\mathrm{T}} & * \\ \Omega - \mathcal{U} + e^{j\theta}\mathcal{V}^{\mathrm{T}}\mathcal{A}_c^{\mathrm{T}} & -\mathcal{V} - \mathcal{V}^{\mathrm{T}} \end{bmatrix} < 0. \tag{7.34}$$

By expanding the matrix Ω as

$$\Omega = \mathcal{P}_c\mathcal{E}_c^{\mathrm{T}} + \mathcal{S}\mathcal{Q}_c,$$

with $\mathcal{S} = \begin{bmatrix} S^{\mathrm{T}} & 0 \end{bmatrix}^{\mathrm{T}}$ and using (7.6) and $2\mathcal{A}_c\mathcal{V}\mathcal{A}_c^{\mathrm{T}} \geq 0$, the inequality (7.34) implies the condition (7.5). This proves the equivalence between (iii) and (iv) and completes the proof. □

Remark 7.2.1 The statements of Theorem 7.2.1 summarizes the existing results on the bounded real lemma for a larger class of dynamic linear systems. Indeed,

- For $\alpha = 1$ and $E = I_n$, the $\mathcal{LMI}$ condition for the $\mathcal{H}_\infty$ norm (7.5) coincides with the $\mathcal{LMI}$ in [4].
- For $\alpha = 1$, both $\mathcal{LMI}s$ conditions for the $\mathcal{H}_\infty$ norm, (7.4) and (7.5), in Theorem 7.2.1 coincide with the condition (5.20) for singular linear systems in [5] and with the condition (19) in [6] respectively.
- For $E = I_n$, the $\mathcal{LMI}$ condition for the $\mathcal{H}_\infty$ norm (7.4) in Theorem 7.2.1 coincides with the condition (14) in [7]

This confirms that results of Theorem 7.2.1 are more general.

By Theorem 7.2.1 and the duality criterion, we obtain the following corollary.

Corollary 7.2.1 *[3] Let the SFOS* (7.1) *be admissible and* $\gamma > 0$. *The transfer function* $G_{wy}(s)$ *in* (7.2) *verifies*

$$\left\| G_{wy}(s) \right\|_{\mathcal{H}_\infty} < \gamma,$$

if and only if the following equivalent statements hold.
(i) There exist matrices $X \in \mathbb{C}^{n\times n}$ such that

$$E^T X = X^* E \geq 0, \tag{7.35a}$$

$$\begin{bmatrix} \mathrm{Her}\{(e^{-j\theta}A)^* X\} & * & * \\ \left(e^{-j\theta}F\right)^* X & -\gamma^2 I_m & * \\ C & 0 & -I_p \end{bmatrix} < 0. \tag{7.35b}$$

(ii) There exist matrices $P \in \mathbb{C}^{n\times n} \succ 0$ and $Q \in \mathbb{C}^{(n-r)\times n}$ such that the following $\mathcal{LMI}$ holds

$$\begin{bmatrix} \mathrm{Her}\{\left(e^{-j\theta}A\right)^* (PE+SQ)\} & * & * \\ \left(e^{-j\theta}F\right)^* (PE+SQ) & -\gamma^2 I_m & * \\ C & 0 & -I_p \end{bmatrix} < 0, \tag{7.36}$$

where $S \in \mathbb{R}^{n\times(n-r)}$ is any matrix with full column rank and satisfies $E^T S = 0$.

(iii) There exist matrices $P > 0$, Q, $\mathcal{V}$ and $\mathcal{U}$ such that

$$\begin{bmatrix} \mathrm{Her}\{\left(e^{-j\theta}\mathcal{A}_F\right)^* \mathcal{U}^*\} + \mathcal{C}^T\mathcal{C} & * \\ \mathcal{P}_F\mathcal{E}_F + \mathcal{S}\mathcal{Q}_F + e^{-j\theta}\mathcal{V}\mathcal{A}_F - \mathcal{U}^* & -\mathcal{V} - \mathcal{V}^T \end{bmatrix} < 0, \tag{7.37}$$

with $\mathcal{E}_F^T\mathcal{S} = 0$, and

$$\mathcal{E}_F = \begin{bmatrix} E & 0_{n\times m} \\ 0_{m\times n} & \frac{\gamma^2}{2}I_m \end{bmatrix}, \mathcal{A}_F = \begin{bmatrix} A & B \\ 0_{m\times n} & -I_m \end{bmatrix}, \tag{7.38a}$$

$$\mathcal{P}_F = \begin{bmatrix} P & 0_{n\times m} \\ 0_{m\times n} & e^{j\theta}I_m \end{bmatrix}, \mathcal{Q}_F = \begin{bmatrix} Q & 0_{(n-r)\times m} \end{bmatrix}, \tag{7.38b}$$

$$\mathcal{C} = \begin{bmatrix} C & 0_{p\times m} \end{bmatrix}. \tag{7.38c}$$

Example 7.2.1 Consider the system (7.1) with the fractional order derivative $\alpha = 1.2$ and system matrices

$$E = \begin{bmatrix} 1 & 2 & 3 \\ 1 & 2 & 3 \\ 1 & 2 & 3 \end{bmatrix}, A = \begin{bmatrix} -1 & 2 & 0 \\ -3 & -4 & 1 \\ 0 & 0 & -2 \end{bmatrix},$$

$$F = \begin{bmatrix} 1 & 2 \\ 0 & 1 \\ 2 & 3 \end{bmatrix}, C = \begin{bmatrix} 1 & 0 & -1 \end{bmatrix}.$$

The unforced system is regular and impulse free, since $\det(\lambda E - A) = 38\lambda + 20$, so $\det(\lambda E - A)$ is not identically zero and verifies $\mathrm{rank}(E) = \deg(\det(\lambda E - A))$.

Its eigenvalue, $\lambda = -\frac{19}{10}$, is located in the stable region because the condition $|\arg(\lambda)| > \alpha\frac{\pi}{2}$ is satisfied. These three criteria show that the system is admissible.

Solving the $\mathcal{LMI}$ (7.4) in Theorem 7.2.1, we obtain

$$P = \begin{bmatrix} 327.1931 & -60.4637 & -25.6958 \\ -60.4637 & 102.2384 & -56.7679 \\ -25.6958 & -56.7679 & 46.9763 \end{bmatrix},$$

$$Q = \begin{bmatrix} 17.2931 & 35.1381 & 23.2448 \\ 22.4470 & 13.7534 & 20.3702 \end{bmatrix}.$$

The transfer function is $G(s) = \left[\frac{-14s^{1.2}-4}{19s^{1.2}+10}, \frac{-14s^{1.2}-2}{19s^{1.2}+10}\right]$. Since $\|G(s)\|_{\mathcal{H}_\infty} < \infty$, according to Lemma 4.2.1, we have

$$\|G(s)\|_{\mathcal{H}_\infty} = \sup_{\omega\geq 0}\overline{\sigma}\left(G(j\omega)\right) = \sup_{\omega\geq 0}\|G(j\omega)\|_2 .$$

The computation of $\|G(j\omega)\|_2$ yields

$$\|G(j\omega)\|_2 = \sqrt{\frac{392\omega^{2.4} - 51.9149\omega^{1.2} + 20}{361\omega^{2.4} - 117.4260\omega^{1.2} + 100}}.$$

As shown in the graph of $\|G(j\omega)\|_2$ in Fig. 7.1,

$$\sup_{\omega\geq 0}\|G(j\omega)\|_2 = 1.0653.$$

Thus, all the maximal singular values of the transfer function are less than γ, which demonstrates validity of the obtained results.

Note that the admissibility of the system in Example 1 in [8] is not guaranteed by solving $\mathcal{LMI}s$ (i) and (ii) of Theorem 1 in [8], as stated in that paper. This point was discussed in several papers (see, for example, [9]).

7.3 $\mathcal{H}_\infty$ Controller Synthesis for SFOS

The bounded real lemma can be applied to the controller synthesis. Since the matrices in $\mathcal{LMI}$ (7.4) are complex and gain control gains must be real, a novel method is proposed to overcome this difficulty.

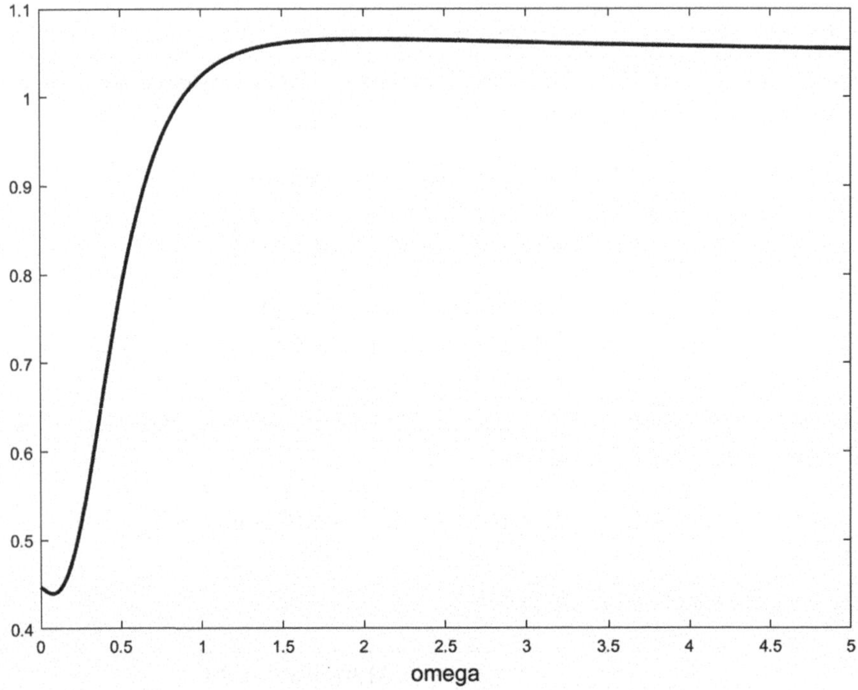

Fig. 7.1 2-norm of $G(j\omega)$ for $\omega \geq 0$

7.3.1 State Feedback $\mathcal{H}_\infty$ Control of SFOS

Let us consider the following SFOS:

$$\begin{cases} ED^{\alpha}x(t) = Ax(t) + Bu(t) + Fw(t), \ 1 \leq \alpha < 2, \\ \quad y(t) = Cx(t), \end{cases} \tag{7.39}$$

where $x(t) \in \mathbb{R}^n$ is the descriptor variable, $u(t) \in \mathbb{R}^m$ is the control input, $w(t) \in \mathbb{R}^q$ is the exogenous input, $y(t) \in \mathbb{R}^p$ is the output, $E, A \in \mathbb{R}^{n\times n}$, $B \in \mathbb{R}^{n\times m}$, $F \in \mathbb{R}^{n\times q}$, $C \in \mathbb{R}^{p\times n}$. The control input is given by the pseudo-state feedback

$$u(t) = Kx(t), \tag{7.40}$$

where $K \in \mathbb{R}^{m\times n}$ is the gain matrix to be determined The system (7.39) becomes

$$\begin{cases} ED^{\alpha}x(t) = (A + BK)x(t) + Fw(t), \\ \quad y(t) = Cx(t), \end{cases} \tag{7.41}$$

and its transfer function has the form

$$G_{wy}^K(s) = C(s^\alpha E - (A + BK))^{-1} F. \tag{7.42}$$

The problem is to design the gain matrix K, such that $\|G_{\omega y}^K(s)\|_{\mathcal{H}_\infty} < \gamma$, where γ is a known positive scalar.

Theorem 7.3.1 *[3] Let the singular fractional order continuous system* (7.41) *with* $w(t) = 0$ *be admissible and* $\gamma > 0$. *For* $w(t) \neq 0$, $\|G_{\omega y}^K\|_{\mathcal{H}_\infty} < \gamma$ *if and only if there exist matrices* $P \in \mathbb{C}^{n\times n} > 0$, $Q \in \mathbb{C}^{(n-r)\times n}$, $\mathcal{U} \in \mathbb{C}^{(n+p)\times(n+p)}$, $\mathcal{V} \in \mathbb{R}^{(n+p)\times(n+p)}$ *and* $K \in \mathbb{R}^{m\times n}$ *such that*

$$\begin{bmatrix} \text{Her}\left\{e^{-j\theta}\mathcal{A}_c\mathcal{U}\right\} + \mathcal{F}\mathcal{F}^T & * \\ \mathcal{P}_c\mathcal{E}_c^T + \mathcal{S}\mathcal{Q}_c + e^{j\theta}\mathcal{V}^T\mathcal{A}_c^T - \mathcal{U} & -\mathcal{V} - \mathcal{V}^T \end{bmatrix} < 0, \tag{7.43}$$

with $\theta = (1-\alpha)\frac{\pi}{2}$, $\mathcal{E}_c\mathcal{S} = 0$, $\mathcal{F} = \begin{bmatrix} F \\ 0_{p\times q} \end{bmatrix}$, *and definitions*

$$\mathcal{E}_c = \begin{bmatrix} E & 0_{n\times p} \\ 0_{p\times n} & \frac{\gamma^2}{2} I_p \end{bmatrix}, \mathcal{A}_c = \begin{bmatrix} A + BK & 0_{n\times p} \\ C & -I_p \end{bmatrix}, \mathcal{F} = \begin{bmatrix} F \\ 0_{p\times q} \end{bmatrix} \tag{7.44a}$$

$$\mathcal{P}_c = \begin{bmatrix} P & 0_{n\times p} \\ 0_{p\times n} & e^{j\theta} I_p \end{bmatrix}, \mathcal{Q} = \begin{bmatrix} Q & 0_{(n-r)\times p} \end{bmatrix} \tag{7.44b}$$

Proof The proof is obtained directly by applying Theorem 7.2.1 to the system (7.41).

In the following, an alternative result is provided.

Theorem 7.3.2 *Let the SFOS* (7.41) *is admissible and* $\gamma > 0$.*The transfer function* $G_{\omega y}^K(s)$ *in* (7.42) *satisfies*

$$\|G_{\omega y}^K(s)\|_{\mathcal{H}_\infty} < \gamma$$

if and only if there exist matrices $P_0 \in \mathbb{R}^{n\times n} > 0$, $P \in \text{H}_n > 0$, Q_0, $Q \in \mathbb{R}^{(n-r)\times n}$ *and scalar* $\varepsilon > 0$ *such that*

$$\begin{bmatrix} \Sigma & * & * \\ e^{j\theta} F^T(PE + SQ) & -\gamma^2 I_q & * \\ C & 0 & -I_p \end{bmatrix} < 0 \tag{7.45}$$

where

$$\Sigma = Her\{e^{j\theta}(A^T(PE + SQ)) - \frac{1}{\varepsilon}e^{j\theta}\sin\left(\alpha\frac{\pi}{2}\right)\Phi^T BB^T(PE + SQ)\},$$

$$\Phi = P_0 E + SQ_0,$$

and S *is any matrix of full column rank and satisfies* $E^T S = 0$. *The controller gain* K *can be designed as*

$$K = -\frac{1}{\varepsilon}\sin\left(\alpha\frac{\pi}{2}\right) B^T\Phi \tag{7.46}$$

Proof (Necessary) Suppose that Since $\left\|G_{\omega y}^K(s)\right\|_{\mathcal{H}_\infty} < \gamma$, then according to (7.36) in Corollary 7.2.1, there exists matrices $P \in H_n > 0$ and $Q \in \mathbb{R}^{(n-r)\times n}$ such that

$$\begin{bmatrix} Her\{(e^{-j\theta}(A+BK))^*(PE+SQ)\} & * & * \\ (e^{-j\theta}F)^*(PE+SQ) & -\gamma^2 I_q & * \\ C & 0 & -I_p \end{bmatrix} < 0, \tag{7.47}$$

Since the system (7.41) is admissible and according to Theorem 6.3.9, it will exist matrices $P_0 \in \mathbb{R}^{n\times n} \succ 0$, $Q_0 \in \mathbb{R}^{(n-r)\times n}$ and scalar $\varepsilon > 0$ such that the gain matrix is given by (6.101). By substitution of $K = -\frac{1}{\varepsilon}\sin(\alpha\frac{\pi}{2})B^{\mathrm{T}}((P_0E+SQ_0))$ into (7.47), we can easily verify that (7.45) holds.

(Sufficiency) The Corollary7.2.1 guarantees the $\mathcal{H}_\infty$ norm condition by simple substitution of the formula (7.46) into (7.45). With this we arrive to the end of the proof. □

Remark 7.3.1 Both Theorems 7.3.1 and 7.3.2 provide necessary and sufficient condition for the system (7.41) to satisfy an $\mathcal{H}_\infty$ performance index γ. However, conditions (7.43) and (7.45) are given by Bilinear Matrix Inequalities ($\mathcal{BMI}$). The following result proposes a sufficient condition in terms of $\mathcal{LMI}$.

Theorem 7.3.3 *[3] The singular fractional order continuous system* (7.41) *is admissible with an $\mathcal{H}_\infty$ performance index γ if there exist matrices $P \in \mathbb{R}^{n\times n} > 0$, $Q \in \mathbb{R}^{(n-r)\times n}$, $U_1 \in \mathbb{R}^{n\times n}$, $U_3 \in \mathbb{C}^{p\times n}$, $V_3 \in \mathbb{R}^{p\times n}$, $U_4 \in \mathbb{C}^{p\times p}$, and $V_4 \in \mathbb{R}^{p\times p}$ such that the following $\mathcal{LMI}$ holds:*

$$\begin{bmatrix} \Psi_{11} & \Psi_{12} & \Psi_{13} & \Psi_{14} \\ * & \Psi_{22} & \Psi_{23} & \Psi_{24} \\ * & * & \Psi_{33} & \Psi_{34} \\ * & * & * & \Psi_{44} \end{bmatrix} < 0, \tag{7.48}$$

where $S \in \mathbb{R}^{n\times(n-r)}$ is any matrix of full column rank satisfying $ES = 0_{n\times(n-r)}$, $\mathcal{J} \in \mathbb{C}^{n\times p}$ is known matrix, $\theta = (1-\alpha)\frac{\pi}{2}$ and

$\Psi_{11} = \mathrm{Her}\left\{e^{-j\theta}(AU_1+BM)\right\} + FF^T,$

$\Psi_{12} = e^{j\theta}(U_1^TC^T - U_3^*) + e^{-j\theta}(AU_1\mathcal{J} + BM\mathcal{J}),$

$\Psi_{13} = (PE^T+SQ)^T + e^{-j\theta}(AU_1+BM) - U_1^T,$

$\Psi_{14} = e^{-j\theta}(AU_1\mathcal{J} + BM\mathcal{J}) - U_3^*,$

$\Psi_{22} = \mathrm{Her}\left\{e^{-j\theta}(CU_1\mathcal{J} - U_4)\right\},$

$\Psi_{23} = e^{-j\theta}(CU_1 - V_3) - \mathcal{J}^TU_1^T,$

$\Psi_{24} = e^{-j\theta}(\frac{\gamma^2}{2}I_p + CU_1\mathcal{J} - V_4) - U_4^*,$

$$\Psi_{33} = -U_1 - U_1^T,$$
$$\Psi_{34} = -U_1\mathcal{J} - V_3^T,$$
$$\Psi_{44} = -V_4 - V_4^T.$$

The gain matrix K of the controller (7.40) *can be designed as*

$$K = MU_1^{-1}.$$

Proof Assume that the $\mathcal{LMI}$ (7.48) holds for some matrices P, Q, U_1, U_3, U_4, V_3, V_4, M satisfying the conditions in the statement of Theorem 7.3.3. Set the matrices $\mathcal{U}$, $\mathcal{V}$ in Theorem 7.3.1 to

$$\mathcal{U} = \begin{bmatrix} U_1 & U_1\mathcal{J} \\ U_3 & U_4 \end{bmatrix}, \mathcal{V} = \begin{bmatrix} U_1 & U_1\mathcal{J} \\ V_3 & V_4 \end{bmatrix}. \tag{7.49}$$

Then, let $M = KU_1$, which yields the inequality (7.43).

Remark 7.3.2 • The matrix $\mathcal{J}$ is usually real but can also be assigned complex or null.
- It is possible to assign distinct matrices $\mathcal{J}_u$ and $\mathcal{J}_v$ in place of J, i.e., define the matrices U and V in (7.49) as $U_2 = U_1\mathcal{J}_u$ and $V_2 = U_1\mathcal{J}_v$.
- Note that choosing the matrix U_1 real leads to a real control gain $K \in \mathbb{R}^{m\times n}$.
- Taking P real allows one to ensure the admissibility and the $\mathcal{H}_\infty$ performance.

Example 7.3.1 This example deals with the $\mathcal{H}_\infty$ state feedback control for the admissible singular system (7.39) with the following data:

$$E = \begin{bmatrix} 1 & 2 \\ 1 & 2 \end{bmatrix}, A = \begin{bmatrix} -5 & 3 \\ 2 & -3 \end{bmatrix}, B = \begin{bmatrix} 2 \\ 3 \end{bmatrix}, F = \begin{bmatrix} 1 \\ 2 \end{bmatrix}, \alpha = 1.12,$$

The transfer function of the open-loop system is equal to

$$G_o(s) = (s^\alpha E - A)^{-1}F,$$

and the transfer function of the closed-loop system (7.41) is given by

$$G_c(s) = (s^\alpha E - A - BK)^{-1}F.$$

Since $\det(s^\alpha E - A) = 20s^\alpha + 9$, the pair (E,A) is admissible.

As observed from Fig. 7.2,

$$\|G_o(j\omega)\|_{\mathcal{H}_\infty} = 2.7759,$$

which means that $\|G_o(j\omega)\|_{\mathcal{H}_\infty}$ is bounded. The objective of this example is to minimize the influence of $w(t)$ on the system (7.41), satisfying the condition

$$\|G_c(j\omega)\|_{\mathcal{H}_\infty} < \|G_o(j\omega)\|_{\mathcal{H}_\infty}.$$

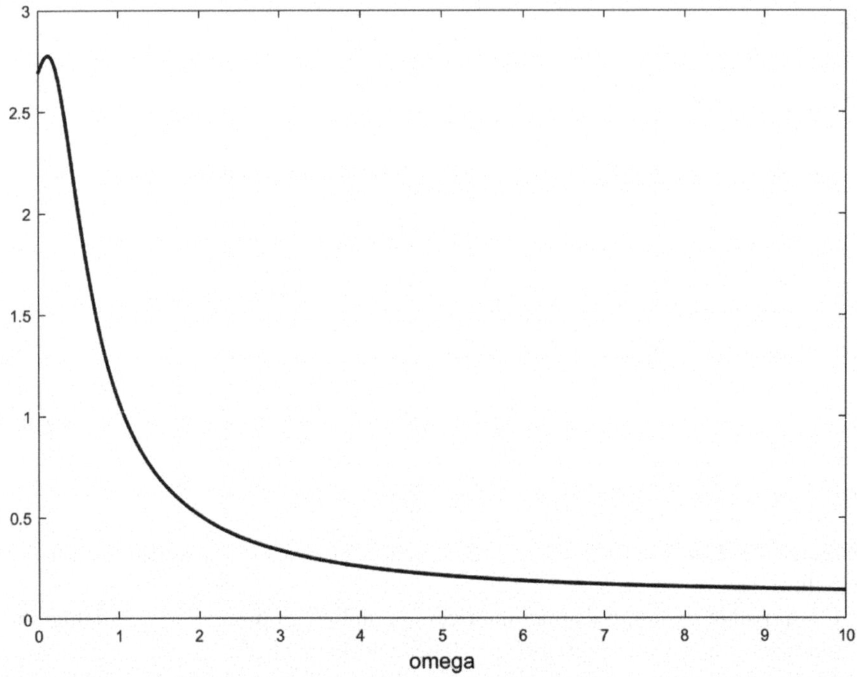

Fig. 7.2 Modulus of $G_o(j\omega)$

Applying Theorem 7.3.3 with the matrix $\mathcal{J} = \begin{bmatrix} 1 & 0 \\ 0 & 1 \end{bmatrix}$, the following results are obtained:

$$P = \begin{bmatrix} 15.4960 & -6.9670 \\ -6.9670 & 5.0455 \end{bmatrix}, \; Q = \begin{bmatrix} -1.3319 & 0.3691 \end{bmatrix},$$

$$U_1 = \begin{bmatrix} 0.5091 & 0.0259 \\ 0.0259 & 0.2143 \end{bmatrix}, \; U_3 = \begin{bmatrix} -3.3351 & -1.0150 \\ -1.0150 & -3.3024 \end{bmatrix},$$

$$U_4 = \begin{bmatrix} 1.9084 + 0.2921i & 0.0092 + 0.0916i \\ 0.0092 + 0.0916i & 1.7072 + 0.2181i \end{bmatrix},$$

$$V_3 = \begin{bmatrix} 0.1566 & 0.1010 \\ 0.1010 & 0.2553 \end{bmatrix}, \; V_4 = \begin{bmatrix} 0.6501 & -0.1242 \\ -0.1242 & 0.8189 \end{bmatrix},$$

and

$$M = \begin{bmatrix} -0.6510 & -1.0317 \end{bmatrix}.$$

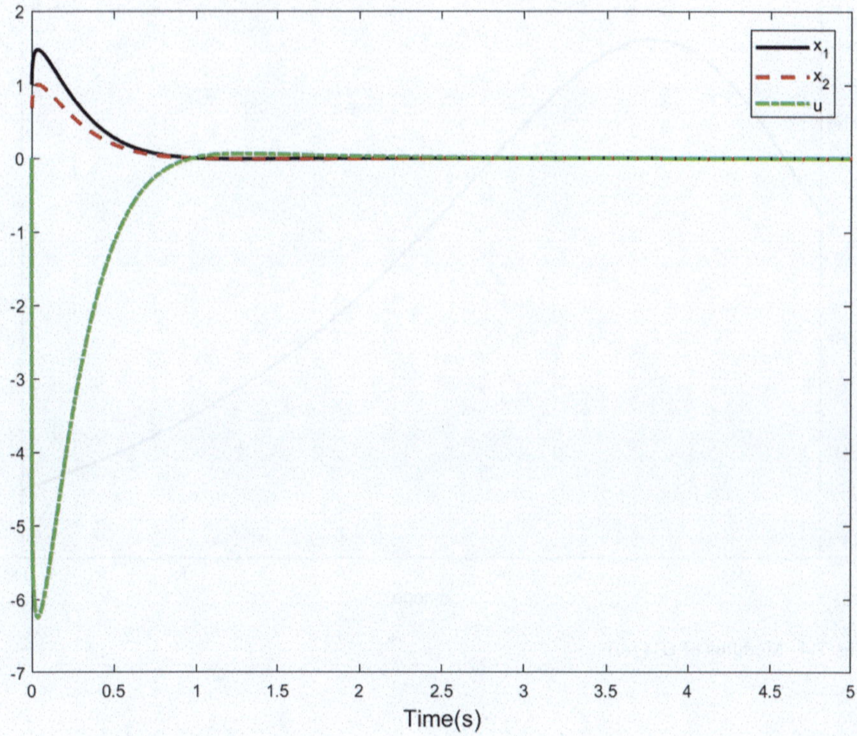

Fig. 7.3 Time histories of system states with $u(t) = Kx(t)$

The obtained $\mathcal{H}_\infty$ controller gain is

$$K = \begin{bmatrix} -1.0400 & -4.6891 \end{bmatrix}.$$

Therefore, the system $(E, A + BK)$ is admissible, its controlled states are plotted in Fig. 7.3, and $\|G_c(s)\|_{\mathcal{H}_\infty}$ is bounded.

Thus, it can be obtained that

$$\|G_c(s)\|_{\mathcal{H}_\infty} = \sup_{\omega \geq 0} \|G_c(j\omega)\|_2 = 0.1286,$$

as shown in Fig. 7.4.

Example 7.3.2 The following example deals with the $\mathcal{H}_\infty$ state feedback control for a no-admissible singular fractional order system. Consider the system (7.39) with the following data

$$E = \begin{bmatrix} 1 & 2 \\ 1 & 2 \end{bmatrix}, \ A = \begin{bmatrix} -5 & 3 \\ 2 & 0 \end{bmatrix}, \ B = \begin{bmatrix} 2 \\ 3 \end{bmatrix}, \ F = \begin{bmatrix} 1 \\ 2 \end{bmatrix}, \ C = \begin{bmatrix} 1 & 0 \\ 0 & 1 \end{bmatrix}$$

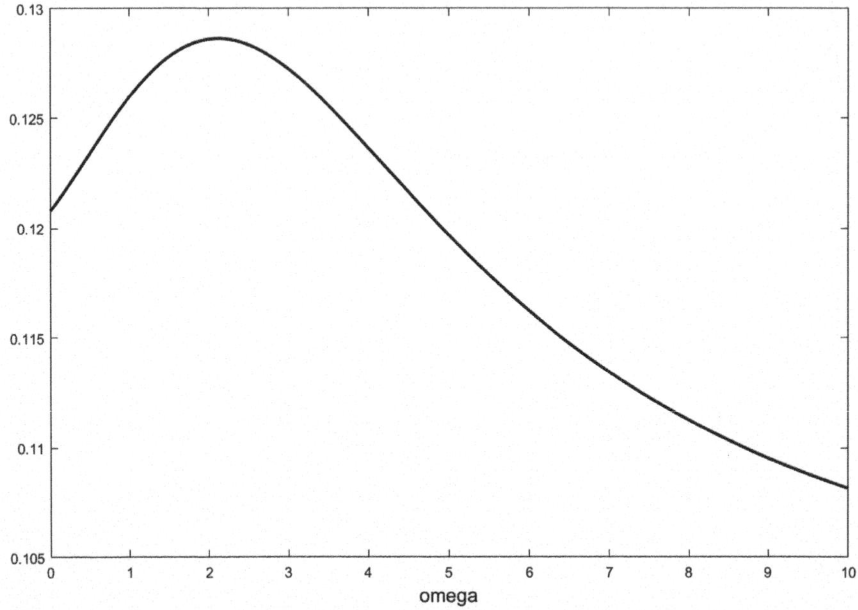

Fig. 7.4 Modulus of $G_c(j\omega)$

and the fractional order $\alpha = 1.5$.

The transfer function of the open-loop system, $u(t) = 0$, is

$$G_o(s) = \begin{bmatrix} \frac{-2s^{1.5}+6}{17s^{1.5}-6} \\ \frac{s^{1.5}+12}{17s^{1.5}-6} \end{bmatrix}$$

It is clear that the pair (E, A) is regular, impulse free but unstable as can be confirmed in Fig. 7.5. In this case $\|G_o(s)\|_{\mathcal{H}_\infty}$ is unbounded. Indeed, since for a complex number s with $\mathrm{Re}(s) > 0$, i.e. $s = |s|\, e^{j\theta}$ and $|\theta| < \frac{\pi}{2}$, s^α has the analytic value $s^\alpha = |s|^\alpha\, e^{j\alpha\theta}$, so

$$\|G_o(s)\|_{\mathcal{H}_\infty} = \sup_{\mathrm{Re}(s)\geq 0} \sqrt{\frac{\left\|-2\,|s|^{1.5}\, e^{(j1.5\theta)} + 6\right\|^2 + \left\||s|^{1.5}\, e^{(j1.5\theta)} + 12\right\|^2}{\left\|17\,|s|^{1.5}\, e^{j1.5\theta} - 6\right\|^2}}$$

The maximum value is obtained for $\theta = 0$ and $|s|^{1.5} = \frac{6}{17}$, which means that $\|G_o(s)\|_{\mathcal{H}_\infty}$ is unbounded. Therefore, we shall find a state feedback law control (7.40) which make the resulting closed-loop system admissible with $\mathcal{H}_\infty$ norm less than some given bound γ. $\mathcal{LMI}$ (7.48) in Theorem 7.3.3 gives for $\gamma = 0.3097$

$$P_1 = \begin{bmatrix} 844.2186 & -844.1857 \\ -844.1857 & 844.2186 \end{bmatrix}, \quad P_2 = 1.6884 \times 10^3,$$

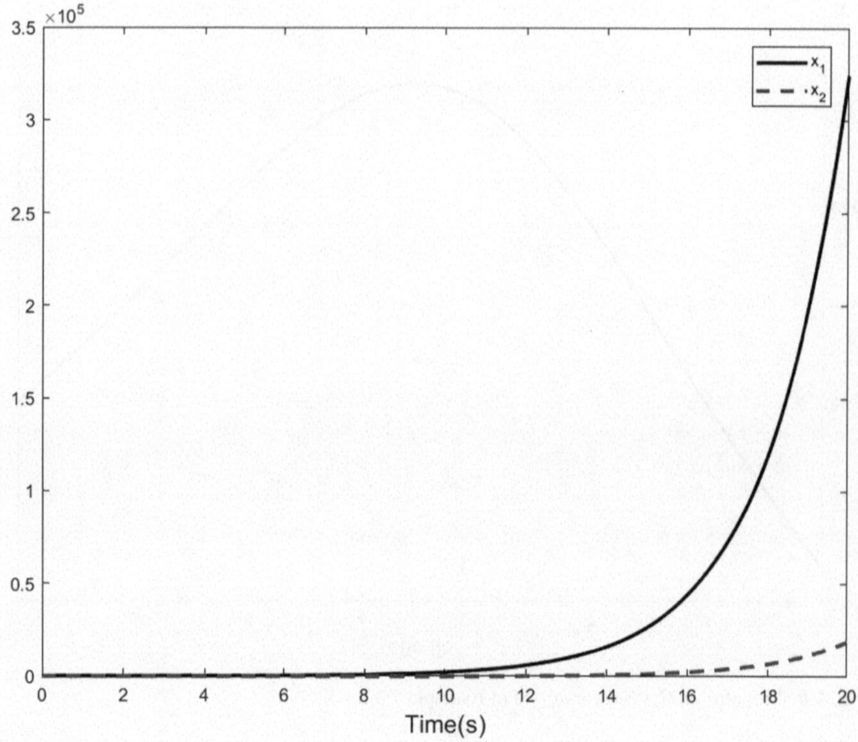

Fig. 7.5 State responses of the open-loop system in Example 7.3.2

$$Q_1 = 10^3 \left[\, 1.0229 \ 1.5347 \,\right], \ Q_3 = 0.0506,$$

$$M = 85.3608, \ N = \left[\, -169.7535 \ -256.7111 \,\right].$$

The control gain matrix obtained is

$$K = \left[\, -1.9887 \ -3.0074 \,\right].$$

Then, The transfer function of the closed-loop selected system is

$$G_c(s) = \begin{bmatrix} \frac{-2s^{1.5}+2.9926}{16.03s^{1.5}+69.0389} \\ \frac{s^{1.5}+13.9887}{16.03s^{1.5}+69.0389} \end{bmatrix}.$$

The system $(E, A + BK)$ is now admissible and $\|G_c(s)\|_{\mathcal{H}_\infty} = 0.3048$. The improvement for $\mathcal{H}_\infty$ performance of the control system can also be shown by the state responses of the closed-loop system and the modulus of its transfer function represented in Figs. 7.6 and 7.7 respectively.

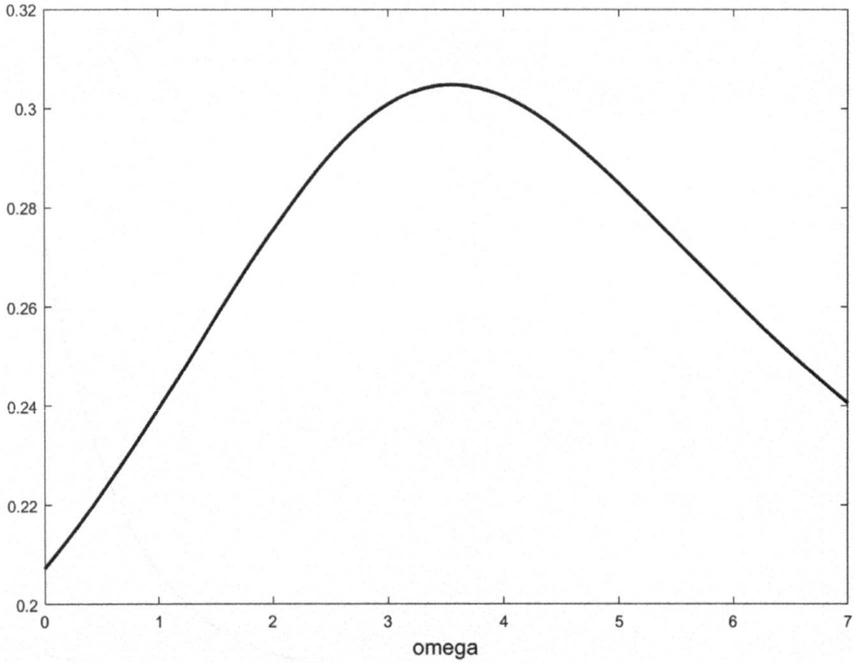

Fig. 7.6 Modulus of $G_c(j\omega)$ for $\omega \geq 0$ in Example 7.3.2

7.3.2 *Static Output Feedback $\mathcal{H}_\infty$ Control of SFOS*

In this section, the problem of static output feedback $\mathcal{H}_\infty$ control for singular continuous-time fractional-order systems with the fractional-order derivative α belonging to $1 \leq \alpha < 2$ will be investigated. All required conditions are characterized in the form of strict linear matrix inequalities whose feasible solutions can be obtained easily and directly.

Consider the continuous-time singular fractional-order system:

$$\begin{cases} ED^\alpha x(t) = Ax(t) + Bu(t) + F\omega(t), \ 1 \leq \alpha < 2, \\ \quad y(t) = Cx(t), \end{cases} \tag{7.50}$$

where $x(t) \in \mathbb{R}^n$ is the descriptor variable, $u(t) \in \mathbb{R}^m$ is the control input, $\omega(t) \in \mathbb{R}^q$ is the exogenous input, $y(t) \in \mathbb{R}^p$ is the output. E, $A \in \mathbb{R}^{n\times n}$, $B \in \mathbb{R}^{n\times m}$, $F \in \mathbb{R}^{n\times q}$, $C \in \mathbb{R}^{p\times n}$. The output control is given by the static output feedback controller

$$u(t) = Ky(t), \tag{7.51}$$

where K is a gain matrix of appropriate dimensions.

The system (7.50) becomes

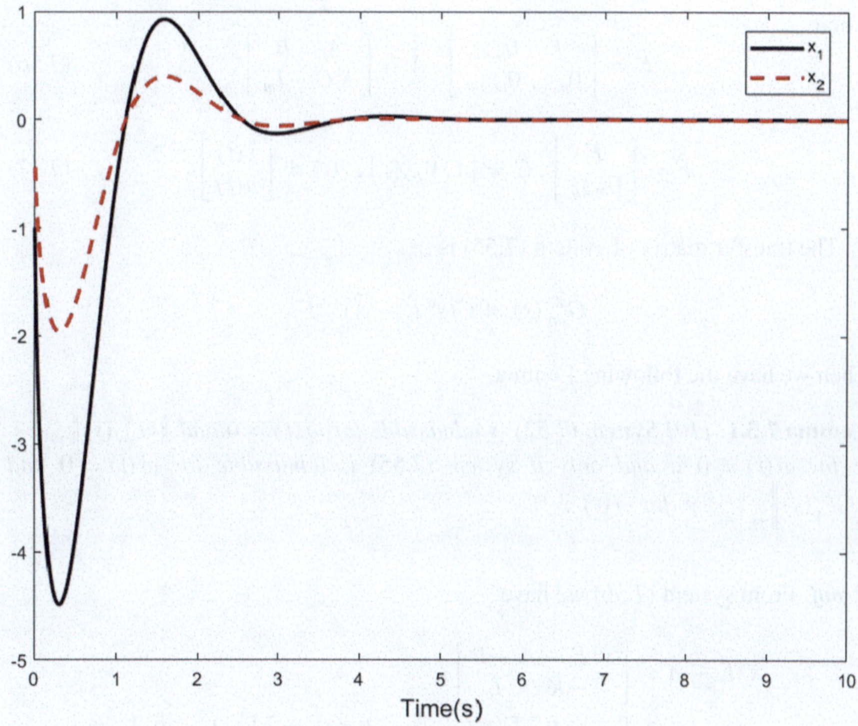

Fig. 7.7 State responses of the closed-loop system in Example 7.3.2

$$\begin{cases} ED^\alpha x(t) = (A + BKC)x(t) + F\omega(t), \\ y(t) = Cx(t). \end{cases} \tag{7.52}$$

and its transfer function has the form

$$G^K_{\omega y}(s) = C(s^\alpha E - (A + BKC))^{-1} F. \tag{7.53}$$

The problem is to design the gain matrix K, such that the system (7.52) is admissible when $\omega(t) = 0$ and $\left\| G^K_{\omega y}(s) \right\|_{\mathcal{H}_\infty} < \gamma$ with γ is a known positive scalar.

To solve the problem of $\mathcal{H}_\infty$ controller, we rewrite the system (7.52) as

$$\begin{cases} ED^\alpha x(t) = Ax(t) + Bu(t) + F\omega(t), \\ 0 \quad = KCx(t) - u(t), \\ y(t) \quad = Cx(t). \end{cases} \tag{7.54}$$

The system (7.54) is also the same as

$$\begin{cases} \hat{E} D^\alpha z(t) = \hat{A} z(t) + \hat{F}\omega(t), \\ y(t) \quad = \hat{C} z(t), \end{cases} \tag{7.55}$$

where

$$\hat{E} = \begin{bmatrix} E & 0_{n\times m} \\ 0_{m\times n} & 0_{m\times m} \end{bmatrix}, \hat{A} = \begin{bmatrix} A & B \\ KC & -I_m \end{bmatrix}, \tag{7.56}$$

$$\hat{F} = \begin{bmatrix} F \\ 0_{m\times q} \end{bmatrix}, \hat{C} = \begin{bmatrix} C & 0_{p\times m} \end{bmatrix}, z(t) = \begin{bmatrix} x(t) \\ u(t) \end{bmatrix}. \tag{7.57}$$

The transfer matrix of system (7.55) is

$$\hat{G}^K_{\omega y}(s) = \hat{C}(s^\alpha \hat{E} - \hat{A})^{-1}\hat{F}.$$

Then we have the following Lemma.

Lemma 7.3.1 *[10] System (7.52) is admissible for $\omega(t) = 0$ and $\left\|G^K_{\omega y}(s)\right\|_{\mathcal{H}_\infty} < \gamma$ for $\omega(t) \neq 0$ if and only if system (7.55) is admissible for $\omega(t) = 0$ and $\left\|\hat{G}^K_{\omega y}(s)\right\|_{\mathcal{H}_\infty} < \gamma$ for $\omega(t) \neq 0$.*

Proof From system (7.56) we have

$$\begin{aligned} s^\alpha \hat{E} - \hat{A} &= \begin{bmatrix} s^\alpha E - A & -B \\ -K & I_m \end{bmatrix}, \\ &= \begin{bmatrix} I_n & -B \\ 0 & I_m \end{bmatrix} \begin{bmatrix} s^\alpha E - A - BKC & 0 \\ 0 & I_m \end{bmatrix} \begin{bmatrix} I_n & 0 \\ -KC & I_m \end{bmatrix}. \end{aligned}$$

Since the matrices $\begin{bmatrix} I_n & -B \\ 0 & I_m \end{bmatrix}$ and $\begin{bmatrix} I_n & 0 \\ -KC & I_m \end{bmatrix}$ are invertible and their determinants are equal to 1, then $\det(s^\alpha \hat{E} - \hat{A}) = \det(s^\alpha E - (A + BKC))$. Therefore the regularity and the admissibility of system (7.52) are equivalent to those of system (7.55). We have also $\operatorname{rank}\left(\hat{E}\right) = \operatorname{rank}(E)$, then the causality of system (7.52) is equivalent to that of the system (7.55). In the other hand, we have

$$\begin{aligned} \hat{G}^K_{\omega y}(s) &= \begin{bmatrix} C & 0 \end{bmatrix} \begin{bmatrix} s^\alpha E - A & -B \\ -KC & I_m \end{bmatrix}^{-1} \begin{bmatrix} F \\ 0 \end{bmatrix} \\ &= \begin{bmatrix} C & 0 \end{bmatrix} \left(\begin{bmatrix} I_n & -B \\ 0 & I_m \end{bmatrix} \begin{bmatrix} s^\alpha E - A - BKC & 0 \\ 0 & I_m \end{bmatrix} \begin{bmatrix} I_n & 0 \\ -KC & I_m \end{bmatrix} \right)^{-1} \begin{bmatrix} F \\ 0 \end{bmatrix} \\ &= \begin{bmatrix} C & 0 \end{bmatrix} \begin{bmatrix} I_n & 0 \\ -KC & I_m \end{bmatrix}^{-1} \begin{bmatrix} s^\alpha E - A - BKC & 0 \\ 0 & I_m \end{bmatrix}^{-1} \begin{bmatrix} I_n & -B \\ 0 & I_m \end{bmatrix}^{-1} \begin{bmatrix} F \\ 0 \end{bmatrix} \\ &= \begin{bmatrix} C & 0 \end{bmatrix} \begin{bmatrix} I_n & 0 \\ KC & I_m \end{bmatrix} \begin{bmatrix} (s^\alpha E - A - BKC)^{-1} & 0 \\ 0 & I_m \end{bmatrix} \begin{bmatrix} I_n & B \\ 0 & I_m \end{bmatrix} \begin{bmatrix} F \\ 0 \end{bmatrix} \end{aligned}$$

$$
\begin{aligned}
&= [C\ 0]\begin{bmatrix}(s^\alpha E - A - BKC)^{-1} & 0\\ 0 & I_m\end{bmatrix}\begin{bmatrix}F\\0\end{bmatrix}\\
&= \left[C(s^\alpha E - A - BKC)^{-1}\ 0\right]\begin{bmatrix}F\\0\end{bmatrix}\\
&= C(s^\alpha E - A - BKC)^{-1}F\\
&= G^K_{\omega y}(s).
\end{aligned}
$$

Then the Lemma is proved. □

Now, by using the results of Lemma 7.3.1 and Theorems 5.4.2 and 7.2.1, we can give the following theorem.

Theorem 7.3.4 *[10] Let the singular fractional order continuous system (7.55) be admissible and $\gamma > 0$. For $\theta = (1-\alpha)\frac{\pi}{2}$, $\left\|\hat{G}^K_{\omega y}\right\|_{\mathcal{H}_\infty} < \gamma$ if and only if one of the following equivalent statements holds:*
(i) There exist matrices $\hat{P} \in \mathrm{H}_{n+m} > 0$ and $\hat{Q} \in \mathbb{R}^{(n+m-r)\times(n+m)}$ such that:

$$
\begin{bmatrix}
Her\{e^{-j\theta}\hat{A}\left(\hat{P}\hat{E}^T + \hat{S}\hat{Q}\right)\} & * & *\\
e^{-j\theta}\hat{C}\left(\hat{P}\hat{E}^T + \hat{S}\hat{Q}\right) & -\gamma^2 I_p & *\\
\hat{F}^T & 0 & -I_q
\end{bmatrix} < 0, \tag{7.58}
$$

where $\hat{S} \in \mathbb{R}^{(n+m)\times(n+m-r)}$ is any matrix with full column rank and satisfies $\hat{E}\hat{S} = 0$.

(ii) There exist matrices $\hat{P} \in \mathrm{H}_{n+m} > 0$ and $\hat{Q} \in \mathbb{R}^{(n+m-r)\times(n+m)}$ such that:

$$
\Sigma = \begin{bmatrix}
Her\{\left(e^{-j\theta}\hat{A}\right)^*\left(\hat{P}\hat{E} + \hat{S}\hat{Q}\right)\} & * & *\\
\left(e^{-j\theta}\hat{F}\right)^*\left(\hat{P}\hat{E} + \hat{S}\hat{Q}\right) & -\gamma^2 I_q & *\\
\hat{C} & 0 & -I_p
\end{bmatrix} < 0, \tag{7.59}
$$

where $\hat{S} \in \mathbb{R}^{(n+m)\times(n+m-r)}$ is any matrix with full column rank and satisfies

$$\hat{E}^T\hat{S} = 0.$$

Proof The proof is obtained directly by replacing A by $\hat{A}$, B by $\hat{F}$ and C by $\hat{C}$ in (7.4) and (7.36) in Theorem 7.2.1, and Corollary 7.2.1 respectively.

Remark 7.3.3 Theorem 7.3.4 gives necessary and sufficient condition for the system (7.55) to get an $\mathcal{H}_\infty$ performance γ. However, it is noted that the condition in (7.59) is in the Bilinear Matrix Inequalities ($\mathcal{BMI}s$) which can be solved by using the existing numerical tools. Nevertheless, strict $\mathcal{LMI}$ formulation is proposed in the following.

Theorem 7.3.5 *[10] The singular fractional order continuous system* (7.55) *is admissible with an $\mathcal{H}_\infty$ performance γ if there exist matrices $P_1 \in \mathbb{R}^{n\times n} > 0$, $Q_1 \in \mathbb{R}^{(n-r)\times n}$, $Q_2 \in \mathbb{R}^{(n-r)\times m}$, $N \in \mathbb{R}^{m\times n}$, and nonsingular matrix $Q_3 \in \mathbb{R}^{m\times m}$ such that the following $\mathcal{LMI}$ holds*

$$\begin{bmatrix} \Psi_{11} & \Psi_{21}^* & \Psi_{31}^* & C^T \\ \Psi_{21} & \Psi_{22} & \Psi_{32}^* & 0_{m\times p} \\ \Psi_{31} & \Psi_{32} & -\gamma^2 I_q & 0_{q\times p} \\ C & 0_{p\times m} & 0_{p\times q} & -I_p \end{bmatrix} < 0, \tag{7.60}$$

where $S_1 \in \mathbb{R}^{n\times(n-r)}$, $S_2 \in \mathbb{R}^{n\times m}$ are any matrices satisfying $E^T[S_1\ S_2] = 0_{n\times(n+m-r)}$, $J \in \mathbb{R}^{m\times n}$ is known matrix and

$$\begin{aligned}
\Psi_{11} &= Her\left\{e^{j\theta}(A^T P_1 E + A^T S_1 Q_1 + A^T S_2 Q_3 J + C^T N^T J)\right\},\\
\Psi_{21} &= e^{j\theta}\left(B^T P_1 E + B^T S_1 Q_1 + B^T S_2 Q_3 J - Q_3 J\right) + e^{-j\theta}\left(Q_2^T S_1^T A + \right.\\
&\quad \left. Q_3^T S_2^T A + NC\right),\\
\Psi_{22} &= Her\left\{e^{j\theta}\left(B^T S_1 Q_2 + B^T S_2 Q_3 - Q_3\right)\right\},\\
\Psi_{31} &= e^{j\theta}\left(F^T P_1 E + F^T S_1 Q_1 + F^T S_2 Q_3 J\right),\\
\Psi_{32} &= e^{j\theta}\left(F^T S_1 Q_2 + F^T S_2 Q_3\right).
\end{aligned}$$

The gain matrix K of the controller (7.51) *can be designed as*

$$K = Q_3^{-T} N.$$

Proof we will assume that the $\mathcal{LMI}$ (7.60) holds for some matrices P_1, Q_1, Q_2, Q_3, N, checking the conditions cited in the statement of the theorem and we will look for matrices $\hat{P} \succ 0$ and $\hat{Q}$ such that the system (7.55) is admissible and condition (7.59) in Theorem 7.3.4 holds.

Let

$$\hat{P} = \begin{bmatrix} P_1 & 0 \\ 0 & I_m \end{bmatrix}, \hat{S} = \begin{bmatrix} S_1 & S_2 \\ 0_{m\times(n-r)} & I_m \end{bmatrix}, \hat{Q} = \begin{bmatrix} Q_1 & Q_2 \\ Q_3 J & Q_3 \end{bmatrix}. \tag{7.61}$$

It is easy to see that $\hat{P} > 0$ since $P_1 > 0$ and $I_m > 0$ and the matrix $\hat{S}$ is of full column rank since S_1 is of full column rank. We have also

$$\begin{aligned}
\hat{E}^T\hat{S} &= \begin{bmatrix} E & 0_{n\times m} \\ 0_{m\times n} & 0_{m\times m} \end{bmatrix}^T \times \begin{bmatrix} S_1 & S_2 \\ 0_{m\times(n-r)} & I_m \end{bmatrix},\\
&= \begin{bmatrix} E^T S_1 & E^T S_2 \\ 0_{m\times(n-r)} & 0_{m\times m} \end{bmatrix} = \begin{bmatrix} E^T\left[S_1\ S_2\right] \\ 0_{m\times(n+m-r)} \end{bmatrix},
\end{aligned}$$

and since $E^T\left[S_1\ S_2\right] = 0_{n\times(n+m-r)}$, then $\hat{E}^T\hat{S} = 0_{(n+m)\times(n+m-r)}$.

By substituting (7.56), (7.57) into the matrix

$$\Sigma = \begin{bmatrix} \Sigma_{11} & \Sigma_{21}^* & \hat{C}^* \\ \Sigma_{21} & -\gamma^2 I_q & 0 \\ \hat{C} & 0 & -I_p \end{bmatrix}$$

defined in (7.59).

$$\begin{aligned}\Sigma_{11} &= \text{Her}\left\{e^{j\theta}\hat{A}^{\mathrm{T}}\left(\hat{P}\hat{E}+\hat{S}\hat{Q}\right)\right\}, \\ &= \text{Her}\left\{ e^{j\theta}\begin{bmatrix} A^{\mathrm{T}}(P_1E+S_1Q_1+S_2Q_3J)+C^{\mathrm{T}}K^{\mathrm{T}}Q_3J \\ B^{\mathrm{T}}(P_1E+S_1Q_1+S_2Q_3J)-Q_3J \\ A^{\mathrm{T}}(S_1Q_2+S_2Q_3)+C^{\mathrm{T}}K^{\mathrm{T}}Q_3 \\ B^{\mathrm{T}}(S_1Q_2+S_2Q_3)-Q_3 \end{bmatrix}\right\}.\end{aligned}$$

$$\begin{aligned}\Sigma_{21} &= e^{j\theta}\hat{F}^{\mathrm{T}}\left(\hat{P}\hat{E}+\hat{S}\hat{Q}\right), \\ &= e^{j\theta}\left[F^{\mathrm{T}}\ 0\right]\begin{bmatrix} P_1E+S_1Q_1+S_2Q_3J & S_1Q_2+S_2Q_3 \\ Q_3J & Q_3 \end{bmatrix}, \\ &= e^{j\theta}\left[F^{\mathrm{T}}(P_1E+S_1Q_1+S_2Q_3J)\ F^{\mathrm{T}}(S_1Q_2+S_2Q_3)\right].\end{aligned}$$

By noting $N = Q_3^{\mathrm{T}}K$ and taking into account the $\mathcal{LMI}$ (7.60), we deduce that $\Sigma < 0$ which means that (7.59) holds and the system (7.55) is admissible since $\Sigma_{11} < 0$. From Q_3 invertible, we get

$$K = Q_3^{-\mathrm{T}}N.$$

This completes the proof. □

Example 7.3.3 Consider the system (7.50) with the following data

$$E = \begin{bmatrix} 1 & 1 \\ 1 & 1 \end{bmatrix},\ A = \begin{bmatrix} -5 & 3 \\ 2 & 0 \end{bmatrix},\ B = \begin{bmatrix} 1 \\ 2 \end{bmatrix},\ F = \begin{bmatrix} 3 \\ 3 \end{bmatrix},\ C = \begin{bmatrix} 1 & 0 \end{bmatrix},$$

with the fractional order $\alpha = 1.5$.

The transfer function of the open-loop system, $u(t) = 0$, is

$$G_o(s) = \frac{9}{10s^{1.5} - 6}.$$

Therefore, we shall find a static It is clear that the pair (E, A) is regular, impulse free but unstable since $\det(\lambda E - A) = 10\lambda - 6$ as can be confirmed in Fig. 7.8. In this case $\|G_o(s)\|_{\mathcal{H}_\infty}$ is unbounded. Indeed, since for a complex number s with

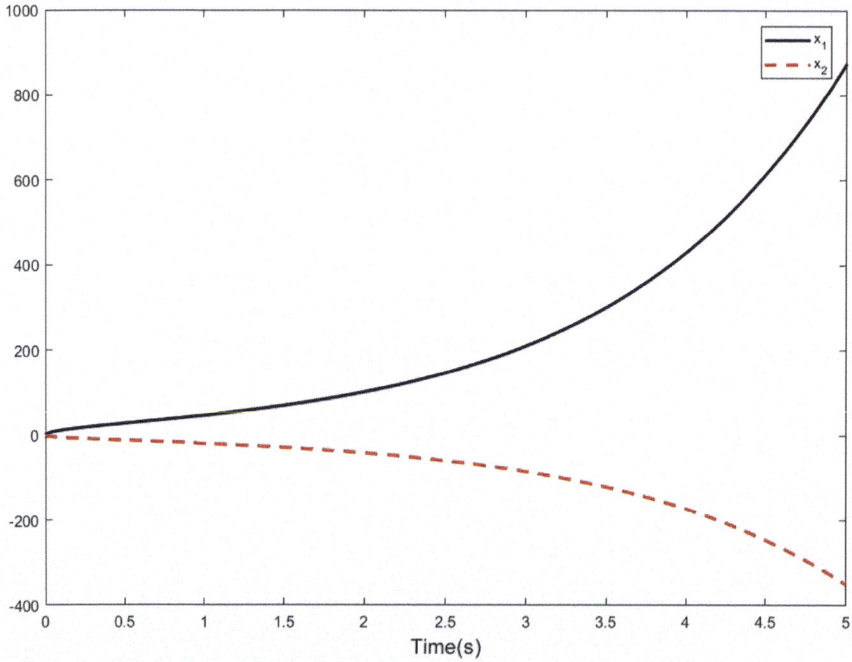

Fig. 7.8 State responses of the open-loop system

$\mathrm{Re}(s) > 0$, i.e. $s = |s| \exp^{j\theta}$ and $|\theta| < \frac{\pi}{2}$, s^α has the analytic value $s^\alpha = |s|^\alpha \exp^{j\alpha\theta}$, so

$$\|G_o(s)\|_{\mathcal{H}_\infty} = \sup_{\mathrm{Re}(s)\geq 0} \sqrt{\frac{9^2}{\left\|10\,|s|^{1.5}\exp^{j1.5\theta} - 6\right\|^2}}.$$

The maximum value is obtained for $\theta = 0$ and $|s|^{1.5} = \frac{6}{10}$, which means that $\|G_o(s)\|_{\mathcal{H}_\infty}$ is unbounded. Therefore, we shall find a static output feedback law control (7.51) which make the resulting closed-loop system admissible with $\mathcal{H}_\infty$ norm less than some given bound γ. $\mathcal{LMI}$ (7.60) in Theorem 7.3.5 gives for $\gamma = 0.5$

$$P_1 = \begin{bmatrix} 0.6718 & -0.6450 \\ -0.6450 & 0.6718 \end{bmatrix}, \; Q_1 = \begin{bmatrix} -0.27984 & -0.0519 \end{bmatrix},$$

$$Q_3 = 0.0972, \;\; Q_2 = -0.0372, \;\; N = -0.5424.$$

The control gain matrix obtained is

$$K = -5.5822.$$

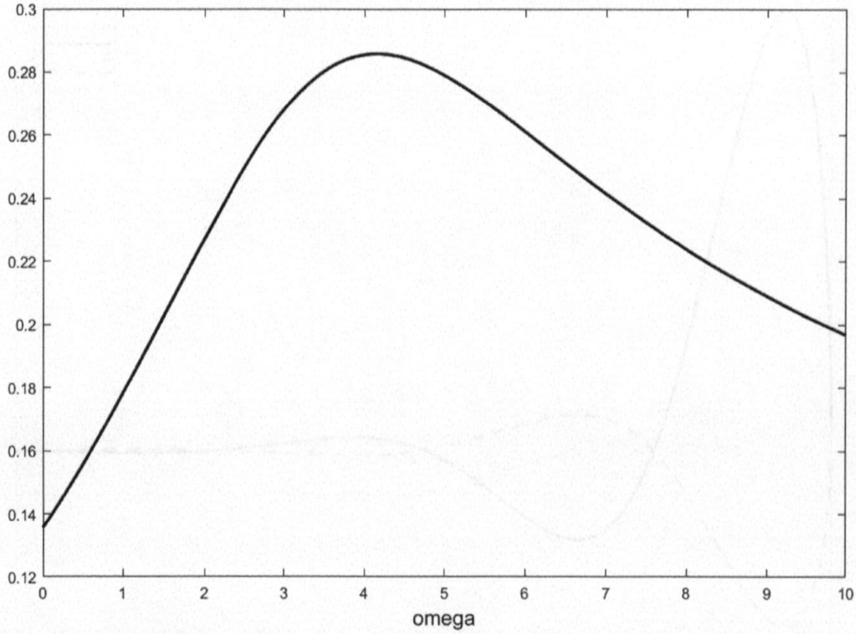

Fig. 7.9 Modulus of $G_c(j\omega)$ for $\omega \geq 0$

Then, The transfer function of the closed-loop selected system is

$$G_c(s) = \frac{-s^{1.5} + 6}{10s^{1.5} + 44.2398}.$$

The system $(E, A + BKC)$ is now admissible and $\|G_c(s)\|_{\mathcal{H}_\infty} = 0.2858$. The improvement for $\mathcal{H}_\infty$ performance of the control system can also be shown by the state responses of the closed-loop system and the modulus of its transfer function represented in Figs. 7.9 and 7.10 respectively.

Example 7.3.4 Consider electrical circuit [11] shown on Fig. 7.11, subject to external perturbation w(t), with given resistances R_1 , R_2, R_3, inductances L_1, L_2, L_3 and source voltages e_1 and e_2, where the voltage $u_L(t)$ on the coil with the inductance L is the α-order derivative of its magnetic flux $\psi(t)$. Taking into account that $\psi(t) = Li_L(t)$, we obtain $u_{L_j}(t) = LD^\alpha i_j$, $j = 1, 2, 3$, where i_j are the current in the coils.

The Kirchhoff's laws give for the electrical circuit the equations

$$e_1 = R_1 i_1 + L_1 D^\alpha i_1 + R_3 i_3 + L_3 D^\alpha i_3 \tag{7.62}$$

$$e_2 = R_2 i_2 + L_2 D^\alpha i_2 + R_3 i_3 + L_3 D^\alpha i_3 \tag{7.63}$$

$$0 = i_1 + i_2 - i_3 \tag{7.64}$$

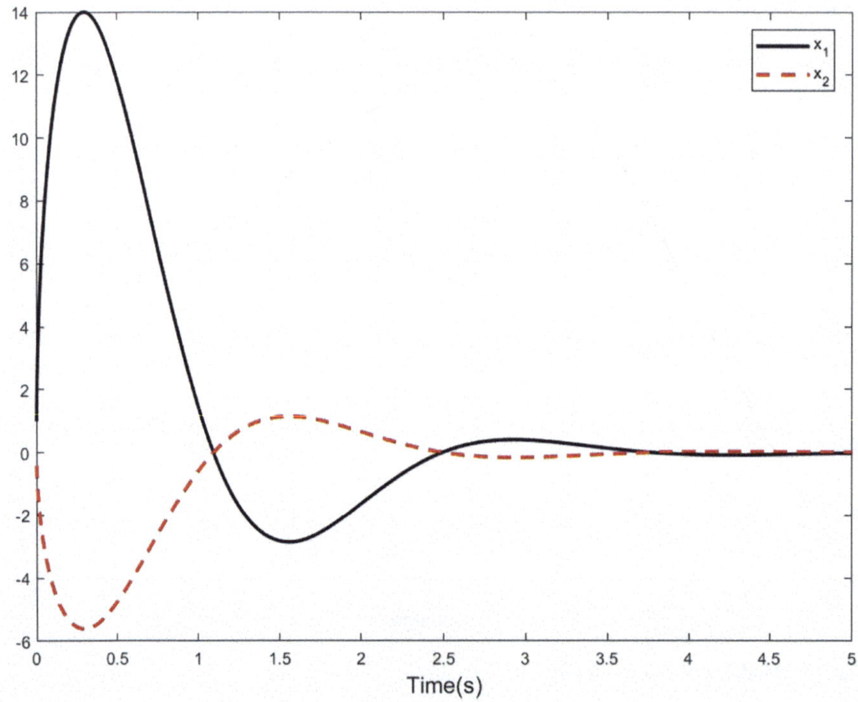

Fig. 7.10 State responses of the closed-loop system

Fig. 7.11 The descriptor fractional-order electrical circuit

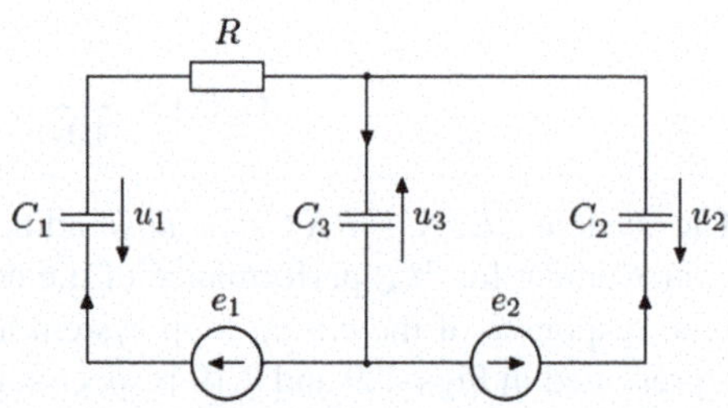

The Eqs. (7.62)–(7.64) can be written in the form

$$ED^{\alpha}x(t) = Ax(t) + Bu(t) + Fw(t) \tag{7.65}$$

with

$$x(t) = \begin{bmatrix} i_1(t) \\ i_2(t) \\ i_3(t) \end{bmatrix}, \; u(t) = \begin{bmatrix} e_1 \\ e_2 \end{bmatrix}, \; E = \begin{bmatrix} L_1 & 0 & L_3 \\ 0 & L_2 & L_3 \\ 0 & 0 & 0 \end{bmatrix},$$

$$A = \begin{bmatrix} -R_1 & 0 & -R_3 \\ 0 & -R_2 & -R_3 \\ 1 & 1 & -1 \end{bmatrix}, \; B = \begin{bmatrix} 1 & 0 \\ 0 & 1 \\ 0 & 0 \end{bmatrix}.$$

Consider the measured output $y(t) = Cx(t)$ and thus system (7.65) is

$$ED^\alpha x(t) = Ax(t) + Bu(t) + Fw(t), \; y(t) = Cx(t). \tag{7.66}$$

Let

$$E = \begin{bmatrix} 1 & 0 & 1 \\ 0 & 1 & 1 \\ 0 & 0 & 0 \end{bmatrix}, \; A = \begin{bmatrix} -10 & 0 & -10 \\ 0 & -5 & -10 \\ 1 & 1 & -1 \end{bmatrix},$$

$$F = \begin{bmatrix} 6 \\ 0 \\ 0 \end{bmatrix}, \; C = \begin{bmatrix} 0 & 1 & 0 \end{bmatrix}, \; \alpha = 1.5.$$

The transfer function of the open-loop system is

$$G_o(s) = \frac{-6}{3s^{1.5} + 20}.$$

One can verifies that $\|G_o(s)\|_{\mathcal{H}_\infty}$ is bounded.

The purpose is to determine a static output feedback law control (7.51) which make the resulting closed-loop system admissible with $\mathcal{H}_\infty$ norm less than some given bound γ. $\mathcal{LMI}$ (7.60) in Theorem 7.3.5 gives for $\gamma = 0.4$

$$P_1 = \begin{bmatrix} 0.0310 & -0.0145 & 0.0000 \\ -0.0145 & 0.2440 & -0.0000 \\ 0.0000 & -0.0000 & 14.0547 \end{bmatrix},$$

$$Q_3 = \begin{bmatrix} 1.4353 & 0.2967 \\ 0.2967 & 5.7075 \end{bmatrix}, \; N = \begin{bmatrix} 1.9174 \\ -5.2962 \end{bmatrix},$$

$$Q_1 = \begin{bmatrix} -4.5598 & -3.0439 & 2.7921 \end{bmatrix},$$

$$Q_2 = \begin{bmatrix} -0.7037 & 1.8397 \end{bmatrix}.$$

The control gain matrix obtained is

$$K = \begin{bmatrix} 1.5443 \\ -1.0082 \end{bmatrix}.$$

Then, The transfer function of the closed-loop selected system is

$$G_c(s) = \frac{-1}{0.5s^{1.5} + 3.9268}.$$

The eigenvalues of the system $(E, A + BKC)$ are -10 and -7.8536, which are located in the stable region, also the maximum singular value of $G_c(jw)$ is approximately 0.3611. So, the admissibility and the transfer function satisfying $\|G_c(s)\|_{\mathcal{H}_\infty} < 0.4$ are both verified.

7.3.3 Observer-Based $\mathcal{H}_\infty$ Control for SFOS

Let us consider the following SFOS:

$$\begin{cases} ED^\alpha x(t) = Ax(t) + Bu(t) + Fw(t), \ 1 \le \alpha < 2, \\ \quad y(t) = Cx(t), \end{cases} \tag{7.67}$$

where $x(t) \in \mathbb{R}^n$ is the descriptor variable, $u(t) \in \mathbb{R}^m$ is the control input, $w(t) \in \mathbb{R}^q$ is the exogenous input, $y(t) \in \mathbb{R}^p$ is the output, $E, A \in \mathbb{R}^{n\times n}$, $B \in \mathbb{R}^{n\times m}$, $F \in \mathbb{R}^{n\times q}$, $C \in \mathbb{R}^{p\times n}$. The sequel is devoted to study the $\mathcal{H}_\infty$ problem via dynamic output feedback controllers. There exist various manners for this topic. We will consider for the system (7.67) a Luenberger-type singular fractional-order linear observer coupled with a state estimated feedback controller.

$$\begin{cases} ED^\alpha \hat{x}(t) = A\hat{x}(t) + Bu(t) + L\left(y(t) - \hat{y}(t)\right) \\ \quad \hat{y}(t) = C\hat{x}(t) \end{cases} \tag{7.68}$$

coupled with a state estimated feedback controller $u(t) = K\hat{x}(t)$. Let $e(t) = x(t) - \hat{x}(t)$ denotes the error signal under the controller, the closed-loop system is given by

$$\begin{cases} ED^\alpha \hat{x}(t) = (A + BK)\hat{x}(t) + LCe\,(t)) + F\omega(t) \\ ED^\alpha e(t) = (A - LC)\,e(t) \\ \quad y(t) = Cx(t) \end{cases}$$

which can be rewritten as

$$\begin{cases} \bar{E}D^\alpha v(t) = \bar{A}v(t) + \bar{F}\omega(t) \\ \quad y(t) = \bar{C}v(t) \end{cases} \tag{7.69}$$

with $v(t) = \begin{bmatrix} \hat{x}(t) \\ e(t) \end{bmatrix}$, $\bar{E} = \begin{bmatrix} E & 0 \\ 0 & E \end{bmatrix}$ and

$$\bar{A} = \begin{bmatrix} A+BK & LC \\ 0 & A-LC \end{bmatrix}, \ \bar{F} = \begin{bmatrix} F \\ 0 \end{bmatrix}, \ \bar{C} = \begin{bmatrix} C & C \end{bmatrix}.$$

The goal of this corrector is to ensure the admissibility and the $\mathcal{H}_\infty$ performance γ for the closed-loop system (7.69). Firstly, let recall that the admissibility of the system (7.69) is equivalent to that of both systems $(E, A+BK)$ and $(E, A-LC)$. The transfer function of the system (7.69) is $G_{\omega y}^{KL}$ defined by

$$G_{\omega y}^{KL} = \bar{C}(s^\alpha \bar{E} - \bar{A})^{-1}\bar{F}. \tag{7.70}$$

i.e.,

$$G_{\omega y}^{KL} = \begin{bmatrix} C & C \end{bmatrix} \tag{7.71}$$

$$\times \begin{bmatrix} s^\alpha E - (A+BK) & -LC \\ 0 & s^\alpha E - (A-LC) \end{bmatrix}^{-1} \begin{bmatrix} F \\ 0 \end{bmatrix} \tag{7.72}$$

Using the following result that for any invertible matrices X, Z and any matrix Y we have

$$\begin{bmatrix} X & Y \\ 0 & Z \end{bmatrix}^{-1} = \begin{bmatrix} X^{-1} & -X^{-1}YZ^{-1} \\ 0 & Z^{-1} \end{bmatrix}$$

then we get

$$G_{\omega y}^{KL} = C(s^\alpha E - (A+BK))^{-1}F \tag{7.73}$$

i.e.,

$$G_{\omega y}^{KL} = G_{\omega y}^{K}, \tag{7.74}$$

then

$$\|G_{\omega y}^{KL}\|_{\mathcal{H}_\infty} = \|G_{\omega y}^{K}\|_{\mathcal{H}_\infty}. \tag{7.75}$$

Now, based on previous results, we can establish the following theorem.

Theorem 7.3.6 *Given a scalar $\gamma > 0$, the closed-loop system* (7.69) *is admissible and its transfer function in* (7.70) *satisfies*

$$\|G_{\omega y}^{KL}\|_{\mathcal{H}_\infty} < \gamma$$

if and only if there exist matrices $P_0, P_1 \in \mathbb{R}^{n\times n} > 0$, $Q_0, Q_1, Y \in \mathbb{R}^{(n-r)\times n}$, $X \in \mathrm{H}_n > 0$ and scalars $\varepsilon_0, \varepsilon_1 > 0$ such that

$$\begin{bmatrix} \Sigma_1 & * \\ (\Phi^T A - A^T \Phi)\cos\left(\alpha\frac{\pi}{2}\right) & \Sigma_1 \end{bmatrix} < 0 \tag{7.76}$$

$$\begin{bmatrix} \Sigma_2 & * \\ (\Psi^T A^T - A\Psi)\cos\left(\alpha\frac{\pi}{2}\right) & \Sigma_2 \end{bmatrix} < 0 \tag{7.77}$$

$$\begin{bmatrix} \Sigma_3 & * & * \\ e^{j\theta}F^T(XE+S_0Y) & -\gamma^2 I_q & * \\ C & 0 & -I_p \end{bmatrix} < 0 \tag{7.78}$$

hold, where

$$\Phi = P_0E + S_0Q_0,$$
$$\Psi = P_1E^T + S_1Q_1,$$
$$\Sigma_1 = \left(A^T\Phi + \Phi^T A\right)\sin\left(\alpha\frac{\pi}{2}\right) - \frac{1}{\varepsilon_0}\sin^2\left(\alpha\frac{\pi}{2}\right)\Phi^T BB^T\Phi,$$
$$\Sigma_2 = \left(A\Psi + \Psi^T A^T\right)\sin\left(\alpha\frac{\pi}{2}\right) - \frac{1}{\varepsilon_1}\sin^2\left(\alpha\frac{\pi}{2}\right)\Psi^T C^T C\Psi,$$
$$\Sigma_3 = Her\{e^{j\theta}(A^T(XE+S_0Y)) - e^{j\theta}\frac{1}{\varepsilon_0}\sin\left(\alpha\frac{\pi}{2}\right)\Phi^T BB^T(XE+S_0Y)\},$$

and S_0, S_1 are any matrices of full column rank satisfying $E^TS_0 = 0$ and $ES_1 = 0$ respectively. The controller gain matrices K, L can be designed by

$$K = -\frac{1}{\varepsilon_0}\sin\left(\alpha\frac{\pi}{2}\right)B^T\Phi. \tag{7.79}$$

and

$$L = \frac{1}{\varepsilon_1}\sin\left(\alpha\frac{\pi}{2}\right)\Psi^T C^T. \tag{7.80}$$

Proof The proof of (7.76), (7.77), (7.79) and (7.80) is exactly similar to the one of (6.161), (6.162), (6.163) and (6.164). Basing on the relationship (7.74) and the results of Theorem 7.3.2, The condition (7.78) is deduced directly. □

Example 7.3.5 Consider the SFOS (7.67) with the following parameters

$$E = \begin{bmatrix} -1 & -2 \\ 1 & 2 \end{bmatrix}, A = \begin{bmatrix} 5 & 3 \\ -5 & -3 \end{bmatrix}, B = \begin{bmatrix} 2 \\ 3 \end{bmatrix}, F = \begin{bmatrix} 1 \\ 1 \end{bmatrix}, C = \begin{bmatrix} 1 & 0 \end{bmatrix},$$

with the fractional order $\alpha = 1.5$.

Obviously, the pair (E, A) is not admissible because $\det(s^\alpha E - A) = 0$ which means also that the $\mathcal{H}_\infty$ performance for $u(t) = 0$ is unbounded. The purpose of this example is the design of an observer-based control law (7.68) to get an admissible closed-loop system defined in (7.69) with γ $\mathcal{H}_\infty$ performance where γ is a given positive scalar.

The resolution of the conditions (7.76)–(7.77) of Theorem 7.3.6 gives

$$P_0 = \begin{bmatrix} 0.1195 & 0.1141 \\ 0.1195 & 0.1141 \end{bmatrix}, Q_0 = \begin{bmatrix} 0.0461 \\ -0.0690 \end{bmatrix}, \epsilon_0 = 0.2195,$$

$$P_1 = \begin{bmatrix} 7.5137 & -2.9625 \\ -2.9625 & 2.8716 \end{bmatrix}, Q_1 = \begin{bmatrix} -0.7405 \\ 0.7405 \end{bmatrix}, \epsilon_1 = 0.01.$$

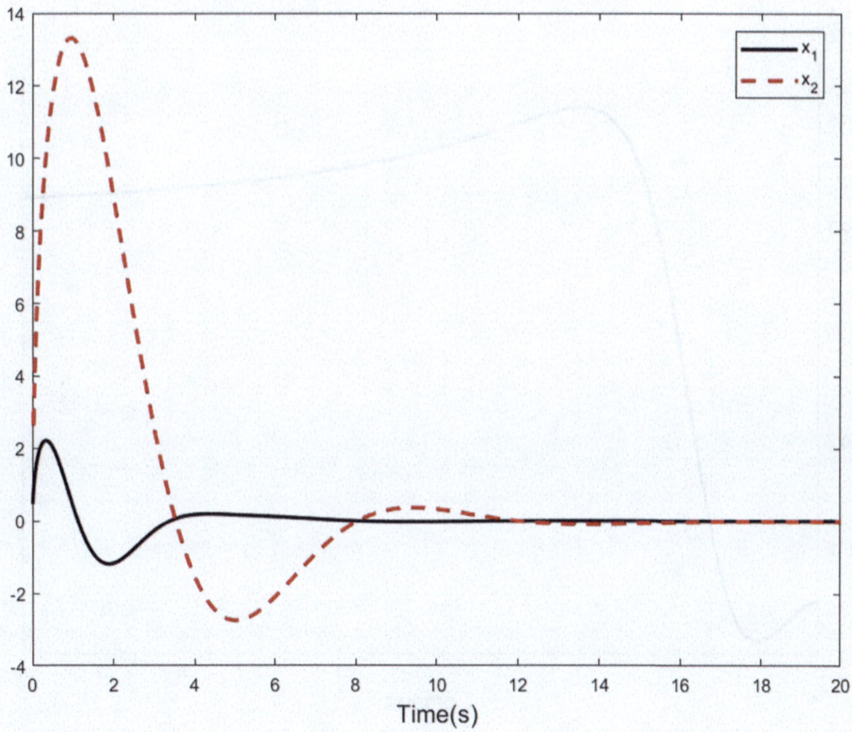

Fig. 7.12 State responses of the closed-loop system

The obtained matrices gains K, L are designed by

$$K = \left[\, -0.7776 \ 1.1010 \,\right], L = \begin{bmatrix} -7.6089 \\ 7.6089 \end{bmatrix}. \tag{7.81}$$

Then the system (7.69) is now admissible since $\text{spec}(\bar{E}, \bar{A}) = \{-0.1248, -1.5\}$ which means that its finite modes lie in the stable region and consequently its state responses are convergent as can be seen in Fig. 7.12 in which the histories of the system states and the observation errors for chosen initial conditions $[\, x_1(0) \ x_2(0) \,] = [\, 0.4637 \ 3 \,]$ and $[\, \hat{x}_1(0) \ \hat{x}_2(0) \,] = [\, 1 \ 1 \,]$.

To verify the $\mathcal{H}_\infty$ performance, we solve the $\mathcal{LMI}$ (7.78) in Theorem 7.3.6 for $\gamma = 0.43$ and obtain a feasible solution

$$X = \begin{bmatrix} 151.3421 & 151.2270 - \text{i}0.0568 \\ 151.2270 + \text{i}0.0568 & 151.3421 \end{bmatrix},$$

$$Y = \left[\, -0.0136 \ -0.4607 \,\right].$$

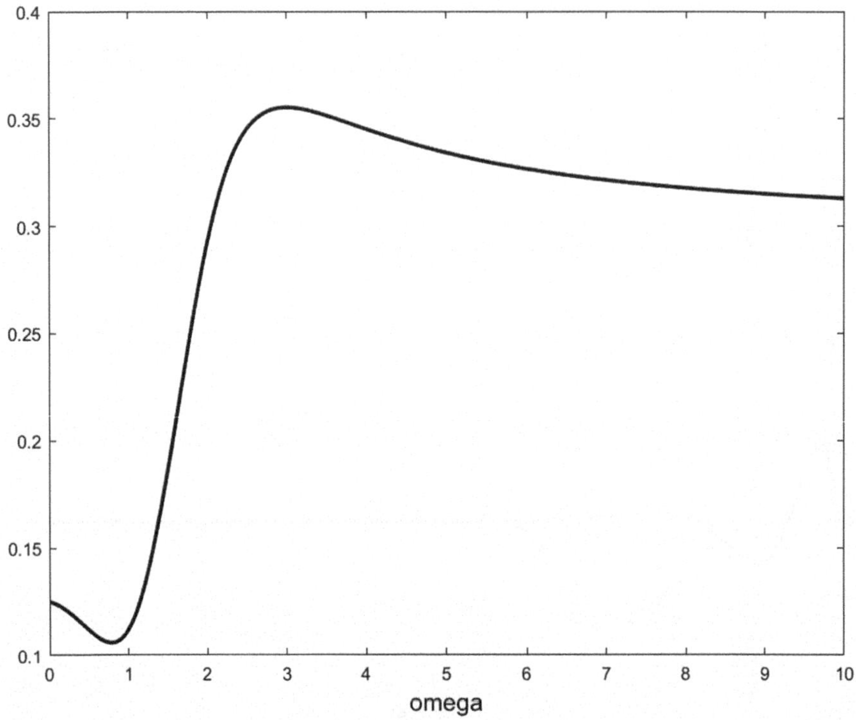

Fig. 7.13 Modulus of the transfer function $G(j\omega)$ of the closed-loop system

This result ensures that the $\mathcal{H}_\infty$ norm of the system (7.69) is now bounded. To show the correctness, we consider the transfer function $G(s)$of the system (7.69).

$$G(s) = \bar{C}(s^\alpha \bar{E} - \bar{A})^{-1}\bar{F} = \frac{4000s^\alpha + 4899}{13281s^\alpha + 39189}.$$

Using the above result, we are content to compute $\|G(j\omega)\|_{\mathcal{H}_\infty}$. After few computation, we get

$$\|G(j\omega)\|_{\mathcal{H}_\infty} = \max_{\omega \geq 0} \sqrt{\frac{1610^6\omega^{2\alpha} + 3919210^3\omega^\alpha \cos(\frac{3\pi}{4}) + 24000201}{176384961\omega^{2\alpha} + 1.040910^9\omega^\alpha \cos(\frac{3\pi}{4}) + 1.535810^9}}.$$

It can be obtained that $\|G(j\omega)\|_{\mathcal{H}_\infty} = 0.3552$. This can be shown in Fig. 7.13.

7.4 Summary

In this chapter, the bounded real lemma (BRL) corresponding to the $\mathcal{H}_\infty$ norm for singular fractional-order systems with fractional derivative α satisfying $1 \leq \alpha < 2$. This lemma transforms the initial computations of the $\mathcal{H}_\infty$ norm into Linear Matrix Inequality ($\mathcal{LMI}$) problems, offering an efficient resolution method. Furthermore, the bounded real lemma approach contribute to addressing the comprehensive challenge of $\mathcal{H}_\infty$ controller design for SFOS.

References

1. Francis, B.A.: A Course in $\mathcal{H}_\infty$ Control Theory. Springer, Heidelberg, Berlin, Germany (1987)
2. Green, M., Limebeer, D.J.N.: Linear Robust Control. Upper Saddle River, NJ, USA, Prentice-Hall, Inc (1995)
3. Marir, S., Chadli, M., Basin, M.V.: Bounded real lemma for singular linear continuous-time fractional-order systems. Automatica **135** (2022). https://doi.org/10.1016/j.automatica.2021.109962
4. Peaucelle, D., Arzelier, D., Bachelier, O., Bernussou, J.: A new robust d-stability condition for real convex polytopic uncertainty. Syst. Control Lett. **40**, 21–30 (2000)
5. Xu, S.Y., Lam, J.: Robust Control and Filtering of Singular Systems. Springer, Berlin (2006)
6. Chadli, M., Shi, P., Feng, Z., Lam, J.: New bounded real lemma formulation and $\mathcal{H}_\infty$ control for continuous-time descriptor systems. Asian J. Control **20**(1), 1–7 (2018). https://doi.org/10.1002lasjc.1606
7. Liang, S., Peng, C., Liao, Z., Wang, Y.: Robust analysis and synthesis for a class of fractional order systems with coupling uncertainties. Int. J. Control Autom. Syst. **13**, 494–500 (2015)
8. Zhang, Q.H., Lu, J.G.: Bounded real lemmas for singular fractional-order systems: the $1<\alpha<2$. IEEE Trans. Circuits Syst. II: Express Briefs (2020). https://doi.org/10.1109/TCSII.2020.3007996
9. Shen, J., Lam, J.: State feedback $\mathcal{H}_\infty$ control of commensurate fractional-order systems. Int. J. Syst. Sci. **45**(3), 363–372 (2014)
10. Marir, S., Chadli, M., Basin, M.V.: $\mathcal{H}_\infty$ static output feedback controller design for singular fractional-order systems. In: 2022 European Control Conference (ECC), London, United Kingdom, July 2022, pp. 1–6. https://doi.org/10.23919/ECC55457.2022.9838112
11. Kaczorek, T.: Selected Problems of Fractional Systems Theory. Lecture Notes in Control and Information Sciences. Springer, Heidelberg, Berlin, Germany (2011). https://doi.org/10.1007/978-3-642-20502-6

Appendix A
Linear Algebra Recall

A.1 Positive Definite Matrices

A.1.1 Definitions

The study of matrices is quite old. Leibnitz is one of the founders of the analysis which developed the theory of determinants in 1693 to facilitate the resolution of differential equations. The matrices are now used for multiple applications and serve in particular to represent the coefficients of linear equations.The matrices are now used for multiple applications and serve in particular to represent the coefficients of linear equations [1–3].

Definition A.1.1 Let A a square matrix.

- A matrix $A \in \mathbb{R}^{n\times n}$ is called symmetric if and only if

$$A^{\mathrm{T}} = A \tag{A.1}$$

- A matrix $A \in \mathbb{C}^{n\times n}$ is called hermitian if and only if

$$A^{*} = A \tag{A.2}$$

Definition A.1.2 1. If A is a symmetric matrix.

- A is called positive definite if and only if

$$\forall v \in \mathbb{R}^{n} (v \neq 0), v^{\mathrm{T}} A v > 0 \tag{A.3}$$

S. Marir and M. Chadli, *Control of Singular Fractional Order Systems: LMI Approach*, Studies in Systems, Decision and Control 292,
https://doi.org/10.1007/978-3-031-87382-9

- A is called positive if and only if

$$\forall v \in \mathbb{R}^n, v^{\mathrm{T}} A v \geq 0 \tag{A.4}$$

2. If A is a hermitian matrix.

- A is called positive definite if and only if

$$\forall v \in \mathbb{C}^n (v \neq 0), v^{\mathrm{T}} A v > 0 \tag{A.5}$$

- A is called positive semidefinite if and only if

$$\forall v \in \mathbb{C}^n, v^* A v \geq 0 \tag{A.6}$$

Here is a simple but fundamental fact.

Theorem A.1.1 *A Hermitian matrix is positive definite if and only if all its eigenvalues are positive.*

Proof (Necessity) Suppose that $v \in \mathbb{C}^n$ is an eigenvector of the Hermitian matrix $A \in \mathbb{C}^{n \times n}$ corresponding to the eigenvalue λ. Then

$$v^* A v = \lambda v^* v \tag{A.7}$$

Since the eigenvector v is non-zero, it follows that $v^* v = \|v_2\| > 0$.
If A is positive definite, then

$$\lambda = \frac{v^* A v}{v^* v} > 0 \tag{A.8}$$

Hence, all eigenvalues of a hermitian positive definite matrix must be positive.

(Sufficiency) suppose $A \in \mathbb{C}^{n \times n}$ is a hermitian matrix whose eigenvalues $\lambda_1 \leq \lambda_2 \leq \cdots \leq \lambda_n$ are all positive. Then let $u_1, \ldots, u_n$ denote an orthonormal basis of eigenvectors, so that any $v \in \mathbb{C}^n$ can be written as

$$v = \sum_{i=1}^{n} \alpha_i u_i \tag{A.9}$$

then

$$Av = \sum_{i=1}^{n} \alpha_i \lambda_i u_i$$

It follows that

$$v^*Av = (\sum_{j=1}^{n} \bar{\alpha}_j u_j^*)(\sum_{i=1}^{n} \alpha_i \lambda_i u_i)$$

which is the same as

$$v^*Av = \sum_{j=1}^{n}\sum_{i=1}^{n} \bar{\alpha}_j \alpha_i \lambda_i u_j^* u_i$$

Since $u_j^* u_i = 0, \forall i \neq j$ and $u_j^* u_i = 1, i = j$, we obtain

$$v^*Av = \sum_{j=1}^{n} \lambda_j \left\|\alpha_j\right\|^2 \geq \lambda_1 \sum_{j=1}^{n} \left\|\alpha_j\right\|^2$$

If $v \neq 0$, then $0 \neq \|v\|_2^2 = \sum_{i=1}^{n} \|\alpha_i\|^2$, and since all the eigenvalues are positive, we must have

$$v^*Av > 0$$

which means that the matrix A is positive definite. □

A.1.2 Properties

- If A is a hermitian matrix then

$$A > 0 \Leftrightarrow A^* > 0 \tag{A.10}$$

  Indeed,
  As $A = A^*$ then spec$(A) =$ spec$(A^{*)}$ and since $A > 0$, all eigenvalues of A are positive (see Theorem A.1.1) which implies that all eigenvalues of A^* are also positive.
  To prove the other implication, it is sufficient to use the same reasoning as in the necessity.
- If A is a hermitian matrix then

$$A > 0 \Leftrightarrow \bar{A} > 0 \tag{A.11}$$

Proof (Necessity) Assume that $A > 0$, then $A = A^*$. This implies

$$\overline{A} = \overline{A^*} = \overline{A}^*$$

which means that $\bar{A}$ is hermitian.

In the other hand, for all $v \in \mathbb{C}^n$, $v \neq 0$, we have

$$v^* \bar{A} v = \overline{\overline{v^* A \bar{v}}} = \overline{\bar{v}^* A \bar{v}} > 0$$

since $A > 0$. We conclude that $\bar{A} > 0$.

(Sufficiency) From the necessity, if $\overline{A} > 0$ then $\overline{\overline{A}} = A > 0$.

A.2 Proof of Lemma 2.4.3

(Necessity). A hermitian matrix $M \in \mathbb{C}^{n \times n}$ is said to be positive definite if and only if for every non-zero column vector $v \in \mathbb{C}^n$

$$v^* M v > 0 \tag{A.12}$$

Without loss of generality, we can define

$$v = x + jy$$

such that $x, y \in \mathbb{R}^n$ and $\|x\|_2 + \|y\|_2 \neq 0$.

Therefore from $M = A + jB > 0$, we can get

$$\begin{aligned} v^* M v &= (x - jy)^{\mathrm{T}}(A + jB)(x + jy) \\ &= (x^{\mathrm{T}} A x - x^T B y + y^{\mathrm{T}} A y + y^{\mathrm{T}} B x) \\ &+ j(x^{\mathrm{T}} A y + x^{\mathrm{T}} B x - y^{\mathrm{T}} A x + y^{\mathrm{T}} B y) > 0 \end{aligned}$$

Then, we can deduce that

$$x^{\mathrm{T}} A x - x^T B y + y^{\mathrm{T}} A y + y^{\mathrm{T}} B x > 0 \tag{a}$$

$$x^{\mathrm{T}} A y + x^{\mathrm{T}} B x - y^{\mathrm{T}} A x + y^{\mathrm{T}} B y = 0 \tag{b}$$

The inequality (a) is the same as

$$\begin{bmatrix} x^{\mathrm{T}} & y^{\mathrm{T}} \end{bmatrix} \begin{bmatrix} A & -B \\ B & A \end{bmatrix} \begin{bmatrix} x \\ y \end{bmatrix} > 0 \tag{A.13}$$

and note that the inequality (A.13) is established for all non-zero real vector

$$z = \begin{bmatrix} x^{\mathrm{T}} & y^{\mathrm{T}} \end{bmatrix}^{\mathrm{T}}$$

then we obtain

$$\begin{bmatrix} A & -B \\ B & A \end{bmatrix} > 0$$

which is exactly (2.26).

To obtain (2.25), it suffises to see that (a) can be rewritten as

$$\begin{bmatrix} y^{\mathrm{T}} & x^{\mathrm{T}} \end{bmatrix} \begin{bmatrix} A & B \\ -B & A \end{bmatrix} \begin{bmatrix} y \\ x \end{bmatrix} > 0$$

(Sufficiency) Assume that the inequality (2.26) is verified. Then we have

$$\begin{cases} A^{\mathrm{T}} = A \\ B^{\mathrm{T}} = -B \end{cases} \tag{A.14}$$

This leads to

$$M^{*} = (A + jB)^{*} = A^{\mathrm{T}} - jB^{\mathrm{T}} = M$$

So, we conclude that the matrix M is hermitian.

For any non-zero complex vector $v = x + jy$ with $x, y \in \mathbb{R}^{n}$, we have

$$v^{*}Mv = (x^{\mathrm{T}}Ax - x^{T}By + y^{\mathrm{T}}Ay + y^{\mathrm{T}}Bx) + j(x^{\mathrm{T}}Ay + x^{\mathrm{T}}Bx - y^{\mathrm{T}}Ax + y^{\mathrm{T}}By)$$

According to (A.14), we have

$$(x^{\mathrm{T}}Ay)^{\mathrm{T}} = y^{\mathrm{T}}Ax$$

and

$$\begin{cases} (x^{\mathrm{T}}Bx)^{\mathrm{T}} = -x^{\mathrm{T}}Bx \\ (y^{\mathrm{T}}By)^{\mathrm{T}} = -y^{\mathrm{T}}By \end{cases} \Rightarrow \begin{cases} x^{\mathrm{T}}Bx = 0 \\ y^{\mathrm{T}}By = 0 \end{cases}$$

then we get that

$$x^{\mathrm{T}}Ay + x^{\mathrm{T}}Bx - y^{\mathrm{T}}Ax + y^{\mathrm{T}}By = 0$$

Therefore

$$\begin{aligned} v^{*}Mv &= (x^{\mathrm{T}}Ax - x^{T}By + y^{\mathrm{T}}Ay + y^{\mathrm{T}}Bx) \\ &= \begin{bmatrix} x^{\mathrm{T}} & y^{\mathrm{T}} \end{bmatrix} \begin{bmatrix} A & -B \\ B & A \end{bmatrix} \begin{bmatrix} x \\ y \end{bmatrix} > 0 \end{aligned}$$

Thus one complete the proof.

Appendix B
$\mathcal{LMI}$ Regions

B.1 A Brief History of LMIs in Control Theory

The history of $\mathcal{LMI}$s in the analysis of dynamical systems goes back more than 100 years, in about 1890 by the early work of Lyapunov [4], which remains to this day one of the most powerful methodologies for stability analysis. The story begins when Lyapunov published his seminal work introducing what we now call Lyapunov theory or the Lyapunov second method that can be summarized by the following statement.

Theorem B.1.1 (Second Lyapunov Method) *System $\dot{x}(t) = f(t, x(t))/$ is stable (globally asymptotically stable around the origin) if there exists a real-valued function $V(x, t)$ such that:*

$$V(0, t) = 0, \;\; \forall t \geq 0$$
$$V(x, t) > 0, \;\; \forall x \neq 0, \;\; \forall t \geq 0$$
$$\lim_{\|x\| \to \infty} V(x, t) = \infty$$
$$\dot{V}(x, t) < 0, \;\; \forall x \neq 0, \;\; \forall t \geq 0$$

$V(x, t)$ is called the Lyapunov function.

In the case of a linear system of the form

$$\dot{x}(t) = Ax(t) \tag{B.1}$$

with arbitrary matrix A, the existence of a quadratic Lyapunov function

$$V(x, t) = x^{\mathrm{T}} P x, \quad P > 0$$

S. Marir and M. Chadli, *Control of Singular Fractional Order Systems: LMI Approach*, Studies in Systems, Decision and Control 292,
https://doi.org/10.1007/978-3-031-87382-9

is a necessary and sufficient condition for asymptotic stability. Finding a positive definite matrix P ensuring that Lyapunov function $V(x, t)$ decreases along the trajectories of the system can be achieved via the following theorem.

Theorem B.1.2 *Integer order system* (B.1) *is asymptotically stable if and only if there exists a positive definite matrix P such that:*

$$A^T P + PA < 0 \tag{B.2}$$

Note that the Theorem B.1.2 is satisfied if and only if the eigenvalues of A lie in the open left half plane. This Lyapunov characterization of stability has been extended to a variety of regions by Gutman [5]. More details can be seen in the next few sections.

B.2 Linear Matrix Inequalities

Definition B.2.1 A strict linear matrix inequality $\mathcal{LMI}$ has the form

$$M(x) = M_0 + \sum_{k=1}^{n} x_k M_k > 0 \tag{B.3}$$

where $x \in \mathbb{R}^n$ is the variable and the symmetric matrices $M_k = M_k^{\mathrm{T}} \in \mathbb{R}^{n \times n}, \ k = 0, 1, \ldots, n$ are given.

Remark B.2.1

- The term linear matrix inequality is used in the literature on systems and control, but the terminology is not consistent with the expression $F(x) > 0$ since F is not a linear function. The term affine matrix inequality may better correspond to the formulation.

- A nonstrict $\mathcal{LMI}$ has the form

$$M(x) = M_0 + \sum_{k=1}^{n} x_k M_k \geq 0$$

- The linear matrix inequalities $F(x) < 0$ and $F(x) < G(x)$ where F, G are affine functions, are special cases of (B.3) since they can be reformulated in the form $\mathcal{LMI}$

$$-F(x) > 0 \tag{B.4}$$
$$G(x) - F(x) > 0 \tag{B.5}$$

- The objective is to find $x \in \mathbb{R}^n$ satisfying the inequality (B.3). This choice of x is called "**feasibility problem**".
  The $\mathcal{LMI}$ (B.3) is a convex constraint on x, i.e., the set $\{x/M(x) > 0\}$ is convex.

- Multiple $\mathcal{LMI}$s

$$M^{(1)}(x) > 0, \cdots, M^{(p)}(x) > 0 \tag{B.6}$$

  can be expressed as the single $\mathcal{LMI}$

$$\text{diag}(M^{(1)}(x), \ldots, M^{(p)}(x)) > 0 \tag{B.7}$$

- Nonlinear (convex) inequalities are converted to $\mathcal{LMI}$ form using Schur complements. The basic idea is as follows: the $\mathcal{LMI}$

$$\begin{bmatrix} Q(x) & S(x) \\ S^{\mathrm{T}}(x) & R(x) \end{bmatrix} > 0 \tag{B.8}$$

  where $Q(x) = Q^{\mathrm{T}}(x)$ and $R(x) = R^{\mathrm{T}}(x)$ and $S(x)$ depend affinely on x, is equivalent to

$$\begin{cases} R(x) > 0 \\ Q(x) - S(x)R^{-1}(x)S^{\mathrm{T}}(x) > 0 \end{cases} \tag{B.9}$$

  or equivalently

$$\begin{cases} Q(x) > 0 \\ R(x) - S^{\mathrm{T}}(x)Q^{-1}(x)S(x) > 0 \end{cases} \tag{B.10}$$

  Indeed, the proof is easily done by multiplying (B.8) to the right by:

$$\begin{bmatrix} I & 0 \\ -R^{-1}(x)S^{\mathrm{T}}(x) & I \end{bmatrix} \tag{B.11}$$

  And to the left by the transpose of this last matrix. These two matrices being defined, then an equivalent condition is obtained:

$$\begin{bmatrix} Q(x) - S(x)R^{-1}(x)S^{\mathrm{T}}(x) & 0 \\ 0 & R \end{bmatrix} > 0 \tag{B.12}$$

B.3 Definition of $\mathcal{LMI}$ Region

The class of LMI regions defined below turns out to be suitable for LMI-based synthesis.

Definition B.3.1 A subset $\mathcal{D}$ of the complex plane is called an $\mathcal{LMI}$ region if there exist a symmetric matrix $\alpha \in \mathbb{R}^{n\times n}$ and a matrix $\beta \in \mathbb{R}^{n\times n}$ such that

$$\mathcal{D} = \left\{z \in \mathbb{C}/f_{\mathcal{D}}(z) = \alpha + z\beta + \bar{z}\beta^{\mathrm{T}} < 0\right\} \tag{B.13}$$

It can be noted that the characteristic function $f_{\mathcal{D}}$ of variable z takes values in the space of $n \times n$ Hermitian matrices. As a result, $\mathcal{LMI}$ regions are convex. Moreover, $\mathcal{LMI}$ regions are symmetric with respect to the real axis since for any $z \in \mathcal{D}$, $\overline{f_{\mathcal{D}}(z)} = f_{\mathcal{D}}(\bar{z})$. This last property is often verified by the regions used for the study of the $\mathcal{D}$-stability[1] of a matrix A since the spectrum of a matrix is auto-conjugate.

Theorem B.3.1 *[6, 7] Let* $A \in \mathbb{R}^{n\times n}$ *and* $\mathcal{D}$ *an* $\mathcal{LMI}$ *region defined by* (B.13). *The matrix* A *is* $\mathcal{D}$*-stable if and only if there exists a symmetric positive definite matrix* $X \in \mathbb{R}^{n\times n}$ *such that*

$$M_{\mathcal{D}}(A, X) = \alpha \otimes X + \beta \otimes (AX) + \beta^T \otimes (AX)^T < 0 \tag{B.14}$$

In practical applications, $\mathcal{LMI}$ regions are often specified as the intersection of elementary regions, such as conic sectors, disks, or vertical half-planes.
Given $\mathcal{LMI}$ regions $\mathcal{D}_1, \mathcal{D}_2, ..., \mathcal{D}_N$ and their associated characteristic functions $f_{\mathcal{D}_i}, i = 1, 2, ..., N$ such that

$$f_{\mathcal{D}_i}(z) = \alpha_i + z\beta_i + \bar{z}\beta_i^{\mathrm{T}}, \qquad i = 1, 2, ..., N$$

The intersection

$$\mathcal{D} = \mathcal{D}_1 \cap \mathcal{D}_2 \cap \ldots \cap \mathcal{D}_N \tag{B.15}$$

is also an $\mathcal{LMI}$ region with characteristic function

$$f_{\mathcal{D}}(z) = \operatorname*{diag}_{i=1}^{N} f_{\mathcal{D}_i}(z) \tag{B.16}$$

If $\mathcal{D}$-stability of a matrix A in region $\mathcal{D}$ defined in (B.15) is of interest, Theorem B.3.1 should be applied to the overall characteristic function $f_{\mathcal{D}}(z)$ defined by (B.16). It has shown (see Corollary 2.3 in [6]) that the eigenvalues of the matrix A belong to

[1] A matrix A is called $\mathcal{D}$-stable if and only if all its eigenvalues lie in a subregion $\mathcal{D}$ of the complex plane.

the region $\mathcal{D}$ if and only if there exists a positive definite matrix $X \in \mathbb{R}^{n\times n}$ such that for any $i = 1, 2, ..., N$

$$M_{\mathcal{D}_i}(A, X) = \alpha_i \otimes X + \beta_i \otimes (AX) + \beta_i^{\mathrm{T}} \otimes (AX)^{\mathrm{T}} < 0 \tag{B.17}$$

i.e.,

$$M_{\mathcal{D}}(A, X) = \alpha \otimes X + \beta \otimes (AX) + \beta^{\mathrm{T}} \otimes (AX)^{\mathrm{T}} = \underset{i=1}{\overset{N}{\mathrm{diag}}} M_{\mathcal{D}_i}(A, X) < 0 \tag{B.18}$$

where

$$\alpha = \underset{i=1}{\overset{N}{\mathrm{diag}}}(\alpha_i), \quad \beta = \underset{i=1}{\overset{N}{\mathrm{diag}}}(\beta_i)$$

This last inequality clearly shows that the region $\mathcal{D}$ can itself be formulated as an $\mathcal{LMI}$ region. This shows that the $\mathcal{LMI}$ approach makes it possible to consider the same matrix X for all subregions of the intersection while preserving the necessity of the $\mathcal{D}$-stability condition.

B.4 Examples of $\mathcal{LMI}$ Regions

The inequality (B.14) can also be written in the following form

$$\begin{bmatrix} \alpha_{11}X + \beta_{11}AX + \beta_{11}(AX)^{\mathrm{T}} & \dots & \alpha_{1n}X + \beta_{1n}AX + \beta_{n1}(AX)^{\mathrm{T}} \\ \vdots & \ddots & \vdots \\ \alpha_{n1}X + \beta_{n1}AX + \beta_{1n}(AX)^{\mathrm{T}} & \dots & \alpha_{nn}X + \beta_{nn}AX + \beta_{nn}(AX)^{\mathrm{T}} \end{bmatrix} < 0 \tag{B.19}$$

where $\alpha = \left(\alpha_{ij}\right)_{1\le i,j\le n}$ and $\beta = \left(\beta_{ij}\right)_{1\le i,j\le n}$.
Below are a few examples of $\mathcal{LMI}$ regions:

- **Open left half-plane**

$$\mathrm{Re}(z) < 0 \Leftrightarrow z + \bar{z} < 0 \tag{B.20}$$

It is enough to take $\alpha = 0$ and $\beta = 1$. From the expression (B.20), we deduce the following $\mathcal{LMI}$

$$AX + (AX)^{\mathrm{T}} < 0, \; X > 0 \tag{B.21}$$

The $\mathcal{LMI}$ (B.21) is the characterization of the asymptotic stability of a dynamical system $\dot{x}(t) = Ax(t)$ introduced in terms of $\mathcal{LMI}$ by the Lyapunov theorem.

- **a-Stability**

$$\mathrm{Re}(z) < -a \Leftrightarrow z + \bar{z} + 2a < 0 \tag{B.22}$$

It is enough to take $\alpha = 2a$ and $\beta = 1$, the following $\mathcal{LMI}$ is obtained

$$2aX + AX + (AX)^{\mathrm{T}} < 0, \quad X > 0 \tag{B.23}$$

- **Vertical strip**

$$a_1 < \mathrm{Re}(z) < a_2 \Leftrightarrow 2a_1 < z + \bar{z} < 2a_2 \tag{B.24}$$

Using the intersection property (B.16), we obtain

$$\begin{bmatrix} -2a_2 + z + \bar{z} & 0 \\ 0 & 2a_1 - z - \bar{z} \end{bmatrix} < 0 \tag{B.25}$$

which leads to the $\mathcal{LMI}$ (B.18)

$$\begin{bmatrix} -2a_2X + AX + (AX)^{\mathrm{T}} & 0 \\ 0 & -2a_1X - (AX + (AX)^{\mathrm{T})} \end{bmatrix} < 0, \quad X > 0 \tag{B.26}$$

with

$$\alpha = \begin{bmatrix} -2a_2 & 0 \\ 0 & 2a_1 \end{bmatrix}, \quad \beta = \begin{bmatrix} 1 & 0 \\ 0 & -1 \end{bmatrix}$$

- **Horizontal strip**

$$|\mathrm{Im}(z)| < a \Leftrightarrow \|z - \bar{z}\| < 2a \tag{B.27}$$

According to the following complex numbers property. For any complex number z

$$\|z\|^2 = z\bar{z} \tag{B.28}$$

The inequality (B.27) is equivalent to

$$-4a^2 - (z - \bar{z})(z - \bar{z}) < 0 \tag{B.29}$$

The application of the Shur complement to (B.29) leads to

$$\begin{bmatrix} -2a & z-\bar{z} \\ -z+\bar{z} & -2a \end{bmatrix} < 0 \tag{B.30}$$

It is enough to take $\alpha = \begin{bmatrix} -2a & 0 \\ 0 & -2a \end{bmatrix}$ and $\beta = \begin{bmatrix} 0 & 1 \\ -1 & 0 \end{bmatrix}$, the following $\mathcal{LMI}$ is then obtained

$$\begin{bmatrix} -2aX & AX-(AX)^{\mathrm{T}} \\ -AX+(AX)^{\mathrm{T}} & -2aX \end{bmatrix} < 0,\ X > 0 \tag{B.31}$$

- **The disk $D(q,r)$ with center $(-q,0)$ and radius r**

The disk $D(q,r)$ is characterized by

$$\|z+q\| < r \Leftrightarrow \|z+q\|^2 < r^2 \tag{B.32}$$

According to the complex numbers property (B.28) the inequality (B.32) is the same as

$$-r^2 + (z+q)(\bar{z}+q) < 0 \tag{B.33}$$

Applying the Shur complement, (B.33) is equivalent to

$$\begin{bmatrix} -r & q+z \\ q+\bar{z} & -r \end{bmatrix} < 0 \tag{B.34}$$

which corresponds to

$$\alpha = \begin{bmatrix} -r & q \\ q & -r \end{bmatrix},\ \beta = \begin{bmatrix} 0 & 1 \\ 0 & 0 \end{bmatrix} \tag{B.35}$$

consequently the following $\mathcal{LMI}$ is obtained

$$\begin{bmatrix} -rX & qX+AX \\ qX+(AX)^{\mathrm{T}} & -rX \end{bmatrix} < 0, \quad X > 0 \tag{B.36}$$

- **Conic sector with apex at the origin and inner angle 2θ**

The conic sector with apex at the origin and inner angle 2θ, $0 < \theta < \frac{\pi}{2}$ caught in the left half plane is resulting from the intersection of the half complex plane defined by

$$\mathrm{Re}(z)\sin(\theta) + \mathrm{Im}(z)\cos(\theta) < 0 \tag{B.37}$$

with the half complex plane defined by

$$\mathrm{Re}(z)\sin(\theta) - \mathrm{Im}(z)\cos(\theta) < 0 \tag{B.38}$$

Then, the conic sector is characterized by

$$\tan(\theta) > \frac{|\mathrm{Im}(z)|}{-\mathrm{Re}(z)} \tag{B.39}$$

which is the same as

$$\mathrm{Re}(z)\tan(\theta) < -|\mathrm{Im}(z)| \tag{B.40}$$

Since $0 < \theta < \frac{\pi}{2}$, so

$$\mathrm{Re}(z) < 0,\ \tan(\theta) > 0,$$

then the inequality (B.40) is equivalent to

$$(\mathrm{Re}(z))^2\tan^2(\theta) > |\mathrm{Im}(z)|^2 \tag{B.41}$$

i.e.,

$$(\mathrm{Re}(z))^2\sin^2(\theta) - \cos^2(\theta)\,|\mathrm{Im}(z)|^2 > 0$$

According to the property (B.28), the last inequality is equivalent to

$$(z+\bar{z})^2\sin^2(\theta) + (z-\bar{z})(z-\bar{z})\cos^2(\theta) > 0 \tag{B.42}$$

According to the Shur complement, (B.42) is equivalent to

$$\begin{bmatrix} \sin(\theta)(z+\bar{z}) & \cos(\theta)(z-\bar{z}) \\ -\cos(\theta)(z-\bar{z}) & \sin(\theta)(z+\bar{z}) \end{bmatrix} < 0 \tag{B.43}$$

It is enough to take

$$\alpha = \begin{bmatrix} 0\,0 \\ 0\,0 \end{bmatrix},\quad \beta = \begin{bmatrix} \sin(\theta) & \cos(\theta) \\ -cos(\theta) & \sin(\theta) \end{bmatrix}$$

and then, we obtain the $\mathcal{LMI}$

$$\begin{bmatrix} \sin(\theta)(AX+(AX)^{\mathrm{T}}) & \cos(\theta)(AX-(AX)^{\mathrm{T}}) \\ -\cos(\theta)(AX-(AX)^{\mathrm{T}}) & \sin(\theta)(AX+(AX)^{\mathrm{T}}) \end{bmatrix} < 0 \tag{B.44}$$

Appendix C
$\mathcal{GLMI}$ Regions

The concept of Generalized LMI ($\mathcal{GLMI}$) regions is recalled here.

C.1 Definition of a $\mathcal{GLMI}$ region

Definition C.1.1 [7, 8] A region $\mathcal{D}$ of the complex plane is a $\mathcal{GLMI}$ region of order l if there exist square complex matrices $\theta_k \in \mathbb{C}^{l\times l}$, $\psi_k \in \mathbb{C}^{l\times l}$, $H_k \in \mathbb{C}^{l\times l}$ and $J_k \in \mathbb{C}^{l\times l}$, $\forall k \in \{1, 2, ..., m\}$, such that

$$\mathcal{D} = \left\{z \in \mathbb{C} : \exists[\omega_1 \,...\, \omega_m]^{\mathrm{T}} \in \mathbb{C}^m s.t. f_{\mathcal{D}}(z,\omega) < 0, g_{\mathcal{D}}(\omega) = 0_{l\times l}\right\} \tag{C.1}$$

where

$$f_{\mathcal{D}}(z,\omega) = \sum_{k=1}^{m}(\theta_k\omega_k + \theta_k^*\bar{\omega}_k + \psi_k z\omega_k + \psi_k^*\bar{\omega}_k\bar{z}) \tag{C.2}$$

and

$$g_{\mathcal{D}}(\omega) = \sum_{k=1}^{m}(H_k\omega_k + J_k\bar{\omega}_k) \tag{C.3}$$

C.2 Stability in a $\mathcal{GLMI}$ Region

Definition C.2.1 A matrix A is said to be $\mathcal{D}$-stable if and only if its eigenvalues are strictly located in region $\mathcal{D}$ of the complex plane.

S. Marir and M. Chadli, *Control of Singular Fractional Order Systems: LMI Approach*, Studies in Systems, Decision and Control 292,
https://doi.org/10.1007/978-3-031-87382-9

In the case where $\mathcal{D}$ is a $\mathcal{GLMI}$ region of the form (C.1), the following theorem presents a necessary and sufficient $\mathcal{LMI}$ condition for matrix A to be $\mathcal{D}$-stable. A proof can be found in [7, 8].

Theorem C.2.1 *Let $A \in \mathbb{C}^{n\times n}$ and $\mathcal{D}$ a $\mathcal{GLMI}$ region. A is $\mathcal{D}$-stable if and only if there exist m matrices $X_k \in \mathbb{C}^{n\times n}$, $\forall k \in \{1, 2, ..., m\}$, such that:*

$$\sum_{k=1}^{m}(\theta_k \otimes X_k + \theta_k^* \otimes X_k^* + \psi_k \otimes (AX_k) + \psi_k^* \otimes (AX)^*) < 0 \tag{C.4}$$

and

$$\sum_{k=1}^{m}(H_k \otimes X_k + J_k \otimes X_k^*) = 0_{nl\times nl} \tag{C.5}$$

C.3 $\mathcal{D}$-Stability in the Union of Convex Sub-regions

This paragraph is devoted to the means by which many simple polynomial regions can be formulated as $\mathcal{GLMI}$ regions. Indeed, these regions actually result from unions of convex polynomial sub-regions, not necessarily symmetrical with respect to the real axis. We describe here polynomial regions, often used in pole placement problems, and which can also be described by a $\mathcal{GLMI}$ formulation.

C.3.1 *First-Order $\mathcal{GLMI}$ Regions*

$\mathcal{GLMI}$ regions of first order are the half plane (see [8]). Indeed, a half plane in the complex plane is defined by

$$\mathcal{D} = \{z = x + jy \in \mathbb{C}/d_0 + d_1 x + d_2 y < 0\} \tag{C.6}$$

where d_0, d_1, d_2 are real constants. knowing that for any complex number $z = x + jy$, we have

$$x = \frac{z+\bar{z}}{2}, y = \frac{z-\bar{z}}{2j} \tag{C.7}$$

we can deduce that

$$\mathcal{D} = \left\{z = x + jy \in \mathbb{C}/d_0 + d_1\frac{z+\bar{z}}{2} + d_2\frac{z-\bar{z}}{2j} < 0\right\} \tag{C.8}$$

which is the same as

$$\mathcal{D} = \left\{ z = x + jy \in \mathbb{C}/d_0 + \frac{d_1 - jd_2}{2}z + \frac{d_1 + jd_2}{2}\bar{z} < 0 \right\} \tag{C.9}$$

This formulation is identifiable with the polynomial formulation

$$\mathcal{D} = \left\{ z = x + jy \in \mathbb{C}/\alpha + \beta z + \beta^* \bar{z} < 0 \right\} \tag{C.10}$$

with

$$\alpha = d_0, \;\; \beta = \frac{d_1 - jd_2}{2}, \;\; \beta^* = \frac{d_1 + jd_2}{2} \tag{C.11}$$

this polynomial formulation is identifiable with a $\mathcal{GLMI}$ region of first order with $m = 1$. It is enough to see that we can take

$$\theta_1 = \alpha, \;\; \omega = 1, \;\; \psi_1 = \beta, \;\; H_1 = -J_1 = 1 \tag{C.12}$$

C.3.2 $\mathcal{GLMI}$ Formulation of the Union of First Order Sub-regions

Let $\mathcal{D}$ be the region of the complex plane described by

$$\mathcal{D} = \bigcup_{k=1}^{m} \mathcal{D}_k \tag{C.13}$$

where $\mathcal{D}_k$ is a sub-region (We restrict ourselves to regions of the first order) described by (C.10)

$$\mathcal{D}_k = \left\{ z \in \mathbb{C}/\alpha_k + \beta_k z + \beta_k^* \bar{z} < 0 \right\}, \;\; \forall k \in \{1, 2, \ldots, m\} \tag{C.14}$$

As presented in [8], a union of m $\mathcal{GLMI}$ regions of the form (C.14) is also a $\mathcal{GLMI}$ region of the form (C.1) with order

$$l = m + 1 \tag{C.15}$$

and for any $k \in \{1, \ldots, m\}$, we have

$$\theta_k = \frac{1}{2} \begin{bmatrix} \Theta_k & 0_{1\times m} \\ 0_{m\times 1} & -\varepsilon_k^m \end{bmatrix} \tag{C.16}$$

$$\psi_k = \begin{bmatrix} \Psi_k & 0_{1\times m} \\ 0_{m\times 1} & 0_{m\times m} \end{bmatrix} \tag{C.17}$$

$$H_k = -J_k = \varepsilon_{k+1}^{m+1} \tag{C.18}$$

$$\Theta_k = \begin{bmatrix} \alpha_k & \dots & 0 \\ \vdots & \ddots & \vdots \\ 0 & \dots & 0 \end{bmatrix} \tag{C.19}$$

$$\Psi_k = \begin{bmatrix} \beta_k & \dots & 0 \\ \vdots & \ddots & \vdots \\ 0 & \dots & 0 \end{bmatrix} \tag{C.20}$$

and

$$\varepsilon_p^q \in \mathbb{R}^{q \times q} \text{ and } \begin{cases} \varepsilon_p^q(\rho, \sigma) = 1 \text{ if } \rho = \sigma = p \\ \varepsilon_p^q(\rho, \sigma) = 0 \text{ else} \end{cases} \tag{C.21}$$

Appendix D
Some Proofs

D.1 Proof of Lemma 4.2.3

Proof Suppose that there exist positive definite matrices P, $Q \in \mathbb{H}_n$ such that $\mathcal{LMI}$ (4.6) holds. Then:

$$\begin{aligned}&\begin{bmatrix} H(\lambda) \\ I_m \end{bmatrix}^* \begin{bmatrix} A & B \\ I_n & 0 \end{bmatrix}^* (\Phi \otimes P + \Psi \otimes Q) \begin{bmatrix} A & B \\ I_n & 0 \end{bmatrix} \begin{bmatrix} H(\lambda) \\ I_m \end{bmatrix} \\ &\quad + \begin{bmatrix} H(\lambda) \\ I_m \end{bmatrix}^* \Theta \begin{bmatrix} H(\lambda) \\ I_m \end{bmatrix} \leq 0.\end{aligned} \tag{D.1}$$

We have

$$\begin{aligned}\begin{bmatrix} A & B \\ I_n & 0 \end{bmatrix} \begin{bmatrix} H(\lambda) \\ I_m \end{bmatrix} &= \begin{bmatrix} A(\lambda I_n - A)^{-1}B + B \\ (\lambda I_n - A)^{-1}B \end{bmatrix}, \\ &= \begin{bmatrix} (A(\lambda I_n - A)^{-1} + I_n)B \\ (\lambda I_n - A)^{-1}B \end{bmatrix}, \\ &= \begin{bmatrix} (A(\lambda I_n - A)^{-1} + (\lambda I_n - A)(\lambda I_n - A)^{-1})B \\ (\lambda I_n - A)^{-1}B \end{bmatrix}, \\ &= \begin{bmatrix} (A + \lambda I_n - A)(\lambda I_n - A)^{-1})B \\ (\lambda I_n - A)^{-1}B \end{bmatrix}, \\ &= \begin{bmatrix} \lambda(\lambda I_n - A)^{-1})B \\ (\lambda I_n - A)^{-1}B \end{bmatrix}, \\ &= \begin{bmatrix} \lambda H(\lambda) \\ H(\lambda) \end{bmatrix}, \\ &= \begin{bmatrix} \lambda \\ 1 \end{bmatrix} \otimes H(\lambda).\end{aligned}$$

S. Marir and M. Chadli, *Control of Singular Fractional Order Systems: LMI Approach*, Studies in Systems, Decision and Control 292,
https://doi.org/10.1007/978-3-031-87382-9

Therefore, (D.1) is equivalent to

$$\left(\begin{bmatrix}\lambda\\1\end{bmatrix}\otimes H(\lambda)\right)^*(\Phi\otimes P+\Psi\otimes Q)\left(\begin{bmatrix}\lambda\\1\end{bmatrix}\otimes H(\lambda)\right) + \begin{bmatrix}H(\lambda)\\I_m\end{bmatrix}^*\Theta\begin{bmatrix}H(\lambda)\\I_m\end{bmatrix}\leq 0. \tag{D.2}$$

Using Kronecker's product properties, (D.2) can be written as

$$\left(\begin{bmatrix}H(\lambda)\\I_m\end{bmatrix}^*\Phi\begin{bmatrix}H(\lambda)\\I_m\end{bmatrix}\right)\otimes(H(\lambda)^*PH(\lambda))+\left(\begin{bmatrix}H(\lambda)\\I_m\end{bmatrix}^*\Psi\begin{bmatrix}H(\lambda)\\I_m\end{bmatrix}\right)\otimes(H(\lambda)^*QH(\lambda)) + \begin{bmatrix}H(\lambda)\\I_m\end{bmatrix}^*\Theta\begin{bmatrix}H(\lambda)\\I_m\end{bmatrix}\leq 0. \tag{D.3}$$

Any $\lambda\in\Upsilon(\Phi,\Psi)$ verifies $\begin{bmatrix}\lambda\\1\end{bmatrix}^*\Phi\begin{bmatrix}\lambda\\1\end{bmatrix}\geq 0$ and $\begin{bmatrix}\lambda\\1\end{bmatrix}^*\Psi\begin{bmatrix}\lambda\\1\end{bmatrix}\geq 0$. In addition, since $P, Q>0$, we also have $(H(\lambda)^*PH(\lambda))\geq 0$ and $(H(\lambda)^*QH(\lambda))\geq 0$. Thus, it follows from these inequalities together with (D.3) that $\mathcal{LMI}$ (4.5) holds $\forall\lambda\in\Upsilon(\Phi,\Psi)$.

D.2 Proof of Lemma 5.4.5

Proof Rewrite (5.60) as

$$\text{sym}\left\{\begin{bmatrix} aX_1 & aX_2 & bX_1 & bX_2\\ aX_3 & aX_4 & bX_3 & bX_4\\ -bX_1 & -bX_2 & aX_1 & aX_2\\ -bX_3 & -bX_4 & aX_3 & aX_1\end{bmatrix}\right\}<0, \tag{D.4}$$

with $a=\sin\left(\alpha\frac{\pi}{2}\right)$, $b=\cos\left(\alpha\frac{\pi}{2}\right)$. By the congruence transformation of

$$\begin{bmatrix} I & 0 & 0 & 0\\ 0 & 0 & I & 0\\ 0 & I & 0 & 0\\ 0 & 0 & 0 & I\end{bmatrix},$$

the inequality (D.4) is equivalent to

$$\text{sym}\left\{\begin{bmatrix} aX_1 & bX_1 & aX_2 & bX_2 \\ -bX_1 & aX_1 & -bX_2 & aX_2 \\ aX_3 & bX_3 & aX_4 & bX_4 \\ -bX_3 & aX_3 & -bX_4 & aX_4 \end{bmatrix}\right\} < 0. \tag{D.5}$$

Thus, it is easy to see that (5.61) holds. This completes the proof. □

References

1. Achdou, Y.: Algèbre linéaire et analyse numérique matricielle. 23 Nov 2005
2. Gantmacher, F.R.: The Theory of Matrices. Chelsea, New York (1974)
3. Kaczorek, T.: Vectors and Matrix in Automatics and Electrotechnics. WNT, Warszawa (1998)
4. Lyapunov, A.M.: Problème général de la stabilité du mouvement. Ann. Frac. Sci. **9**, 203–474, Toulouse (1907)
5. Gutman, S., Jury, E.I.: A general theory for matrix root clustering in subregions of the complex plane. IEEE Trans. Automat. Contr. **AC-26**, 853–863 (1981)
6. Chilali, M., Gahinet, P.: h_∞ design with pole placement constraints: an $\mathcal{LMI}$ approach. IEEE Trans. Automat. Contr. **41**, 358–367 (1996)
7. Chilali, M.: Méthodes $\mathcal{LMI}$ pour l'analyse et la synthèse multi-critère. PhD thesis, Université Paris IX-Dauphine, U.F.R. Mathématiques de la Décision, Paris, France (1999)
8. Bachelier, O.: Commande des systèmes linéaires incertains: Placement de pôles robuste en d-stabilité. Ph.D. Thesis, INSA, Toulouse, France (1998)

Index

S. Marir and M. Chadli, *Control of Singular Fractional Order Systems: LMI Approach*, Studies in Systems, Decision and Control 292,
https://doi.org/10.1007/978-3-031-87382-9

The manufacturer's authorised representative in the EU is Springer Nature Customer Service Centre GmbH, Europaplatz 3, 69115 Heidelberg, Germany. If you have any concerns regarding our products, please contact ProductSafety@springernature.com

Printed and bound by CPI Group (UK) Ltd, Croydon, CR0 4YY
30/09/2026
02225592-0014